Ecosystem-Based Fisheries Management in the Western Pacific

Ecosystem-Based Fisheries Management in the Western Pacific

Edward Glazier, Editor

WILEY-BLACKWELL
A John Wiley & Sons, Ltd., Publication

WESTERN PACIFIC REGIONAL FISHERY MANAGEMENT COUNCIL

Wiley-Blackwell is an imprint of John Wiley & Sons, formed by the merger of Wiley's global Scientific, Technical and Medical business with Blackwell Publishing.

Registered office: John Wiley & Sons Ltd, The Atrium, Southern Gate, Chichester, West Sussex, PO19 8SQ, UK

Editorial offices: 2121 State Avenue, Ames, Iowa 50014-8300, USA
The Atrium, Southern Gate, Chichester, West Sussex, PO19 8SQ, UK
9600 Garsington Road, Oxford, OX4 2DQ, UK

For details of our global editorial offices, for customer services and for information about how to apply for permission to reuse the copyright material in this book please see our website at www.wiley.com/wiley-blackwell.

Library of Congress Cataloging-in-Publication Data

Glazier, Edward W. (Editor)
Ecosystem-based fisheries management in the western pacific / Edward
Glazier.
p. cm.
Includes bibliographical references and index.
ISBN 978-0-8138-2154-2 (hardback)
1. Fishery management–Oceania. 2. Ecosystem management–Oceania. 3. Fishery management–Pacific Ocean. 4. Ecosystem management–Pacific
Ocean. I. Title.
SH319.A2G53 2011
639.20995–dc22 2010043545

A catalogue record for this book is available from the British Library.

This book is published in the following electronic formats: ePDF 9780470959459; Wiley Online Library 9780470959480; ePub 9780470959466

Set in 10/12 pt Dutch801BT by Aptara® Inc., New Delhi, India

The left bottom cover image courtesy of D.J. Struntz, Senior Staff Photographer, Surfing Magazine, GLOBE. Middle bottom cover image courtesy of Kimberlee Harding.

1 2011

Contents

Foreword

In late 2004 my old friend and colleague Kitty Simonds, the Executive Director of the Western Pacific Regional Fishery Management Council (the Council), asked me to attend and synthesize discussions during a workshop on ecosystem-based fishery management in the Western Pacific. Having worked with Kitty and the Council for many years, I was happy to accept the invitation.

When I arrived at the workshop I noted an interesting phenomenon: David Fluharty of the University of Washington and I—two social scientists—were responsible for synthesizing the many and various ecosystem concepts brought forth during the meeting. This was fine, except that virtually all of the rest of the invitees were biological or physical scientists whose works, comments, and interactions we were supposed to synthesize at the end of the meeting. Dave and I joked about this, but since we both were appointed in interdisciplinary schools at our respective universities, we were comfortable in these roles.

In the middle of the second day of the meeting, I began to be uncomfortable about the fact that the workshop attendees were discussing not only biophysical science issues but also issues involving the human dimensions of fishery ecosystems, and related policy and management issues. I approached Kitty about this and said, offhandedly, "You need two more workshops: One with social scientists focused on the human dimensions, and one that brings both biophysical and social scientists together with stakeholders, policy-makers, and managers." Kitty, always quick to recognize an opportunity, said, "If you'll organize the second one and moderate the third one, we'll do it!" Thus was born the three workshop series detailed in this book.

Two notable developments occurred during the subsequent workshops. The first was the idea of three separate but related "ecological systems"—the biophysical, the human, and the institutional. The biophysical ecological system includes the nonhuman biophysical resources and environments related to the Western Pacific fisheries. The human ecological system was defined as "those humans and human behaviors that affect, are affected by, or are otherwise concerned with Western Pacific fisheries." The institutional ecology was defined as "those institutions that govern or affect the behavior of those people in the human ecological system, and include governance bodies at all levels of governance." These three ecological systems together—the "total ecology"—were taken to comprise the Western Pacific fishery ecosystems and their biophysical, human, and institutional characteristics. The first workshop had focused primarily on the biophysical ecology. The second focused on the human ecology. The third brought the biophysical and human ecologies together within the institutional framework of governance for Western Pacific fishery ecosystems.

The second development was the emergence of the *ahupua'a* as a conceptual model for modern fishery ecosystem management. In ancient Hawaii, *ahupua'a* were sections of the islands typically pie-shaped within which all activities were governed by the nobility. Management of all lands and nearshore reefs and seas was integrated into a seamless composite of compatible, interrelated activities. This is, essentially, the modern concept of integrated coastal management, a companion concept to ecosystem-based management. It was suggested that our task might be not to invent something completely new, but to take advantage of concepts embodied in the ancient Hawaiian system that had governed “fishery ecosystems” for centuries.

The implication of using such a concept was that we could not speak only of resources, activities, and environments under the direct jurisdiction of the Western Pacific Council, which extends generally from 3 to 200 nautical miles offshore of the US Pacific Islands. Tracing the biophysical, human, and institutional ecologies of Western Pacific fishery ecosystems makes it clear that these systems extend from the highest reaches of the mountains to the deep sea, areas under the jurisdiction of scores of different local, state, and federal agencies. In order to be compatible with the *ahupua'a*/fishery ecosystem concept, we would have to discuss many things that were the official “jurisdiction” of people and institutions other than the Western Pacific Council.

Kitty and the workshop participants made the courageous decision to go down this path. Our thinking was that we must identify the true extent of the ecological systems even if the official responsibility for the policy-makers and managers was fragmented among different agency jurisdictions. Once the true extent of the ecological systems and the attendant fishery ecosystem issues were identified, the process of actually constructing policy and management measures would have to involve not only the Western Pacific Council but also many other policy and management agencies.

These two developments set the stage for the success of the workshops. Without the organizing concept of the “total ecology” of Western Pacific fishery ecosystems and the *ahupua'a* model of holistic resource management, the discussions would not have been as complete, fulfilling, and productive as they were. With those concepts in hand, the workshops produced an ideal lens through which to view and deliberate upon ecosystem-based fisheries management in the Western Pacific.

Michael K. Orbach
Professor of Marine Affairs and Policy
Duke University Marine Laboratory

Preface

The biophysical and human processes addressed by fishery scientists and managers are challengingly complex. Marine organisms are complex in genetic and behavioral terms, and the nature of their evolution, distribution, and population dynamics continues to challenge scientists worldwide. Such organisms are interactive components of ocean systems of varying scope and complexity, the dynamics of which further complicate scientific inquiry into the nature of the sea and its myriad creatures.

With great significance in a world of rapidly increasing coastal populations and growing reliance on marine resources, humans play crucial roles in the dynamics of ocean systems. It is also significant that the effects of human activities on the ocean and its organisms can be regulated through various social and institutional means. As such, the human dimension is the principal vector through which change in marine systems can be effected.

Achieving understanding of complex marine and human systems and the manner in which they interact is a challenging endeavor. From the perspective of the biophysical sciences, many questions about specific fish populations and how they are affected by fishing pressure cannot be sufficiently answered without understanding the nature of their relationships with other species and with the physical properties of the ocean itself.

From a managerial perspective, understanding marine systems and fisheries interactions is vital to decisions that have the potential to affect the well-being of people involved in the pursuit, distribution, and consumption of seafood. Societies of such persons are complex in themselves, and adequate understanding of regulatory implications requires an understanding of human motivations for pursuing seafood and the social, economic, and cultural factors that condition life along the margin of the world's oceans.

Many fisheries scientists and managers increasingly relate their thinking to *marine ecosystems and human ecology*. Despite the complexities of marine systems and an historical tendency for scientists and managers to focus on single species or relatively limited interactions between species, there is growing attention to *whole marine systems and the physical, biological, and human relationships that comprise those systems.* While the practical challenges of addressing such complexities are daunting to some, advocates are moving forward with the understanding that the intricate realities of marine fisheries call for a more encompassing strategy than has been used in years past.

The Western Pacific Regional Fishery Management Council (referred to in this text alternately as the Western Pacific Council, the Council, or WPRFMC) is one such entity. Recognizing the need for improved understanding of relationships between open-ocean and island-specific biophysical systems, and the human groups that use or inhabit such areas, the Council has enthusiastically undertaken the planning processes required to formally adopt the ecosystem approach to fishery management, and it is now implementing the approach across the region. This is in line with an international trend toward ecosystem-based science and management, and it is in keeping with the unique attributes of Pacific islands and the deep ocean systems that surround them. Further, and perhaps most significantly, the approach is compatible with the recognition that humans and human institutions are particularly well-established elements of marine systems across the Western Pacific.

The ecosystem approach bears great potential for enhancing scientific understanding and effective management of dynamic fishery-environmental interactions. But given the complexities inherent in such interactions, the Council is implementing the ecosystem approach on an incremental and adaptive basis.

One aspect of this strategy has been to seek the advice of experts on matters of ecosystem science and policy. As such, ecosystem principles and the potential challenges and benefits of ecosystem-based fishery management were explored in great depth during a series of three formalized ecosystem workshops convened in Honolulu between 2005 and 2007. The workshops focused on advances in and the problematic nature of ecosystem modeling, data needed to better understand the prospects and implications of the ecosystem approach, and policy matters that will need to be addressed moving forward.

This book describes the three-part Council workshop series by summarizing expert input on the process of incorporating ecological principles into the science and management of marine fisheries across the Western Pacific. The lessons so derived bear value for fisheries managers and scientists worldwide.

The first workshop was held to examine biophysical data and models suitable for transitioning from conventional species-based management to the new ecosystem-based approach. The second workshop was held to examine social, economic, and institutional aspects of the ecosystem approach in the region. The workshop focused on the human ecology of marine ecosystems. The third and final workshop involved the participation of local, regional, and national experts who worked to synthesize output from the first two workshops and to develop policy options and information needed to effectively implement the ecosystem-based management strategy in the Western Pacific in the upcoming years.

One of the benefits of shifting to an ecosystem approach in the Western Pacific is that the new process enables fisheries to be more efficiently addressed as integrated components within archipelago-based units of management. For example, instead of studying and managing bottomfish species and fisheries across island areas that often differ in many ways, bottomfish species and fisheries can now be more appropriately and efficiently examined and managed as they relate to conditions that are unique to each distinct archipelago.

This is critically important because the region managed by the Western Pacific Council is truly vast and varied. It includes the entirety of the Exclusive Economic Zone (EEZ; 3 to 200 miles offshore and 0–200 miles offshore for Guam, CNMI, and PRIA) surrounding the various archipelagos and remote islands possessions of the United States in the Central and Western Pacific. These include the following: (1) the Mariana archipelago (Guam and the Northern Mariana Islands); (2) the Hawaii archipelago (including the Main Hawaiian Islands, the Northwestern Hawaiian Islands, and Midway and Johnston Atolls); (3) American Samoa; and (4) the US Pacific Remote Islands (Howland, Baker, Jarvis, Kingman Reef, Palmyra Atoll, and Wake Island). This area of nearly 1.5 million square nautical miles comprises 48 percent of the nation's EEZ and is by far the largest area addressed by any fishery council in the United States.

In the Western Pacific region, the shift to an ecosystem-based approach to management is being implemented through Fishery Ecosystem Plans (FEPs) for each of the archipelagos. These subsumed the original fishery management plans (FMPs). An FEP for open ocean pelagic resources and fisheries is being implemented separately.

Given extensive environmental variability in physical and social conditions across the Western Pacific region, and the desirability of streamlining introduction of the new ecosystem approach, effective early planning was considered essential. The workshop series described in this book was an important element of that process, and in fact the Western Pacific Council is now far along the road of implementing EAFM in the region.

As provided in the following pages, each workshop event enabled regional and national experts to examine a promising new approach to fisheries management in the region and world. It is hoped that the following discourse and analysis will inform readers about the nature and challenges of that approach and the importance of maintaining healthy marine ecosystems not only in island settings of the Western Pacific but also in island and coastal regions throughout the world.

Summary Overview of the Ecosystem Workshops

The Ecosystem Science and Management Planning Workshop

The first workshop of the three-part series was held during April 2005. Approximately 60 scientists and marine policy experts participated in the meeting. Presentations and discussions were organized around three central topics: (1) data needed to support ecosystem-based science and management, (2) state-of-the-art ecosystem models and modeling, and (3) indicators useful for gauging ecosystem processes and the effects of management on species and ecosystems of interest.

Overview

The overarching goal of the initial workshop was to identify scientific protocol and information needed to support an ecosystem-based approach to marine resource management in the WPRFMC region of jurisdiction. In order to achieve this goal, the workshop was designed to address six basic objectives: (1) review ecosystem models in terms of management utility and application; (2) identify management requirements in the Western Pacific region; (3) identify the best suite of quantitative ecosystem indicators and associated trade-offs to support ecosystem-based management; (4) within the confines of existing mandates, identify the most effective short-term application of ecosystem-based approaches to management that can be implemented based on current data, and in this context address whether a precautionary approach has a role; (5) identify new data or models that would be required to advance ecosystem-based approaches; and (6) identify changes in policy or science that would be needed to effectively implement those approaches in the region.

Meeting participants regarded tasks (1) through (3) above as within the purview of scientists, and tasks (3) through (6) as within the purview of marine policy experts. Participants determined that the role of scientists should be limited to research, and that subjective decision-making about allocation of resources and related issues is the distinct role of managers and policy-makers. Three breakout groups were established during the course of the workshop to enable informed and interactive discussion about the new ecosystem approach, with directed discussion regarding ecosystem data needs, modeling possibilities, and development and use of indicators.

Ecosystem Data

Obviously, the function of research in fisheries management is to provide resource managers with valid information needed to make well-grounded resource decisions. With this in mind, workshop participants identified three imperatives for collection and use of relevant data: (1) it must be appropriate in terms of scale and suitable for purposes of modeling; (2) it must involve a triangulated focus on human, ecological, and environmental dimensions; and (3) it must be well-managed, archived, and accessible. Participants recognized that data needs would depend in part on modeling requirements that inevitably varied across time and space.

Other important workshop findings regarding data and data collection included the following: (1) high-quality data will facilitate development of quality indicators and models which will ultimately enhance management of the resources and fisheries, and local knowledge can enhance the quality of data; (2) new or different data will be needed to support ecosystem models and ecosystem-based management, especially as regards nontarget species; (3) adaptive management experiments involving spatially sensitive comparison of policy options are critical for improving understanding of the ecosystem effects of fishing; and (4) a data expert or clearinghouse will be essential for coordinating the appropriate collection, storage, distribution, and analysis of ecosystem-relevant data.

Participants also recommended formation of a Data Needs Working Group for the Western Pacific and identified a range of interim requirements for new data needed to support ecosystem-based management in the region. These needs included the following: (1) improved commercial, recreational, and subsistence landings and effort data; (2) information regarding by-catch and fishery interactions; (3) trophic interactions data; (4) information regarding habitat-species associations and habitat-fishery interactions; (5) data regarding spatial distribution of stocks; (6) data regarding the life history of relevant species; (7) data regarding marine environmental variability and the consequences of and responses to climate change and oceanic regime shifts; (8) data regarding inherent ecosystem productivity and habitat alteration; (9) information regarding pertinent social and economic dimensions of marine ecosystems; (10) data supporting carrying capacity analysis and forage base interactions; and (11) information revealing ecosystem processes under differing fishery-related scenarios.

Ecosystem Modeling

Extensive discussion about ecosystem models and modeling occurred throughout the workshop. The transferability of models and associated data and indicators across the region and its archipelagos was of particular concern. While management issues and priorities will drive the development and application of specific models in each of the five subregions, the group identified four data layers that would be appropriate for modeling applications across the Western Pacific. These are: (1) hydrodynamics, (2) biological community dynamics, (3) habitats and species-habitat associations, and (4) the behavior of fishers. Regarding the last data layer, participants acknowledged that understanding fishing operations and the involvement of fishers and the interest public in the ecosystem-based management approach would be crucial to its success.

Participants also reached consensus regarding processes fundamental to successful ecosystem modeling efforts in the region. These include (1) identification of salient resource and resource management issues, (2) identification of potentially viable management policies and options, (3) matching the model in question with appropriate management policies and options, (4) identification of data needs for the selected model(s), (5) inventory and collection of the requisite data, and (6) identification of any other biophysical processes that may be important in terms of analytical or experimental control. Participants agreed that the most important aspect of modeling is clear initial delineation of objectives. This includes determining whether predictive or evaluative models will be most useful for the application of interest. Participants recommended that adaptive management considerations should be incorporated into ecosystem modeling in the region.

The modeling discussion concluded with the group achieving consensus on the need to develop appropriate base models. It was agreed that these could be refined and adapted to predict or evaluate environmental or regulatory changes over the course of time.

Ecosystem Indicators

The role and utility of indicators in this context gave rise to vigorous discussion among workshop participants. Participants recognized that no single set of indicators would serve in a functionally holistic manner across the archipelagic subregions of the Western Pacific. Rather, indicators would need to be prioritized and adapted to fit specific places and situations. There was agreement that scientists should (1) distinguish between emergent properties operating in a given ecosystem and measures used for theoretical or experimental control, (2) distinguish between ecosystems properties that are intractable and those which can be manipulated to create a desirable effect or mitigate an adverse effect, (3) develop a mechanistic understanding of how indicators are derived, and (d) exercise caution when using ecosystem indicators as performance measures.

While discussion of caveats and conditions for developing and applying ecosystem indicators was extensive, participants were able to identify a variety of indicators that could be used as to examine the status of ecosystems and various pressures on those systems. These included the following: habitat "quantity" and "quality"; keystone/functional species; sentinel and protected species; assemblage structure; biodiversity; pathogens; harmful events; and fishery measures. There was clear consensus that the final choice of indicators should be clearly linked to management objectives.

The indicators working group generated the following recommendations and priorities for developing valid ecosystem indicators for the Western Pacific. First, it will be necessary to identify and evaluate valid candidate indicators. Second, each priority should be ranked in terms of its applicability in each archipelago or open ocean pelagic zone (which occurs throughout the region). Third, the performance of specific indicators should be assessed by experts in each region. Fourth, indicators should be in keeping with salient management needs and modeling requirements. Finally, indicators should address the status of and pressures on marine ecosystems and be capable of evaluating feedback effects of management actions.

Concluding Summary and Recommendations

In sum, participants identified eight broadly conceived operational objectives for the region's FEPs. These are: (1) conserving and managing the target species, (2) minimizing by-catch, (3) managing trade-offs, (4) accounting for feedback effects, (5) establishing appropriate ecosystem boundaries, (6) maintaining ecosystem productivity and balanced ecosystem structure, (7) accounting for climate variability, and (8) using adaptive approaches to management. It was agreed that consideration of ecosystem management must extend beyond the biophysical components of the region's marine ecosystems. Participants in all three breakout sessions recognized the need for data, models, and indicators of utility for understanding and addressing human dimensions of marine ecosystems in the Western Pacific.

Six recommendations were developed as general policy advice for the Council as it moves forward with implementation of ecosystem-based management in the region. First, it was recommended that the fishing industry and managers should endeavor to be proactive in changing the burden of proof regarding the impacts of fishing. This would be enabled in part by industry taking an active participatory role in research and monitoring and resource conservation and sustainability.

Second, it was recommended that a precautionary approach should be employed when implementing the ecosystem-based approach in the region. This would enable sufficient time for scientific understanding to meet the requirements of the new approach to management.

A third recommendation asserted the need for spatial or other latitude in development and implementation of ecosystem-related policy. The intent of this recommendation is to identify ways and means for scientists and managers to develop sufficient understanding of changing environmental conditions as per the parameters of a truly adaptive approach to managing fisheries and fishery resources.

Fourth, it was recommended that lessons should be drawn from other regions, and an adaptive approach should be employed in the Western Pacific. This is in keeping with the assertions that fisheries management in the United States and elsewhere has involved successes, and that lessons deriving from those successes may well augment current and future management strategies.

Fifth, it was recommended that proper incentives should be used to aid in the achievement of management goals. This reflects an understanding of the historic and potential future needs, interests, and tendencies of fishery participants vis-à-vis the management and regulatory processes.

Finally, it was recommended that the issues of fairness and equity should be duly considered in the ecosystem-based approach to management in the Western Pacific and elsewhere. This relates to concerns for appropriate and ethical balancing of social and economic benefits and liabilities potentially following from implementation of the ecosystem-based approach.

The Ecosystem Social Science Workshop

The Ecosystem Social Science Workshop was held during January 2006. Thus, nationally recognized social scientists and regional experts were convened to examine a range of pertinent issues. These included the following: (1) marine fisheries, fisheries management, and related human and biophysical factors in the Western Pacific; (2) the need for and utility of social science in the context of ecosystem-based management in this region and elsewhere; (3) institutional constraints and opportunities for incorporating social science into ecosystem-based management; (4) relevant information needs, useful types of data, and data collection methods; (5) ecosystem-relevant human behavior and resource modeling; (6) indicators for assessing regulatory effects and the performance of management strategies; and (7) the appropriate scope and scale of social science applications to ecosystem-based management in the Western Pacific.

Overview

The overarching goal of the meeting was to facilitate informed discussion of social science requirements for implementing effective ecosystem-based fisheries management in the Council's region of jurisdiction. Meeting this goal involved a series of in-depth discussions organized around the major functional–analytical components of marine ecosystems.

As asserted at the outset of the meeting, these components include the biophysical ecology of marine ecosystems, and the human ecology of marine ecosystems. The latter has two distinct components of relevance to social scientific inquiry: (a) the human ecology of the constituent groups—the people whose behavior affects or is affected by a defined biophysical ecology, or who are otherwise concerned with the state of that biophysical ecology; and (b) the human ecology of the governance institutions that have authority or responsibility for establishing and/or enforcing formal rules of human behavior with respect to the defined biophysical ecology. It was determined that these components together comprise the ecosystems to be addressed by fishery management agencies in the Western Pacific.

The workshop was organized so that relevant aspects of the Western Pacific region and its archipelagic subregions were discussed at the outset, thereby providing context for meaningful discussion of ecosystem-based fishery management. Presentations and related discussions were both general and specific in scope, and regional experts were on hand to provide their own perspectives and experiences regarding the realities of island life in the Western Pacific, and the various fishery management challenges and solutions that have been encountered and applied in the region.

Presenters made clear that each archipelago in the region is distinct in terms of its sociocultural, socioeconomic, and demographic attributes. Mode and culture of governance, marine environmental conditions, and types and extent of fishing and other pursuits, and uses of marine resources also vary extensively. It was determined that (a) this variation may be effectively addressed for purposes of meeting FEP objectives through appropriate application of social science methods, including those that facilitate public participation in relevant decision-making processes, and (b) selection of social science methods and analytical techniques should be closely tailored to the particular environmental and social conditions and specific information needs and objectives that characterize each archipelago.

Concepts regarding the relationship of humans to marine ecosystems were repeatedly discussed during the course of the workshop. It was argued that scientists and managers must understand that humans are not merely exogenous entities affecting marine systems. Rather, we are integral and pivotally important components of those systems. As such, it was asserted that any institutional mandate promoting sustainability of marine resources would be effective only insofar as it can successfully manage human behavior.

Island Variability and the Critical Importance of Seafood and Fisheries in the Region

Workshop discussions tended to underscore human and environmental variability within and across the island groups that comprise the vast Western Pacific. Participants recognized that social science research must address such variation and translate findings in a manner that is optimally useful for resource managers seeking to make fair and equitable decisions in an increasingly complex and contested sociopolitical environment. Regional variation notwithstanding, pursuit and consumption of seafood and related cultural processes were seen as constant and critically important aspects of life throughout the archipelagos. It was agreed that there was vital need for monitoring the full range of factors that might impinge on these activities and processes, including the potential effects of conservation interests and ecosystem-based management.

The Importance of Traditional Ecological Knowledge

An important outcome of the social science workshop was recognition of the ongoing importance of indigenous fishery practices and traditional and local knowledge of marine resources and ecosystems. Indigenous Pacific islanders draw on lengthy histories and ever-evolving knowledge and traditions of interaction with ocean ecosystems and with each other to successfully draw sustenance from the marine environment. The Council's approach to ecosystem-based management involves, among other strategies, adaptive management, emphasis on indigenous forms of resource management, and opportunities for community involvement in the management process across the archipelagic subregions. There was consensus among workshop participants that this was a valid approach and that it should continue to be emphasized by the Council as it moved forward with the FEPs.

Additional Points of Summary and Recommendation

An assortment of valuable insights, lessons, and pertinent background information about ecosystems, ecosystem social science, and the context of fisheries in the Western Pacific may be derived from the social science workshop and proceedings. Again, individuals and social institutions were clearly recognized as critically important elements of marine ecosystems, and given their place in the trophic hierarchy, human behaviors, beliefs, and values were envisioned as primary considerations in the implementation of ecosystem-based approaches to fishery management.

Workshop participants indicated that the nascent paradigm shift to ecosystem-based management might potentially lead to further institutional complexity in this unique region of multiple jurisdictions. Given its size, extensive diversity in sociodemographic and sociopolitical context, and increasing involvement of international entities in allocation decisions regarding migratory species, participants recommended that an incremental and adaptive strategy coupled with appropriate incentives might augment the success of ecosystem-based management in the Western Pacific.

Valid social and economic indicators were seen as particularly useful for assessing and monitoring direct and indirect interactions between humans and the environment, and as a basis for adjusting resource use policy under the new mode of management. Although specific variables were not identified, there was agreement that social indicators of utility for ecosystem-based management in the region should articulate with a wide range of climatic, macroeconomic, sociodemographic, regulatory, and community-related factors. It was determined that such indicators would ideally be based on: (a) their potential utility for meeting Council objectives, (b) extant and readily obtainable data regarding the social and biophysical contexts in question, and (c) relevant elements of the social indicators literature and associated theory of social and economic change.

As indicated during the course of the workshop, a well-formulated social science approach to ecosystem-based management could enhance Council efforts to meet its FEP objectives and to administer the new form of management over the long term. The approach would ideally include a series of related elements: (1) a venue or venues for choosing high-priority FEP objectives, (2) design of research to meet prioritized objectives and related information needs, (3) implementation of a research strategy to gather and analyze requisite information, (4) development of an indicators-based archipelagic monitoring system through which to gauge and analytically parse social change potentially associated with Council actions, and (5) implementation of a liaison and performance evaluation program to ensure the validity and effectiveness of the social science approach to ecosystem-based management in the region.

The ecosystem approach calls for greater attention to relationships between components of marine ecosystems, including relationships between marine fisheries and the broader social communities of the islands. But participants noted that social science could not be equated with community development per se. Rather, application of social science may further understanding of community context and the potential for community input, local receptivity to or need for fisheries-related development programs, and the potential or actual social and economic costs and benefits of such processes and programs. Therefore, social science may be used to help identify ways in which communities and individuals may participate in the abundance of positive ocean opportunities available throughout the Western Pacific region.

Given that a number of fisheries or fisheries-relevant social science research and monitoring programs have been undertaken in the United States and abroad in recent years, participants indicated that a strong social science approach supporting the Council FEPs would ideally articulate with these, both drawing upon and contributing to the base of knowledge regarding human interaction with the marine environment and the many related aspects of human behavior discussed during the course of the workshop.

The Ecosystem Policy Workshop

The final workshop in the series was convened during January 2007. The event was held to synthesize the results of the first two meetings and to deliberate on policy options for the Council's fishery ecosystem planning process. The final workshop provided essential context for understanding policy dimensions of fisheries management in both the national and international contexts and for refining the ecosystem approach in the Western Pacific. As was the case for the previous meetings, local, regional, national, and international experts representing a variety of relevant disciplines were involved in the final event.

Overview

Three basic topics were examined during the course of the policy workshop, with the intent of supporting the FEP process. Deliberations drew from the findings of the previous biophysical and social science workshops and from facilitated discussion of policy-specific matters pertinent to biophysical and social conditions across the archipelagos. Participants focused on the following topics: (1) opportunities for fishery ecosystem research and monitoring across the Western Pacific; (2) governance, institutional ecology, and social connectivity as core issues in efforts to maximize the benefits of ecosystem-based fishery management in the cross-jurisdictional and cross-cultural settings that typify the archipelagos; and (3) traditional ecological knowledge, customary fishing practices, and community participation as means for enhancing fishery management across the region.

Ecosystem Research and Monitoring

Policy workshop participants agreed that the ecosystem-based approach to management of fisheries in the Western Pacific would require expanded scientific attention to a larger field of physical–environmental, social, and political factors and issues than had been addressed in years past. But a complete shift in existing science programs was by no means indicated. Rather, because the Council is incrementally and adaptively shifting to a more comprehensive approach, the intent is to build upon and complement existing research and monitoring programs. The new approach will continue to support the sustainability, productivity, and conservation goals engaged by the Council and NOAA Fisheries.

Workshop participants worked to identify options for funding an expanded program of research and monitoring. It was determined that the value of available human and fiscal resources could be maximized through administration of internship programs, community-based data collection and monitoring programs, and programs designed to incorporate traditional knowledge into the management process.

A range of issues and options were discussed as solutions for challenges typical of community-based research and monitoring programs. Practical solutions in this case included technical assistance from local agencies, reciprocal data arrangements, careful attention to issues of proprietary and confidential information, and the participation of on-site island coordinators, cultural practitioners, and social scientists who were familiar with the cultural and linguistic subtleties of island societies in the Western Pacific. Identified options for funding ecosystem-related science and monitoring included Work Force Training Act funds, various nontraditional sources of federal funds, and funds and partnerships with nongovernmental organizations.

Governance, Institutional Ecology, and Connectivity

An emphasis on sociocultural dimensions of marine fisheries in the Western Pacific emerged during the course of the policy workshop. This resulted from efforts to address challenges inherent in understanding and managing marine resources in the complex jurisdictional settings that characterized the region. While some discussions were focused on formal scientific investigation of physical properties and processes associated with the region's marine ecosystems, deliberation tended to force recognition that all means of acquiring knowledge of marine systems, and that knowledge itself, ultimately related to human objectives.

A central objective of the Council in establishing an ecosystem-based approach is to attend more closely to connectivity within and between biophysical and human elements of marine systems. In terms of the human dimension of those systems, workshop participants agreed that increased attention to the needs and interests of communities across the archipelagos might yield a variety of benefits.

For instance, it was agreed that the ecosystem approach would require that managers work more closely with persons using traditional and local knowledge of marine resources and ecosystems, and with persons engaging in long-standing fishing and shoreline food-collecting practices. This is significant in that each archipelago is home to indigenous peoples who have accumulated extensive knowledge of marine and terrestrial components of island ecosystems. It was agreed that enhanced interaction with island communities and their representatives would improve the potential for identifying and mitigating a wide range of factors impinging on the health of the region's marine ecosystems. Finally, it was agreed that the ecosystem approach might augment community development initiatives both directly and indirectly related to marine fisheries.

Community Participation, Customary Practices, and Traditional Knowledge

The Council's Regional Ecosystem Advisory Committee (REAC) process was the subject of much discussion throughout the course of the ecosystem policy workshop. The REACs, which have been established for each of the archipelagos, are comprised of Council members with expertise in marine fisheries and related issues, and representatives from federal, state, and local government agencies, businesses, and nongovernmental organizations with responsibilities or interests in human activities potentially affecting the marine environment. The committees function as means for gathering and disseminating information about place-specific issues affecting marine fisheries and related aspects of community life.

Workshop participants identified a range of options for increasing the probability that the REAC process would be successful now and in years to come. It was recommended that the Council should clearly determine and communicate its objectives and expectations prior to initiating formal relationships with participating agencies and individuals. Participants also suggested that the REACs would likely coalesce upon identifying and working on a problem of direct and compelling interest to the members and their constituents. Finally, it was recommended that long-term research or monitoring efforts associated with the REACs should be fostered through the participation of individuals committed to the well-being of communities on their respective islands.

The importance of understanding customary fishing practices and the potential value of traditional ecological knowledge was discussed throughout the policy workshop. Participants asserted that traditional ecological knowledge would very likely contribute meaningfully to a place-based approach to fishery management, and it was recommended that the Council continue to seek the wisdom of indigenous practitioners and other knowledgeable persons in each island region.

Benefits notwithstanding, some participants felt that traditional knowledge in solo might be insufficient for understanding certain highly complex ecosystem processes such as interaction between pelagic species and associated trophic systems during oceanic regime shifts. Similarly, it was felt that formalized science might at times be inadequate for developing sufficient understanding of complex environmental processes such as global climate change. Given such uncertainties and gaps in knowledge, participants tended to advocate use of the full suite of knowledge-gathering tools available to resource managers across the region. Trial and error were seen as basic elements of adaptive management, and participants discussed the scientific method in terms of its capacity for generating understanding through failed trials.

Summary and Concluding Recommendations

In sum, the final ecosystem workshop generated valuable insight into factors and issues of critical importance to formulation of fishery management policy in the Western Pacific. The findings may also prove useful for managers undertaking the ecosystem approach in other regions of the Pacific and beyond. Workshop participants generated practical insight into the following: (a) working with communities and governments to undertake place-specific ecosystem-based management of marine fisheries, (b) establishing effective long-term consultation with communities across an area that is vast and complex in terms of its biophysical and sociocultural trends and current conditions, (c) documenting and productively incorporating traditional knowledge and the needs of customary practitioners into management considerations via culturally sensitive collaboration with such persons, and (d) identifying possible venues for funding and human resources needed to establish long-term ecosystem research and monitoring programs in the region.

As was made clear throughout the series of three workshops, an ecosystem approach to management is at once potentially beneficial and challenging. The science and information requirements are challenging in themselves and demand a wide range of formalized research methods, analytical approaches, areas of inquiry, and modes of interaction with persons highly knowledge of and/or directly involved with the ocean, its resources, and factors impinging on the sustainability and productivity of marine systems and adjacent fishing societies.

Given (a) the expanded realm of inquiry under the nascent ecosystem-based approach to fishery management, and (b) the desirability of using the full suite of scientific tools and principles currently available to scientists and managers for maximizing understanding of ecosystems and the effects of ecosystem-based management, a final recommendation of the policy workshop involved development of a comprehensive long-term plan for ecosystem research and monitoring. Participants agreed that an effective plan would necessarily meet the information needs and management goals of the Council, with particular attention to the unique and highly varied attributes of human communities and their integral relationships with marine ecosystems and resources across the Western Pacific.

Ideally, such a plan would guide scientists and managers through critical and overarching objectives. These were identified as follows: (1) inventory existing biophysical, social science, and TEK data and related research programs and projects, (2) identify management objectives specific to implementation of the ecosystem approach across the archipelagos, (3) identify sources of funding to complement existing science programs with new research, analysis, monitoring, and programmatic evaluation, (4) articulate ongoing and new research, data management, and data analysis strategies with specific management objectives, and (5) develop means for disseminating needed information in a manner that would best support implementation of the Council's ecosystem approach and associated projects and programs.

Edward Glazier

Acknowledgments

We would like to acknowledge and thank the hundreds of fishery scientists and managers who collectively participated in the three-part ecosystem workshop series and in the development of the technical reports upon which this text is based.

Incorporation of ecosystem principles into the Western Pacific Regional Fishery Management Council's management processes, and steps along the way, including the ecosystem workshops have been advanced through the energetic efforts of the Council's Executive Director, Kitty Simonds; and those of numerous Council staff members, especially Paul Dalzell, Charlie Kaaiai, Sylvia Spalding, Jared Makaiau, Marcia Hamilton, and others.

NOAA's Pacific Islands Fisheries Science Center (PIFSC) provided extensive support and expertise throughout the course of the workshop series. PIFSC Director Samuel Pooley and numerous Center staff members were particularly helpful in moving the process forward.

David Fluharty of the University of Washington played a key role in organizing the biophysical workshop, and Michael Orbach of Duke University was instrumental in expanding the scope of the workshops to include and emphasize human ecological and policy concerns and dimensions. Dr. John Sibert of the Pelagic Fishery Research Program at UH offered ongoing expertise during the course of the series. Dr. John Petterson of Impact Assessment, Inc., co-organized the social science and policy workshops.

Substantive and editorial input for the social science and policy chapters were graciously provided by Craig Severance of the University of Hawaii at Hilo, Dr. Richard Pollnac of the University of Rhode Island, Dr. Susan Hanna of Oregon State University, Dr. Mike Orbach of Duke University, and Mr. Paul Bartram.

Assistance in developing summaries of speaker dialog for the social science workshop was provided by Janna Shackeroff, then a doctoral student at Duke University, by Mr. Paulo Morais, and by Dr. Laura Stanley of Impact Assessment, Inc. Artistic renderings, editorial assistance, and archival research support were provided by Julia Stevens of Impact Assessment, Inc. Spatial analysis and cartographic products were generated by Russell Scalf of Impact Assessment, Inc.

List of Acronyms

ACOE	Army Corps of Engineers
ADF&G	Alaska Department of Fish and Game
AIMS	Australian Institute of Marine Science
ANU	Australia National University
B/BMSY	Biomass at Maximum Sustainable Yield
CCAMLR	Commission for the Conservation of Antarctic Marine Living Resources
CEA	Comprehensive Ecosystem Amendment
cf	Latin for "compare" or "consult"
CLIOTOP	CLimate Impacts on Oceanic TOp Predators
CMT	Configuration Management Environment
CNMI	Commonwealth of the Northern Mariana Islands
CoML	Census of Marine Life
CRC	Cooperative Research Center
CREIOS	Coral Reef Ecosystem Integrated Observing System
CRM	Coastal Resources Management
CSIRO	Commonwealth Scientific and Industrial Research Organization
CZM	Coastal Zone Management
DAWR	Division of Aquatic and Wildlife Resources (Guam)
DEQ	Division of Environmental Quality
DFW	Division of Fish and Wildlife (CNMI)
DLNR-DAR	Department of Land and Natural Resources—Division of Aquatic Resources
DMWR	American Samoa Department of Marine and Wildlife Resources
DOH	Department of Health
DOI	Department of Interior
DPW	Department of Public Works
EAFM	Ecosystem Approach to Fisheries Management
EAM	Ecosystem Approach to Management
EBA	Ecosystem-Based Approaches
EBFM	Ecosystem-Based Fisheries Management
Ecopath	Modeling program establishing a mass-balanced snapshot of the ecosystem
Ecosim	Time-dynamic simulation module for policy exploration
EEZ	Exclusive Economic Zone
EGT	Ecosystems Goal Team
EIS	Environmental Impact Statement
ENSO	El Niño Southern Oscillation
EPA	Environmental Protection Agency
EPAP	Ecosystem Principles Advisory Panel
ESSW	Ecosystem Social Science Workshop
EwE	Ecopath with Ecosim
F/FMSY	Fishing Mortality at Maximum Sustainable Yield

FAD	Fish Aggregating Device
FFA	Forum Fisheries Agency
FISH	Fishery Information Survey and History
FMP	Fishery Management Plan
FSM	Federated States of Micronesia
GBR	Great Barrier Reef
GBRMPA	Great Barrier Reef Marine Park Authority
GIS	Geographic Information System
GLOBEC	Global Ocean Ecosystem Dynamics
HARP	Hawaii Archipelagic Ecosystem Research Plan
HDAR	Hawaii Division of Aquatic Resources
HRS	Hawaii Revised Statutes
IAI	Impact Assessment, Incorporated
IATTC	Inter-American Tropical Tuna Commission
IBM	Individual-Based Modeling
ICES	International Council for the Exploration of the Seas
ID	Identification
IFQ	Individual Fishery Quota
IIS	Institute of Island Studies (through the University of Prince Edward Island)
IKONOS	Commercial earth observation satellite
IOOS	Integrated Ocean Observing System
ITQ	Individual Transferable Quota
IUUF	Illegal, Unregulated, Unreported Fishing
JCU	James Cook University
JIMAR	Joint Institute for Marine and Atmospheric Research, University of Hawaii
kg	kilogram
lb	pound
LIDAR	Light Detection and Ranging technology
LMEs	Large Marine Ecosystem
m	meter
MHI	Main Hawaiian Islands
MIS	Management Information System
MMA	Marine Managed Areas
MMCA	Managaha Marine Conservation Area (Commonwealth of the Northern Marianas)
MPA	Marine Protected Area
MRFSS	Marine Recreational Fisheries Statistics Survey
MSE	Management Strategy Evaluations
MSFMCA	Magnuson-Stevens Fishery Conservation and Management Act
MSFOR	Multi-Species Forward (projection model)
MSVPA	Multi-Species Virtual Population Analysis
MSY	Maximum Sustainable Yield
NAO	North Atlantic Oscillation
NB	Nota Bene (Latin: Note well)
NEPA	National Environmental Policy Act
NESDIS	National Environmental Satellite, Data, and Information Services

NGO	Nongovernmental Organization
nm	nautical mile
NMFS	National Marine Fisheries Service
NOAA	National Oceanic and Atmospheric Administration
NOS	National Ocean Service
NPFMC	North Pacific Fishery Management Council
NRC	National Research Council
NRDA	National Research Damage Assessment
NS	National Standard (under the Magnuson-Stevens Act)
NSF	National Science Foundation
NWHI	Northwestern Hawaiian Islands
NWS	National Weather Service
OAR	Office of Oceanic and Atmospheric Research
OPC	Ocean Policy Council
OSCAR	Ocean Surface Current Analyses—Real time
PDOI	Pacific Decadal Oscillation Index
PFRP	Pelagic Fisheries Research Program, University of Hawaii
PICES	North Pacific Marine Science Organization
PIFSC	Pacific Island Fisheries Science Center
PIRO	Pacific Islands Regional Office
PIT	Policy Implementation Tool
PNA	Parties to the Nauru Agreement
PPI	Office of Program Planning and Integration
PSR	Pressure-State-Response
QDPI	Queensland Department of Primary Industries
REAC	Regional Ecosystem Advisory Committee
RFMC	Regional Fishery Management Council
RMI	Republic of the Marshall Islands
ROP	Republic of Palau
ROVs	Remotely Operated Vehicles
SAFE	Stock Assessment and Fishery Evaluation
SCUBA	Self-Contained Underwater Breathing Apparatus
SEK	Scientific Ecological Knowledge
SFA	Sustainable Fisheries Act
SIA	Social Impact Assessment
SocMon	Global Socioeconomic Monitoring Initiative for Coastal Management
SPC	Secretariat of the Pacific Community
SSC	Science and Statistical Committee (within each Fishery Management Council)
SSP	Social Sciences Plan (within National Ocean Service)
SST	Social Sciences Team
TEK	Traditional Ecological Knowledge
TOGA	Tropical Ocean Global Atmosphere
TURF	Territorial Use Rights in Fisheries
UCSC	University of California at Santa Cruz
UH	University of Hawaii
UNC	University of North Carolina

USCG	United States Coast Guard
USCOP	United States Commission on Ocean Policy
USDA	United States Department of Agriculture
USFWS	Unites States Fish and Wildlife Service (also abbreviated as FWS)
VMS	Vessel Monitoring System
WCPFC	Western and Central Pacific Fisheries Commission
WG	Working Group (within GLOBEC CLIOTOP)
WPacFIN	Western Pacific Fishery Information Network
WPRFMC	Western Pacific Regional Fishery Management Council
WTFCC	Western and Central Pacific Fisheries Commission

Section 1
Ecosystem Science and Planning

Jared Makaiau, Paul Dalzell, Gerard Dinardo, Charly Alexander, Svein Fougner, and Dirk Zeller

Chapter 1
Introduction to Ecosystem Science and Planning

Much has been written in recent years about the prospective benefits of an ecosystem approach to fisheries management (EAFM) in the United States and abroad. The Reauthorized Magnuson–Stevens Fishery Conservation and Management Act calls for expanded attention to ecosystem principles in fishery conservation and management actions, and such principles have been advocated by the US Commission on Ocean Policy and the Pew Ocean Commission. The potential role of ecosystem theory in assessing environmental effects under the National Environmental Policy Act (NEPA) has been considered for many years.

Recognizing the potential benefits of an ecosystem-based approach to management, the Western Pacific Regional Fishery Management Council (WPRFMC), one of the nation's eight regional fishery management councils, has progressively applied ecosystem principles to management of fisheries under its jurisdiction. As part of its planning process for undertaking EAFM, the Council convened a series of workshops through which to exchange information and learn from experts in various fields related to marine ecosystem research and analysis.

The first such workshop was held in 2005 at the Council offices in Honolulu, Hawaii. The workshop focused on the science and data needed to support application of ecosystem principles to fisheries planning and management across the Western Pacific. Experts among an international community of fishery scientists and managers were invited to present and engage in discussions about their work, experiences, and views on EAFM and related topics. This chapter presents the results of this important event.

Background

Over the past decade, fishery managers have moved away from a single-species approach to management, and toward a more holistic approach that addresses fishery interactions with other components of marine *systems*. This shift was evident in the 2006 reauthorization of the Magnuson–Stevens Act, which included a requirement for fishery management plans (FMPs) to incorporate considerations of essential fish habitat, which was defined as "waters and substrate necessary for spawning, breeding, feeding, or growth to maturity." FMPs were required to "describe and identify essential fish habitat for the fishery, minimize to the extent practicable adverse effects on such habitats caused by fishing, and identify other actions to encourage the conservation and enhancement of such habitat." The Reauthorized Act also contained a new National Standard for bycatch (NS-9), which was defined as "fish which are harvested in a fishery, but which are not sold or kept for personal use, including economic

Ecosystem-Based Fisheries Management in the Western Pacific, First Edition. Edward Glazier.

discards and regulatory discards." FMP conservation and management measures were required to "minimize by-catch and to the extent by-catch cannot be avoided, minimize the mortality of such by-catch."

Moreover, the Reauthorized Act also included the establishment of an Ecosystem Principles Advisory Panel (EPAP) to expand the application of ecosystem principles in fishery conservation and management. Following the directives of the Magnuson–Stevens Act, this Panel completed a report to Congress in 1999. In 2003 and 2004, the Pew Ocean Commission and the US Commission on Ocean Policy both advised the National Oceanic and Atmospheric Administration (NOAA) National Marine Fisheries Service (NMFS) to adopt ecosystem approaches to management. It was clear that the next reauthorization of the Magnuson–Stevens Act would likely include a requirement for the Regional Fishery Management Councils to prepare fishery ecosystem plans (FEPs). The Western Pacific Council prudently began preparations to move from FMPs to FEPs, and, in fact, was experienced in the process, having implemented the nation's first ecosystem-based FMP, (for management of coral reef ecosystems), in 2004.

Workshop Goal and Objectives

The overarching goal of the initial workshop of the series was to identify information that would be needed to support EAFM in the Western Pacific Region. Meeting this goal required a series of objectives and tasks, as follows:

1. Review existing ecosystem models and their applications and management of marine resources around the Pacific.
2. Identify fisheries management requirements in the Western Pacific.
3. Identify the best suite of quantitative ecosystem indicators and associated tradeoffs to support EAFM in the Western Pacific.
4. Within the confines of existing mandates such as the Magnuson–Stevens Act and the National Marine Sanctuaries Act, identify the most effective short-term application of EAFM that can be implemented based on current data; and in this context, address whether the precautionary approach has a role.
5. Identify new data or models that would be required to advance the ecosystem approach in the Western Pacific.
6. Identify changes in policy that would be required to effectively implement the new approach in the region.

Organization

The workshop was organized into three sessions (Figure 1.1): (1) an initial plenary session addressing data, models and ecosystem indicators; (2) a breakout session, in which three working groups would deliberate on the respective topics; and (3) a final plenary session consisting of reports from the three working groups and overall synthesis of workshop findings. In summarizing workshop findings, the working groups were asked to develop both short- and long-term recommendations for implementing EAFM in the Western Pacific.

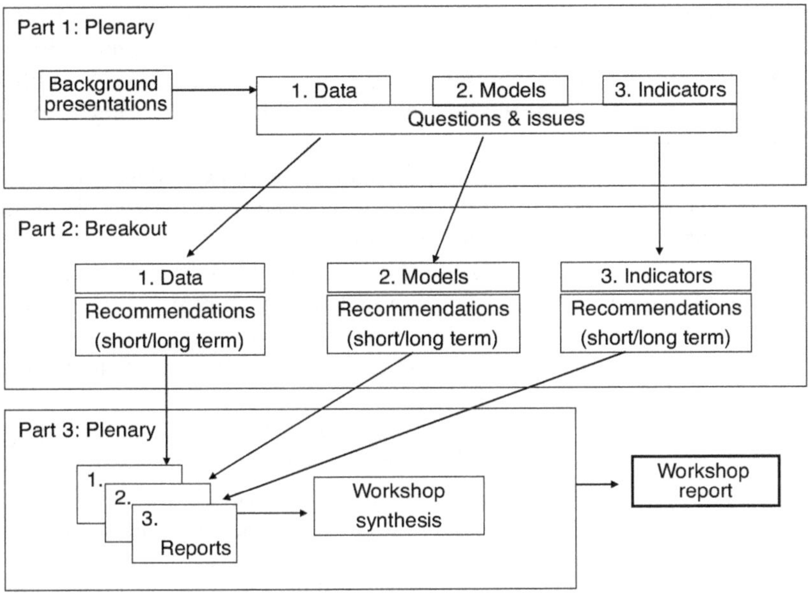

Figure 1.1. Workshop structure and outline.

A panel of experts was tasked to lead discussions in the plenary and working groups. The expert panel involved the following individuals:

Jerald Ault	University of Miami, Rosenstiel School of Marine and Atmospheric Science
Villy Christensen	University of British Columbia, Fisheries Centre
David Fluharty	University of Washington, School of Marine Affairs
Mike Fogarty	NOAA, NMFS, Northeast Fisheries Science Center
Neil Gribble	Queensland Dept of Primary Industry, Northern Fisheries Center
Steve Murawski	NOAA, NMFS, Office of Science and Technology
Carl Walters	University of British Columbia, Fisheries Centre

During the closing session, panelist David Fluharty and workshop participant Michael Orbach of the Duke University Marine Lab presented a synthesis of the workshop and recommended that social science and policy workshops subsequently be convened to further assist the Council in implementing EAFM in the region.

Opening Discussion

Kitty M. Simonds, Executive Director, Western Pacific Regional Fishery Management Council

Ms Simonds welcomed participants and opened the series of workshops with the following discussion:

> "Aloha kakahiaka. E komo mai. Good morning and welcome." We have before us this week a daunting but exciting task. It was almost 20 years ago that National

Marine Fisheries Service (NMFS) convened the first ecosystem workshop with the Regional Councils, producing the NMFS Program Development Plan for Ecosystem Monitoring and Fisheries Management. In 1999, NMFS published ecosystem principles for fisheries management, which several of you present today were responsible for developing. We at the Western Pacific Regional Fishery Management Council began work that year on our Coral Reef Ecosystem Fishery Management Plan, the first fishery management plan in the United States to take an ecosystem approach to management.

All councils have, in one way or another, integrated ecosystem principles in their Fishery Management Plans. This council is fully committed to the ecosystem approach, and we are now in the process of converting all of our existing species-based fisheries management plans to archipelagic fishery ecosystem plans.

Fishery managers, scientists, and policy-makers have been deliberating on the ecosystem approach for years, including talks at the recent "Managing Our Nation's Fisheries II" conference. Support for the concept has remained strong, but so many important factors—such as data needs, modeling requirements, and effective ecosystem indicators—have yet to be determined.

It is our hope that the discussions that take place this week will leave us with some level of consensus on these critical issues. Undoubtedly, we may not all see eye to eye on all of the details. But it is important that we reach some basic agreement so that we can move forward, and through an adaptive approach, improve management of the region's fisheries as our understanding grows.

As we debate these issues, let us be reminded of the Hawaiian proverb: I ka ʻōlelo ka make, i ka ʻōlelo ke ola. That is, "in the word is death, in the word is life." This proverb relates to the importance of speech in old Hawaiian. Everyone should be mindful, speak honestly and compassionately, and truly mean and be responsible for his or her discourse.

In closing, I hope you have an enjoyable and productive workshop and that the outcomes of your discussions will greatly enhance and improve the management of fisheries in our islands. E kuahui like ka hana! That is, "let everybody pitch in and work together!"

Opening Presentations

Principles of an EAFM, with an Update from the Workshop on "Ecosystem-Based Decision Support Tools for Fisheries Management"

Steve Murawski, NOAA, NMFS Headquarters

The foundation of an effective EAFM involves the application of proven approaches for managing the major fisheries in a designated region. Added to this are considerations of the impacts of fisheries on nontarget species, the effects of fishing on habitats and predator–prey dynamics, and relationships between the biota and the physical environment. Given the nature of marine ecosystems, relationships between species and their environment are typically highly complex. Management strategy evaluation (MSE) is a modeling approach that enables scientists to test for the effects of alternative relationships among species in relation to harvesting scenarios.

A problem managers have encountered in implementing EAFM is the lack of a "road map" for integrating an ecosystem approach into a working system of governance. In an integrated EAM, those in the governance system examine various kinds of information to develop management measures which achieve strategic goals. This requires taking into account the perspectives and interests of a variety of stakeholders. An effective EAFM also requires a comprehensive ecosystem observing system to collect data at various spatial and temporal scales, and a management decision support system that can: (a) synthesize the information, (b) develop status indicators for components of the system in question, (c) forecast trends in those components, and (d) evaluate the biological, social, and economic effects of policy decisions.

One of the most important challenges in EAFM is how to link high-level principles, such as maintaining healthy and productive ecosystems, to informative performance indicators (Figure 1.2). Unfortunately, aside from mortality rates and population size, there are no established criteria for determining proper reference levels for ecosystems. Additionally, quantification of the social costs and benefits associated with a given management measure is typically absent.

The literature on EAFM suggests that eight broad categories of operational objectives should be considered in developing FEPs.

1. *Conserving and managing the species:* currently, fishery management policies require managers to conserve target and nontarget species, including endangered and threatened species, and minimize adverse effects on habitat. Moreover, there is increasing interest in how management can be strengthened to protect biodiversity.
2. *Minimizing bycatch:* The principal rationale for minimizing bycatch is to limit waste.
3. *Managing tradeoffs:* tradeoff analysis must examine the costs and benefits of management options among fishing and nonfishing sectors to optimize fishery benefits and to prevent depletion and undue transfer of effort to alternative fisheries. Management actions should also be equitable and transparent, address cumulative impacts, and evaluate impacts on nonfishery sectors.

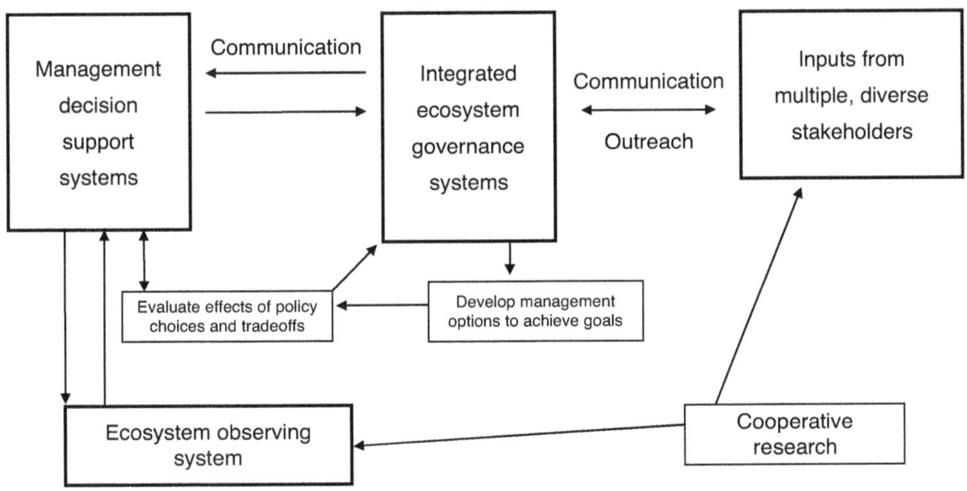

Figure 1.2. Components of an integrated ecosystem approach to management.

4. *Account for feedback effects:* this includes trophic interactions (predator–prey relationships), the effects of fishing on habitat productivity, potential irreversibility of fishing impacts, and fishing-induced habitat effects.
5. *Establish appropriate ecosystem boundaries:* this is a challenging concept for fishery managers, as ecosystems can be defined on multiple scales depending on the problem one is dealing with.
6. *Maintain ecosystem productivity and balance ecosystem structure:* in principle, one should identify the properties of an ecosystem in a balanced state. However, ecosystems are dynamic and even if impacts were managed properly, a "balanced" ecosystem is not a foregone result.
7. *Account for climate variability:* both low-frequency (decadal) and high-frequency (year-to-year) climate variability can exert profound effects on marine ecosystems. Management measures should err on the side of caution to account for this uncertainty.
8. *Use adaptive approaches to management:* currently, our knowledge of marine ecosystems is often quite limited. High degrees of uncertainty require that managers proceed with caution and experiment with different strategies to increase knowledge of the systems in question and the effectiveness of prospective management measures.

In February 2005, NOAA held a workshop in Key Largo, Florida, to examine the status of various ecosystem management approaches. The workshop focused on four topics: (1) the status of ecosystem information; (2) ecosystem components and functional relationships; (3) indicators and reference points; and (4) models and forecasts (Figure 1.3). The workshop also involved discussion of the importance of decision support tools for ecosystem approaches to fisheries management (see www.st.nmfs.gov/st7/ecosystem/workshop/2005).

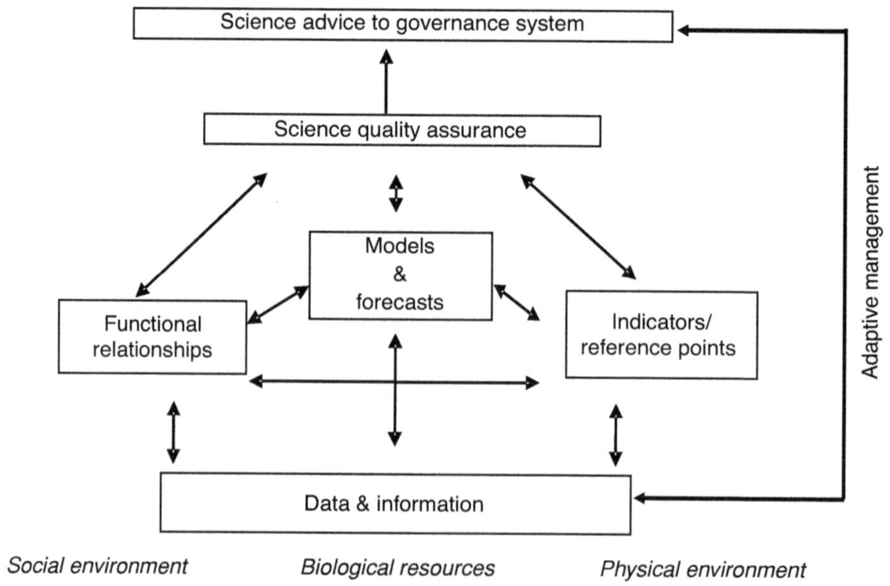

Figure 1.3. Decision tools supporting ecosystem approaches to fisheries management.

Decision support tools are important in that they have the capacity to link governance decisions to real data, indicators, models, and socioeconomic analyses. They can also help diverse stakeholder groups understand the likely consequences of alternative management actions and the tradeoffs of management objectives. With respect to functional relationships in marine ecosystems, participants observed the following:

- Specification of functional relationships among ecosystem components is an essential precursor to the development of predictive ecosystem models.
- Determining the nature of functional relationships among ecosystem components require evaluation of time series information and adaptive management experiments.
- Ecosystem modeling must specify functional forms of bottom–up and top–down forcings in system dynamics.
- Incorporating increased dimensionality leads to changes in management reference points, and increases uncertainty in predicted outcomes.
- Uncertainty regarding the effects of management measures on complex ecosystems requires risk assessment and the application of the precautionary principle.
- While we cannot readily conduct controlled and replicated ecosystem-level studies, the prospects are good for learning about ecosystem control mechanisms from intersystem comparative studies as well as from intrasystem time-series studies.
- Evaluation of functional relationships for key components of an ecosystem may be facilitated using statistical approaches to guide formulation and parameterization of ecosystem-level models.
- Testing of ecosystem models is currently focused on fitting time-series data using log-likelihood criteria while evaluating the relative impact of multiple drivers, such as climatic factors, nutrient loading, habitat modification, and fisheries effects.
- Related activities are in progress worldwide, including research and monitoring at several NOAA agencies and fishery councils. Results indicate that most ecosystems are impacted by multiple drivers.
- Formal evaluation of how assumptions about functional relationships impact the predictive capabilities of ecosystem models for fisheries management is required.
- Reliable evaluation requires information collected over long-time periods of time; data-contrast is essential for testing alternative hypotheses about functional response.
- Such testing calls for back-calculation and typically requires expanding time series data regarding abundance and fishing pressure for multiple ecosystem components, including noncommercial species—this calls for data scavenging activities.
- Multiple modeling tools should be used to test various ecosystem relationships.
- Ecosystem modeling should be designed to generate reliable directional guidelines as part of a strategic management approach and adaptive management scheme.
- Testing alternative hypotheses may call for adaptive management experimentation, notably where time series information shows little contrast. Examples include experiments with: closed areas, exploitation rates, stock enhancement, and habitat modification.

With respect to indicators, Key Largo workshop participants determined that:

- Indicators are measures of an ecosystem characteristic or process of interest that may be selected by scientists, stakeholders, or policy-makers.

- Numerical values change directionally in association with the direction of change in the characteristic or process of interest.
- Indicators should be relatively convenient to measure using available data; they should also be comprehensible by different audiences, such as scientists, decision makers, interest groups, and educators.

With respect to current use of indicators, the Key Largo participants noted the following:

- Indicators are increasingly important aspects of EAFM.
- Biological indicators are more widely used than socioeconomic indicators.
- Lists of indicators have been assembled to address multiple dimensions of EAFM.
- Correlative relationships may be useful in determining the status of ecosystems.
- Indicators should not be treated as management "performance targets" without understanding direct mechanistic linkages.
- Most indicators are not yet usable as reference points.
- Empirical use of status indicators, such as biomass, as a partial function of fishing pressure can help specify thresholds or Limit Reference Points.
- Empirically based indicator thresholds need further development but can be used now to establish intermediate decision criteria.

A working group of the International Commission for the Exploration of the Sea reviewed a variety of indicators that should be applied to EAFM. The group asserted that indicators should be (a) easy to understand, (b) responsive to human activities, (c) linked to specific management actions, (d) easily and accurately measured, (e) relatively unresponsive to other factors, (f) measurable over large areas, and (g) based on existing data that can help in the understanding of historic ecosystem dynamics.

The group also determined that in terms of selection metrics for indicators, considerations should be given to: (a) relevance, (b) cost of monitoring, (c) frequency of need for reevaluation, (d) strength of underlying signal, (e) precision, (f) accuracy, and (g) level of sensitivity to multiple drivers.

The current state of ecosystem modeling appears sufficiently advanced for use in EAFM. Various multispecies and ecosystem models have been used to approximate the complex and dynamic interrelationships among biotic and abiotic processes occurring over a range of spatial and temporal time scales (Table 1.1). Simple models can be used to show the relative consequences of alternative management strategies. For the purposes of ecosystem-based fishery management, a classification scheme is available to characterize operational models that have been used in various locations. This includes multispecies interaction models, bycatch models, and age-structured trophodynamic models. Each require different data, and it is typically the case that data gaps limit applicability in real-time management settings.

Ecosystem modeling requirements are universal. Specific biotic, abiotic, and socioeconomic data are needed to construct useful and relevant ecosystem models. Characterizing uncertainty is vital for interpreting model output and for providing advice on the likely consequences of alternative management actions. Methods from other disciplines may be useful for EAFM modeling. While not typically considered in traditional fisheries management, these include: fuzzy logic, credibility theory, artificial intelligence, gaming theory, network analysis, decision theory, and a variety of theories and methods used in economics.

Table 1.1. Categories of single-species (SS), multispecies (MS), and ecosystem (ES) models.

Model type	Biological interactions	Predator-prey feedback	Environmental and/or lower trophic levels	Age structure	Spatial structure
MS fishery technical interaction models				×	×
SS models with unidirectional drivers	×		×	×	×
MS production models	×	×			
Age/size-structured MS models	×	×		×	×
Aggregate trophodynamic ES models	×	×	×		×
Age/size-structured trophodynamic ES or individual-based models	×	×	×	×	×

The interplay between ecosystem modeling and policy is very important. Research and management actions need to be well coordinated in order to better understand ecosystem dynamics. Factors associated with participatory decision making must also be considered in EAFM, as concerns about the status of ecosystems are increasingly voiced by the general public.

Finally, there is a need to provide adequate support for ecosystem-related research, modeling, and monitoring. There is need for emphasis on assessing ecosystem goods and services, and identifying appropriate metrics to account for nonconsumptive ecosystem services that must have an equal footing in decisions regarding market-based ecosystem goods and services.

What Goes Wrong with Ecosystem Models as Tools for Policy Design?

Carl Walters, Fisheries Centre, University of British Columbia, Canada

A wide range of information is needed to make well-informed marine policy decisions. Maximizing the ultimate efficacy of EAFM requires a variety of data and predictive models. These vary in terms of level of bias, empiricism, and scope of coverage. Such models include the following:

- Intuitive assessment about the likely direction and magnitude of response in ecosystem dynamics;
- Dogmatic and simplistic concepts about "natural" systems being better systems;
- Straightforward models that emphasize major interactions and dependencies; and/or
- Detailed models that seek to capture all relevant factors.

The recent increase in ecosystem modeling activity has produced numerous models that fit historical data and generate reasonable policy predictions. However, habitat and environmental changes, including those associated with fishing pressure and displacement or removal of fishing pressure, are creating novel and highly complex situations for fishery managers working in real-time settings. Current understanding of ecological response is insufficient for predicting the outcomes of such situations. Indeed, "details" that are difficult to incorporate in a given model may have large effects on the modeled ecosystem. For example, factors associated with small particulates can dramatically affect climate models; compensatory behaviors associated with habitat changes can significantly affect fish habitat models; and "minor" prey consumed in small space-time windows can significantly affect trophic dynamics.

Application of ecosystem models to fishery issues has induced many challenges and few great successes. The not-so-wonderful score is subjectively indicated in Table 1.2.

Unfortunately, thus far, the predictive approaches seem to be more or less failing, for reasons that include the following:

- Lack of long-term monitoring data on nontarget species and their life stages;
- Concentration of trophic and habitat interaction effects during early life stages (recruitment); these are difficult to monitor;
- Confounding of fishery, environmental, and trophic effects in historical data;

Table 1.2. Score card for using models to address ecosystem management questions.

Issue	Grade	Comment
Bycatch impacts	A	We are reasonably successful at predicting direct effects of fishing in general.
Top–down effects (predator culling or protection)	C	Trophic effects of fishing can be classified as "top down" or "bottom up" with respect to where management controls are exerted.
-on valued prey	B	Changes in M for prey species already subject to assessment.
-on "rare" prey	F	Outbreaks of previously rare species.
Bottom–up effects (effects of prey harvesting on predator stocks)	C	Uncertainty here is about flexibility of predators to find alternative food sources when prey are fished.
Multiple stable states	B	Cultivation–depensation mechanism appears to be main mechanism that could cause "flips."
Habitat damage	D	Lack of understanding about real habitat dependencies, bottlenecks.
Production regime changes	C	Models look good when fitted to data, but have not stood test of time.
Selective fishing practices/policies	F	We have not yet looked closely at options in this area.
Regime shifts	C	Policy adjustments in response to ecosystem-scale productivity change.

- Failure to anticipate "new problems" (emergent novelty) associated with unpredictable changes in system structure, such as exotic invasions, new fishing technology, and unexpected biotic population changes; and/or
- Unpredictable preadaptations to altered habitats.

Most ecosystem models are age-structured. However, some older models, such as the multispecies virtual population analysis model and the multispecies forward projection model focused on mortality rates for senescent age cohorts, for which data are more readily available, while treating recruitment rates as "black boxes" as in most single-species models. Newer models, such as Ecosim, attempt to represent: interaction effects at all life stages, mortality rates, habitat effects, and other factors affecting the recruitment component of system dynamics. All models inevitably require real data for verification and testing.

One hope has been that ecosystem models will somehow perform better than single-species assessment procedures, presumably by accounting more realistically for sources of population variation. In reality, this has occurred in only a few situations; generally, the single-species prescriptions are very similar, since they are based on the same kinds of data. It has also been hoped that ecosystem models will resolve uncertainties about the importance of fisheries effects, such as those related to harvest of incidental species, habitat damage, displacement of fishing effort to new fisheries and regions, and so forth. Such questions cannot be effectively examined with single-species assessment models.

Attempts to test ecosystem models against historical data have revealed potentially serious weaknesses that will likely be resolved only through adaptive management experiments. Thus, modeling exercises should be considered aids in experimental design, rather than trustworthy prescriptive tools. These weaknesses range from lack of information about major components of the system, such as small mid-trophic level fishes that have not been monitored in the past, to the "emergent" impacts of species that were previously too rare to study. Furthermore, we can typically explain historical data equally well with a set of different models/hypotheses that make widely different policy predictions, especially in cases where historical effects of fishing and environmental change cannot be clearly separated.

Another impediment to development of useful ecosystem models has been the inability or unwillingness of fisheries managers to clearly specify the policy changes that would be needed to establish an ecosystem-based management program. In the absence of clear guidance about what policy options should be compared, modelers either have typically tried to answer the wrong questions or have chosen to model outcomes for certain objectives only, such as protection of nonconsumptive values associated with biodiversity, for instance.

For example, assuming that the catch reconstruction undertaken by Zeller et al. (see later data section) is generating valid results, then governance entities in the Western Pacific will need to address a variety of inshore fishery issues in the years to come. More models, assessments, or data collection may not be the answer. Following are the options for dealing with this problem:

- Ignore it (e.g., wait for bioeconomic equilibrium at a more severely degraded state);
- Band Aid it (e.g., the approach taken in Florida: ongoing size and bag limits to discourage fishing, and numerous small marine protected area (MPAs); or

- Face it (e.g., establish moratoria until stocks recover, reopen small areas/times of particular subsistence concern, initiate development of rights-based fishing systems).

In moving towards ecosystem-based management, some clarity and imagination may supersede the practical utility of models. Ideas include the following:

- Regulations favoring selective fishing, such as numerical quotas that create incentives to avoid taking smaller fish.
- Pulse rotation closures of sufficient spatial and temporal magnitude to prevent erosion of biodiversity.
- Formalized territorial use rights in fisheries (TURF) and localized monitoring and enforcement responsibilities.
- Ecosystem-scale monitoring technologies.

Of special mention should be the observation that, particularly when the behavior of fishermen is included, ecosystem models often predict counterintuitive policy effects, even in terms of the direction of response. Thus, in some ways, policies may be seen as experimental treatments, the understanding of which should be incorporated as an adaptive management aspect of EAFM.

A word of caution to managers: If you want to get useful results from ecosystem models, be very careful to tell scientists exactly what policy options you want to examine otherwise, scientists will answer the wrong questions. In this sense, ecosystem modeling for adaptive management would require an approach in which:

- Modelers deliberately attempt to uncover alternative models that explain historical data but indicate unexpected policy outcomes.
- Policy options involve deliberate, diagnostic management experiments aimed at identifying the utility of alternative models (e.g., using spatial closures to test recovery predictions or selective culling to test trophic interaction effects).

High uncertainty in ecosystem model predictions implies a clear need to treat all ecosystem management initiatives as adaptive management experiments. The problem lies in implementing such experiments because of the following:

- People who fervently believe that they already have the answers (i.e., know what policies to use);
- People who refuse to embrace uncertainty in policy design because they fear loss of credibility or authority;
- People who fear any risk of policy failure or do not understand the need for consideration of alternatives;
- The high cost of monitoring experimental responses, and/or high risks of investing in new technologies and institutional arrangements (e.g., data gathering by fishermen) that could reduce monitoring costs.

Group Discussion

Workshop attendees raised various questions about the kinds of adaptive management experiments that would be needed to explore the utility of models for EAFM. Carl Walters asserted that "real MPAs" should be considered, arguing that MPAs large enough to be ecologically integrated at an archipelagic scale could indicate challenges for which experiments in stakeholder buy-in and participation, enforcement, and monitoring could provide valuable insight.

Steve Murawski added that large MPAs would also be helpful for building scientific knowledge in that if an MPA was established in an overfished area, and a proper monitoring effort was initiated, it was likely that a change in population dynamics would be observed. He noted that experience elsewhere has shown that such contrasts can occasionally be detected relatively quickly (i.e., within a few years) but generally may take many years or decades. Thus, it is prudent to incorporate the long-term time horizon into planning, governance, monitoring, and enforcement, and to ensure stakeholder understanding of the temporal dimension of an EAFM.

The Western Pacific Region and Its Fishery Management Plans: How Far from Ecosystem-Based Fishery Management?

Paul Dalzell, Western Pacific Regional Fishery Management Council

In 1976, the United States Congress passed the Magnuson Fishery Conservation and Management Act (the Magnuson Act), which established eight quasi-federal regional fishery management councils to manage fisheries in the Exclusive Economic Zone (EEZ) surrounding the United States. The Western Pacific Council is one such entity, and it is the policy-making body for the management of fisheries in the EEZ adjacent to the Territory of American Samoa, Territory of Guam, state of Hawaii, the Commonwealth of the Northern Mariana Islands and the US Pacific island possessions of Jarvis, Johnston, Wake, Howland and Baker Islands, Kingman Reef, and Palmyra and Midway Atolls (Figure 1.4). This 1.5 million nm^2 region, is far and away the largest

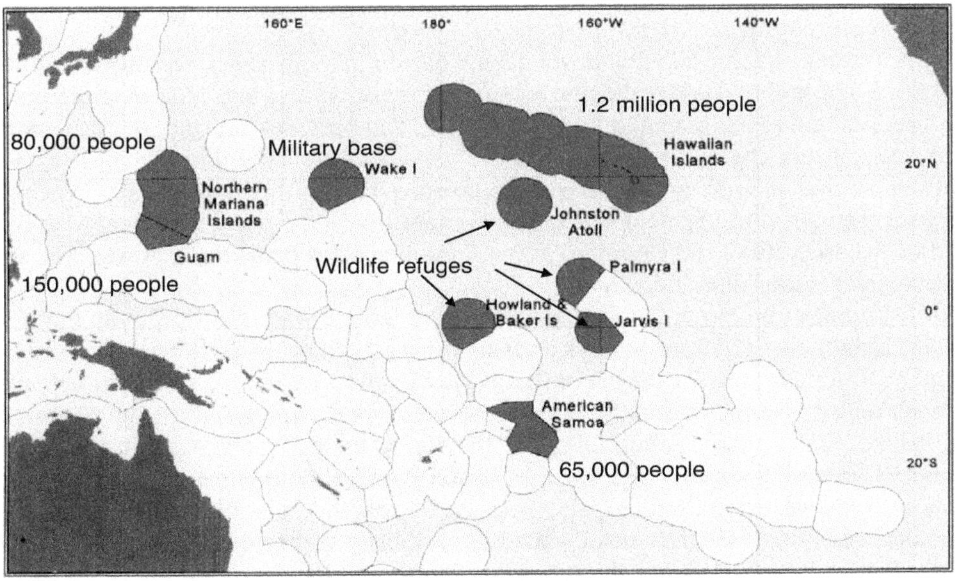

Figure 1.4. US and adjacent EEZs of the Western Pacific. (For a color version of this figure, see Plate 1.)

area managed by any of the regional fishery management councils and comprises about half of the total EEZ of the U. S. The area spans the Pacific dateline and the equator.

The original goals of the Magnuson Act were to phase out foreign fishing in the US EEZ and to replace it with domestic fisheries. The Act has achieved its original goal; however, the expansion of domestic fisheries has in turn led to some of the problems that were faced through the 1980s and 1990s, with the depletion of some fishery resources and interactions with protected species.

The Magnuson Act, reauthorized in 2006, and is now known as the Magnuson–Stevens Fishery Conservation and Management Act. Reauthorization laid the groundwork for ecosystem-based fishery management by reinforcing requirements to end overfishing, rebuild overfished stocks, minimize bycatch, and protect habitat. The Reauthorization also laid out 10 national standards to guide the future of fisheries management in the US EEZ.

Every FMP and plan amendment developed by one of the regional fishery management councils has to address the 10 national standards: (1) prevent overfishing while achieving optimum yield; (2) use the best scientific information available; (3) manage fish stocks as units; (4) when implementing conservation and management measures, avoid discrimination between residents of different states; (5) consider efficiency; (6) take into account variation and contingencies; (7) minimize costs and avoid duplication; (8) consider human communities; (9) minimize bycatch and mortality; and (10) promote safety of human life at sea.

The Reauthorized Act also laid out a process through which primary stakeholders can play a substantial role in managing fisheries and resources in their respective areas. The Councils involve participation of fishermen, scientists, public officials, and others knowledgeable of specific given fisheries. Council members are appointed by the Secretary of Commerce for 3-year terms. The Western Pacific Council has 16 members, three of whom are nonvoting federal agency representatives. Current membership includes 13 voting members, five of whom are local and federal fishery agency representatives.

The Western Pacific Council develops policies to manage fishery resources in the US EEZ across the Western Pacific Region, prepares FMPs and plan amendments for fisheries and resources needing management, and provides a forum for discussion and decision making. The Council has designed and implemented five FMPs: (1) bottomfish and seamount groundfish, (2) crustaceans, (3) precious coral, (4) coral reef ecosystems, and (5) pelagics. Given the nature of the Western Pacific region and its fisheries, all of the Council FMPs are multispecies plans that address species assemblages rather than individual species.

Most domestic commercial fishing activities in federal waters surrounding the US Pacific islands are variations of hook and line fishing. In Hawaii and American Samoa, longline fisheries dominate the fishing industries. Domestic commercial fishing activity in each region also includes handlining for large snappers and groupers on the outer reef slope, and trolling, handlining, and longlining for pelagic species in the open ocean. The now dormant trap fishery for spiny and slipper lobsters based in the remote Northwestern Hawaiian Islands (NWHI) was the only lobster fishery of any significance in the United States. Pacific islands lobster fishing continues in the Main Hawaiian Islands (MHI) state waters.

Mention should also be made of precious coral harvesting, which have historically taken place in federal waters around Hawaii, and currently occurs in state waters.

Harvesting of precious corals in Hawaii is conducted with SCUBA gear in shallow waters, and with submersibles in deeper water.

Bottomfishing is conducted throughout the region, but is only of economic significance in Hawaii, where bottomfish represent a small fraction of total landed value of all catches. Most bottomfish grounds in American Samoa, Guam, the Northern Mariana Islands, and the MHI are within the 0–3 nautical mile (nm) zone and, thus, fall under state or territorial jurisdiction. However, the NWHI represents a substantial area of bottomfishing grounds, and there are significant bottomfishing banks and seamounts in the MHI, which lie within or extend into federal waters. The four island groups also have a variety of small-scale inshore fisheries for reef fish, mostly within the 0–3 nm zone. The Council's Coral Reef Ecosystem FMP addresses reef fishing conducted in federal waters.

The Bottomfish and Seamount Groundfish FMP, implemented in 1983 manages a multispecies complex of snappers (*Lutjanidae*), groupers (*Serranidae*), emperors (*Lethrinidae*), jacks (*Carangidae*), and several species of seamount groundfish. The plan prohibits destructive fishing techniques, including explosives, poisons, trawl nets, and bottom-set gillnets; establishes a moratorium on the commercial harvest of seamount groundfish stocks at Hancock Seamount; and implemented a permit system for fishing for bottomfish around the NWHI.

In 1988, the Council developed regulations that divided the NWHI into two fishing zones: the Mau and Ho'omalu Zones (Figure 1.5). Access to the Mau Zone is limited to 10 permits, two of which are reserved for indigenous communities. Available permits are issued to fishermen based on past participation in the MHI and/or NWHI bottomfish fisheries. Access to the more distant Ho'omalu Zone is limited to seven permits, and entry is through accumulation of points through fishing in the MHI or

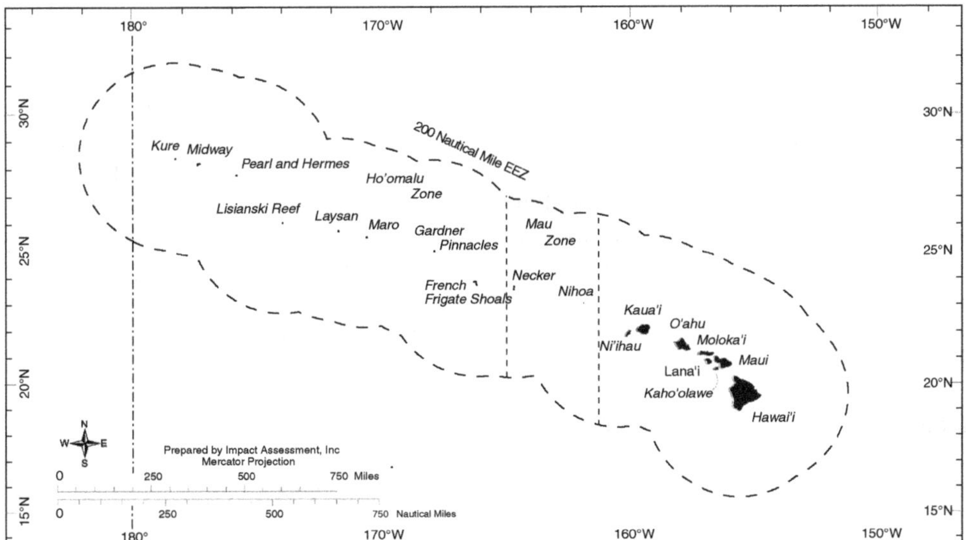

Figure 1.5. The Mau, Ho'omalu, and Main Hawaiian Islands Management Zones.

Mau Zone. Fishermen who have permits to fish in one zone may not fish in the other zone and must meet minimum annual landing requirements to retain their permits.

In addition to the deep-slope fisheries in the MHI and NWHI, a trawl and bottom longline fishery targeting Alfonsin (*Beryx* spp.) at the southeast Hancock Seamount in the NWHI and the Emperor Seamount Chain was initiated by Russian and Japanese fishing vessels in the late 1960s. After 10 years of high levels of productivity, overfishing caused the fishery to collapse. A moratorium on the harvest of Alfonsin on the Hancock Seamounts has been in effect since 1986 in an effort to rebuild the stock. The moratorium was initially in effect until 2004 but has been extended. Periodic reviews of the stock indicate that recovery has not occurred.

The NWHI lobster fishery developed in the late 1970s with several commercial vessels that had relocated from the Pacific Northwest where crustacean overfishing was occurring. By the mid-1980s, the NWHI lobster fishery was Hawaii's most lucrative fishery.

Under the Council's Crustacean FMP, implemented in 1983, 20-mile closed areas around Laysan Island and 0–10 fathoms depth closures throughout the remaining NWHI were established as a spawning refuge zone and to protect monk seals. Regulations required that traps deployed in the NWHI lobster fishery be equipped with juvenile escapement panels. The Council also amended the Crustacean FMP to specify the maximum dimensions of the trap funnel entrance needed to minimize the risk the traps posed to monk seals. In 1990, lobster catch rates declined dramatically, likely due to climate-induced change in oceanic productivity throughout the NWHI, which also affected the abundance of reef fish, seabirds and Hawaiian monk seals. The decrease in lobster catch prompted the Council to establish a limited access program and fleetwide seasonal harvest quotas that significantly altered fishing operations.

Vessels concentrated on trapping lobsters on the banks around Necker Island, Gardner Pinnacles, and Maro Reef during the derby-style fishing season. From 1992 to 1997, Necker Island accounted for 48–64 percent of total effort. In 1998, the quota was allocated among four fishing areas to prevent localized depletion of the lobster population at the most heavily fished banks, and to encourage fishermen to broaden the geographic distribution of their efforts.

The lobster harvest guidelines are an example of implementation of the precautionary approach to fisheries management. The guidelines allowed only a 10 percent risk of overfishing to the total exploitable population. Initially, a minimum size limit of 5 cm tail width for spiny lobsters and 5.6 cm width for slipper lobsters was established, along with a ban on the retention of berried females. However, observations on the mortality of discarded lobsters, both on deck and through predation, led to a Council decision in 1996 to permit a "take all" fishery in which all lobsters brought on deck were counted against the annual quota. The Hawaii lobster fishery landed 261,000 pounds with ex-vessel revenue of $1.2 million in 1999, which was the last year the fishery was active.

The majority of the vessels participating in this fishery voluntarily deployed a satellite vessel monitoring system (VMS) for vessel location tracking and daily catch reporting. This allowed managers to monitor the progress of the fishery through real-time reporting of catch and enabled immediate notice when annual quotas were reached.

While calculating the year 2000 estimates for the exploitable population of lobsters in the NWHI, NMFS scientists expressed alarm at the increasing level of uncertainty in their computations. The scientists also noted a lack of appreciable rebuilding of lobster populations, despite significant reductions in fishing effort throughout the NWHI. Given these concerns, the Council recommended that NMFS close the NWHI lobster fishery as a precautionary measure, until stock status is better understood.

The Council's Precious Corals FMP was approved in 1980 and regulations for the fishery were determined in 1983. The plan established a permit requirement, harvest quotas for separate beds, a minimum size limit for pink coral, gear and area restrictions, and specific fishing seasons. In 1991, an amendment to the FMP defined a bed as overfished with respect to recruitment when the total spawning biomass for all species combined had been reduced to 20 percent of its unfished condition.

The Hawaii precious coral fishery includes two distinct sectors. One sector extracts deepwater (400–1,500 m) pink, gold, and bamboo corals. This fishery historically employed dredges and tangle nets to extract the precious coral, but the Council now requires selective methods, such as remotely operated vehicles and submersibles. Starting in 1973, Maui Divers of Hawaii adopted the use of a manned submersible to commercially extract pink, gold, and bamboo coral at the Makapu'u bed. These operations were discontinued in 1978 due to high operating costs. In 2000, American Deepwater Engineering collected precious corals at the Makapu'u bed and in the Exploratory Area of the EEZ around the MHI. New precious coral beds continue to be discovered in the archipelago, with new beds identified in both the MHI and NWHI during 2002 and 2003.

The second sector of the Hawaii precious coral fishery, which occurs predominantly in state waters, involves hand-collecting black coral using SCUBA gear at depths of 30–100 m. Since the inception of the black coral fishery in Hawaii in the late 1950s, fewer than 10 operations have been active in the fishery at any one time. Harvest levels of black coral have fluctuated widely over the past four decades, reflecting changes in demand. During the 1970s, the state of Hawaii drafted regulations requiring a minimum height of 48 inches, the estimated minimum size for maintaining maximum sustainable yield. In the 1990s, the state advanced regulations to implement the size restriction. Between 1990 and 1997, the annual harvest of black coral in Hawaii ranged between 846 and 6,017 lbs, with an annual average of 3,084 lbs.

Much of the basic biology and distribution of deepwater precious corals is poorly understood. Furthermore, there is controversy over the degree of adverse effects an NWHI precious coral fishery would have on populations of the endangered Hawaiian monk seal. Although monk seals have been observed preying on eels found among precious coral colonies, the importance of eels in the monk seal diet is unknown.

The Council's 2001 FMP for Coral Reef Ecosystems in the Western Pacific Region was the first ecosystem-based fisheries plan developed in the US. It incorporates many of the principles and policies recommended by the NMFS EPAP. The goal of the FMP is to establish a management regime that will maintain sustainable coral reef fisheries for the entire western Pacific Region, while preventing adverse impacts to stocks, habitat, protected species, and the ecosystem. To achieve this goal, the FMP implements several management measures, including: (a) designation of zoned MPAs; (b) permit and reporting requirements to fish in designated low-use MPAs (reporting of fisheries information in non-MPA areas will continue through locally administered monitoring systems), and if needed, a general permit program for all EEZ reef fisheries; and (c) a prohibition on nonselective/destructive use of gear, and conditions on the types and uses of allowable gear.

The central feature of the Coral Reef Ecosystem FMP is adaptive management, which recognizes the uncertainty, changing conditions, and resilience associated with coral reef ecosystems. Strategies for managing for coral reef ecosystems have allowed Pacific Islanders to survive for millennia and are best viewed as adaptive responses that have evolved over time, not as mere traditions.

The Pelagics FMP, implemented in 1986, is designed to manage tuna, billfish, pelagic sharks, wahoo, mahi, mackerel, gempylids, and pomfrets. The FMP banned the use of drift gillnets in the EEZ and established a limited entry permit and reporting program for longliners in Hawaii and American Samoa. The Council has periodically considered actions to limit entry for handline fisheries in Hawaii as a means to reduce overfishing of bigeye tuna. The FMP also prohibits longline fishing within 50 miles of the NWHI to prevent interactions with monk seals and other protected species. The FMP further closes large areas around the MHI and American Samoa to longlining to eliminate gear interactions between longline and small vessels engaged in handline and troll fishing.

A significant component of the pelagics fisheries management regime is the conservation of sea turtles and seabirds. The Council's Protected Species Program includes several turtle conservation projects at nesting beaches and foraging grounds on the Pacific Rim and also testing and implementing turtle and seabird mitigation strategies.

In working toward incorporating ecosystem-based management principles, the Council will manage fishery resources on an archipelagic basis. Under this approach, FEPs are developed for each area under the Council's jurisdiction. Within each FEP, benthic-associated fishery resources (i.e., bottomfish, crustacean, coral reef ecosystems, and certain precious coral management unit species) are managed together. Although pelagic resources are part of the ecosystems of each archipelago, these will be managed independently under a single Pelagic FEP. This is based on considerations of the larger international issues involved in this fishery.

In conclusion, of immediate importance from the Council's perspective, is the need to gain an understanding of the relationship between managing fisheries on an ecosystem interaction level rather than managing fisheries by traditional single species management approaches. Particular emphasis will be placed on the development of limit reference points, indicators, and performance standards for fishery regulations. Recognizing that ecosystems are neither static nor predictable, the basic tenets of ecosystem-based management must be founded on a precautionary basis within an adaptive management approach.

The Western Pacific Council's Coral Reef Stock Assessment and Bottomfish Stock Assessment Workshops

Gerard DiNardo, National Marine Fisheries Service, Pacific Islands Fisheries Science Center

In early 2004, the Western Pacific Council convened two workshops to address the need for stock assessments of coral reef fish and deep-slope bottomfish for the Western Pacific Region. Gerard DiNardo discussed the objectives and outcomes of the workshops and implications for a regional shift to EAFM.

The coral reef and bottomfish stock assessment workshops were convened to develop a plan to improve data collection and assessment methodologies. Expert panels were assembled for each workshop. Short-, medium-, and long-term recommendations were generated during each workshop and are briefly summarized below:

Short-term recommendations

- Define near- and long-term management objectives and appropriate management units.
- Review, synthesize, and analyze extant biological, ecological, and fishery data relative to management needs, and identify data gaps and analytical shortcomings.
- Based on gap analysis, design and implement fishery-dependent data collection programs and fishery-independent research/monitoring programs at appropriate spatiotemporal scales in the Pacific Islands region.
- To ensure compatibility, data should share sufficient information in consistent formats to allow straightforward linking among data sets.
- Complete habitat inventory with measures of habitat suitability, and map the location of habitats and associated biota.
- Review existing stock assessment models and reassess them under a variety of different scenarios.
- Acquire funding and hire or contract the personnel needed to complete the various work tasks.

Mid-term recommendations

- Combine data on common species collected in the bottomfish and coral reef fisheries for purposes of assessment.
- Advance assessment tools through application of spatially structured models.
- Develop operational models for evaluating valid and useful approaches to stock assessment and management.
- Conduct pilot programs to assess the feasibility of using hydroacoustics to survey bottomfish populations.

Long-term recommendations

- Assess metapopulation structure of bottomfish stocks and recruitment connectivity between the MHI and NWHI.
- Investigate possibility that groupers aggregate to spawn in EEZ waters of the Western Pacific region.
- Review existing stock assessment models.

Group Discussion

A question was raised regarding the state of knowledge about interconnectivity between the MHI and NWHI, and whether it was unidirectional. Mr Dalzell stated that the general belief is that there may be an east/west movement, that is, from the MHI to the NWHI. However, efforts to model the dispersal of lobster larvae by Jeff Polovina at the Pacific Islands Fisheries Science Center (PIFSC) have indicated that interconnectivity is bi-directional, and could be influenced by vertical migration and other larval behavior patterns. Connectivity is indicated by taape (*Lutjanus kasmira*), which was introduced into the MHI, but had dispersed as far northwest as Kure Island. With respect to larval dispersal, it appears that in the NWHI, lobster larvae move both northwest and southeast along the chain. This may influence source- and sink-opportunities for stocks exploited in the MHI.

Jeff Polovina added that two factors may influence connectivity between the NWHI and the MHI. First is the surface flow that moves northwest along the Hawaiian ridge in response to Eckman transport and the predominant north-east tradewinds. The deeper geostrophic flow—part of the subtropical gyre, flows predominantly southeast. Also important are eddies generated around the archipelago; these typically move from east to west across the ridge and could retain larvae for some time.

It was suggested that concerns about connectivity between populations should not be restricted to connectivity within the Hawaiian Archipelago. For example, genetic analysis of Hawaiian bottomfish stocks suggests linkages with stocks around Johnston Atoll.

A question was raised about the next steps needed to further an ecosystem perspective within the context of the Council's multispecies FEPs; for example, by taking into account species interactions or regime shifts. Mr Dalzell responded that the Pelagics FMP was the most advanced, since NOAA's PIFSC has traditionally focused on pelagic research, and thus, the transition from the Pelagics FMP to the Pelagics FEP would probably be the most expedient.

It was suggested that one of the major problems encountered in fisheries assessment is the effort required to develop a sufficient knowledge base to undertake a new approach with confidence. Extensive effort is required to undertake archival data-detective work, and this was likely to increase when shifting to an EAFM. Sam Pooley noted that if NOAA was going to undertake ecosystem-based modeling, assessment, and management, then it would be the responsibility of PIFSC and associated scientists to undertake the scientific work needed to move in this direction.

Chapter 2
Data Sources

Fishery-Dependent Data

The biophysical workshop involved discussion of fishery-specific information needed to develop EAFM in the Western Pacific. Both biophysical and economic data were addressed. The discussions focused on gaps in understanding of historic populations of commercial and recreational fish stocks, data and analytical needs, and other important issues.

Fishery-Dependent Data for Ecosystem Management

Kurt Kawamoto, National Marine Fisheries Service (NMFS), Pacific Islands Fisheries Science Center

The NMFS Pacific Island Fisheries Science Center currently monitors four principal fisheries in Hawaii: the pelagic longline fishery, NWHI lobster fishery (currently closed), precious corals, and the bottomfish fishery. The longest running dataset with detailed size data was originally designed for economic data collection purposes. These data are collected through the public fish auction in Honolulu, the primary venue for the commercial transfer of seafood in the state. These data have been collected since at least 1948. In 2001, the State of Hawaii's Commercial Marine Dealer's Report replaced previous economic monitoring instruments.

Data and information (Table 2.1) related to the longline and bottomfish fisheries are provided through four sources: logbooks, the auction house, seafood dealer reports, and the at-sea observer program (see Russell Ito's presentation on pelagic longline fisheries, p. 33). By virtue of a data-sharing agreement between the state of Hawaii and National Oceanic and Atmospheric Administration (NOAA), NMFS analysts also have access to all state datasets through the Western Pacific Fisheries Information Network (WPacFIN, www.pifsc.noaa.gov/wpacfin/), which is housed at the PIFSC in Honolulu.

Monitoring instruments and datasets are provided for both the federally managed NWHI lobster fishery and the state-managed lobster fishery in Figures 2.1 and 2.2 as examples of data collections and monitoring in the region. Similarly, fishery monitoring instruments for the federally managed NWHI bottomfish fishery are presented in Figure 2.3. Approaches for detailed monitoring of biological information are illustrated in Figure 2.4.

Ecosystem-Based Fisheries Management in the Western Pacific, First Edition. Edward Glazier.

Table 2.1. Inventory of fishery-dependent data for Hawaiian archipelago.

Responsible agency	Dataset name	Details
Federal (FMP reporting requirements)	Pelagic longline	Daily longline logbook
		Transshipment logbook
	NWHI lobster fishery	NWHI crustacean daily logbook
		Trip sales report
	Hawaii precious coral fishery	Daily logbook
		Sales report
	NWHI bottomfish fishery	Trip daily logbook[a]
		Trip sales report[a]
Federal (other datasets)	PIRO longline observer (mandatory)	Economic Data Survey (Trip)
	PIRO bottomfish observer (intermittent)	Economic Data Survey (Trip)
	Honolulu fish auction data	NWHI bottomfish (size and sales)
		Pelagic longline (size and sales)
		Other (size and sales)
State of HawaiI	Fisher's catch report (commercial license)	—
	Dealer's report	—
	Longline trip report	—
	Tuna handline trip report	—
	Aku boat report	—
	Baitfish report	—
	Albacore trolling trip report	—
	Crustacean trip report	—
	Deep sea handline trip report	—
	Net, trap, dive activity report	—
	Aquarium fish catch report	—
	Pond operator's monthly report	—
	NWHI bottomfish trip daily report[a]	—
	NWHI trip sales report[a]	—

[a]State form; Federal data requirement.

In summary, a variety of fishery-dependent datasets are available for use in ecosystem modeling and EAFM. However, the manner in which data are collected has varied over time, and data quality varies depending on the dataset in question. Therefore, the data must be used judiciously.

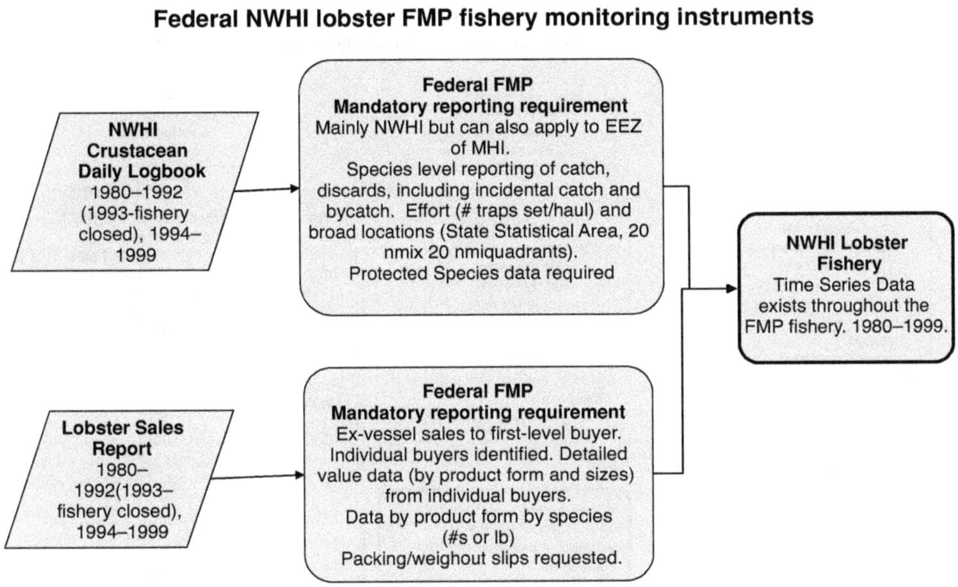

Figure 2.1. Fishery monitoring instruments and resultant datasets for the Federal NWHI Lobster FMP Fishery.

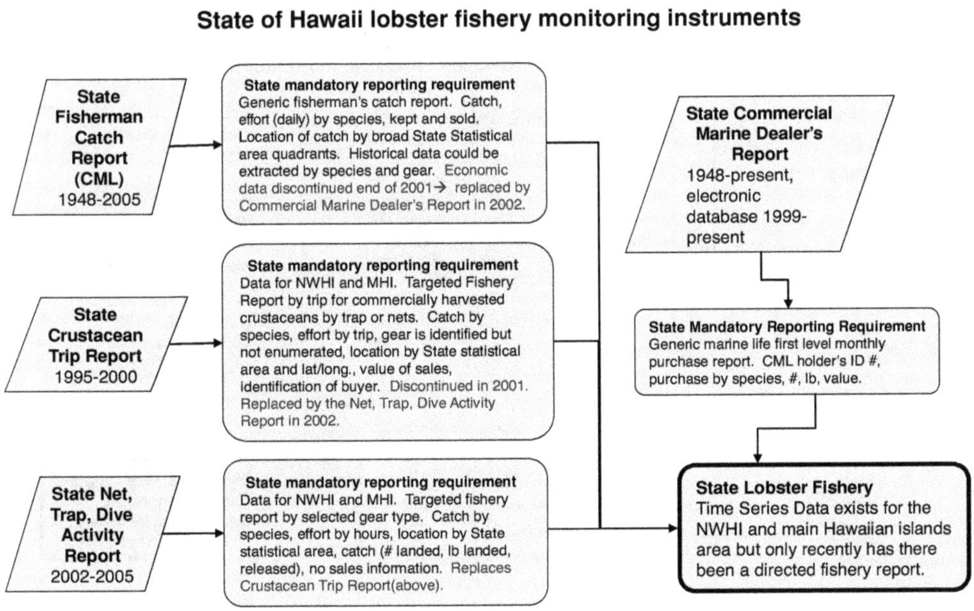

Figure 2.2. Fishery monitoring instruments and resultant datasets for the State of Hawaii Lobster Fishery.

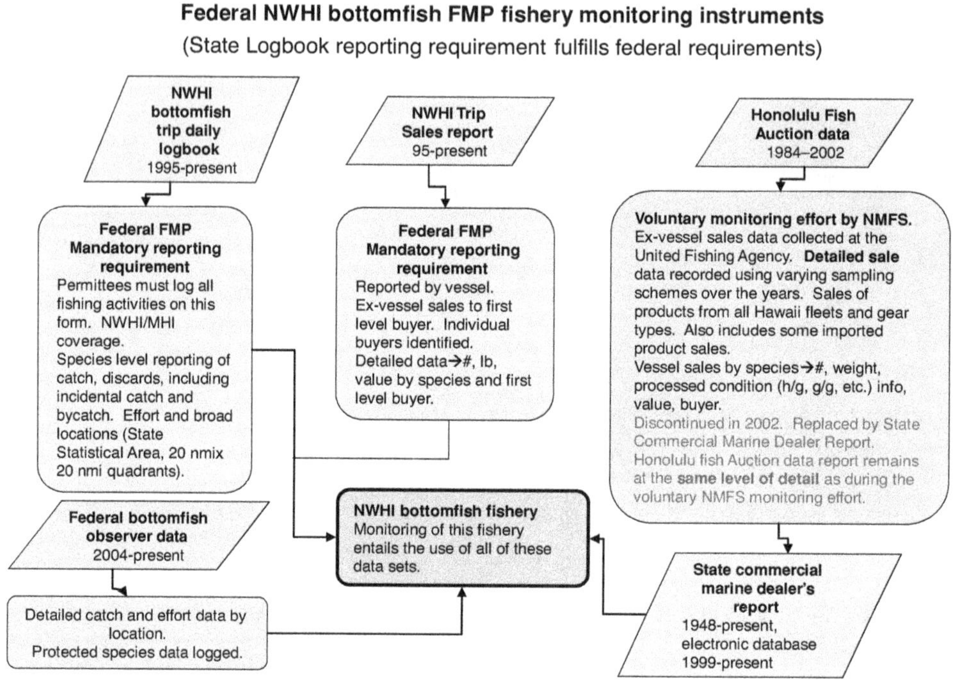

Figure 2.3. Fishery monitoring instruments and resultant datasets for the Federal NWHI Bottomfish FMP Fishery.

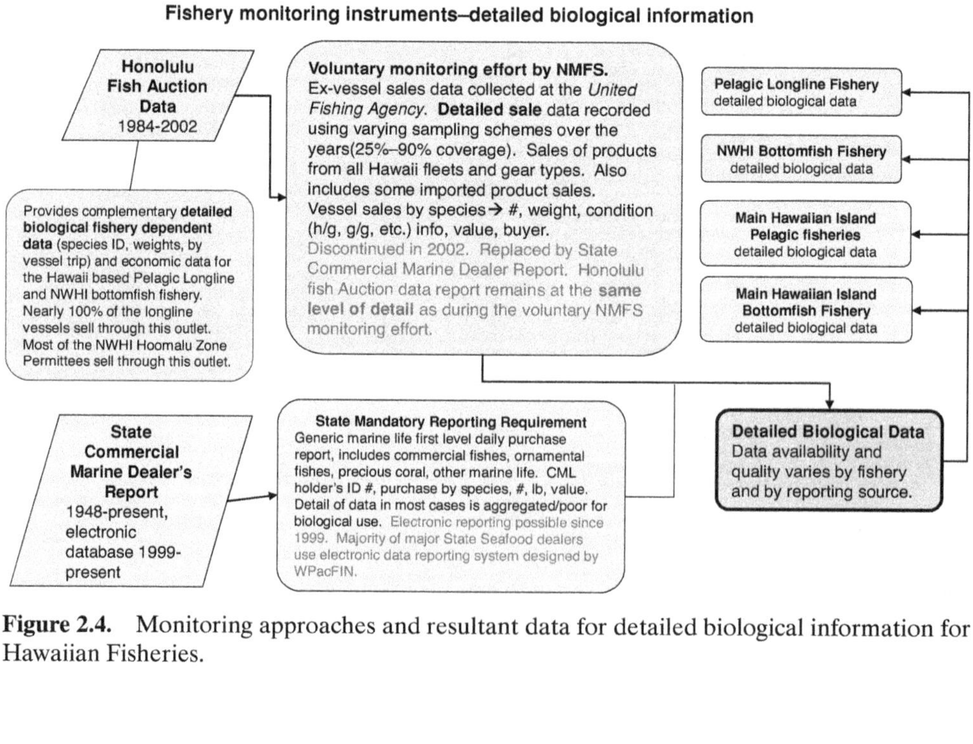

Figure 2.4. Monitoring approaches and resultant data for detailed biological information for Hawaiian Fisheries.

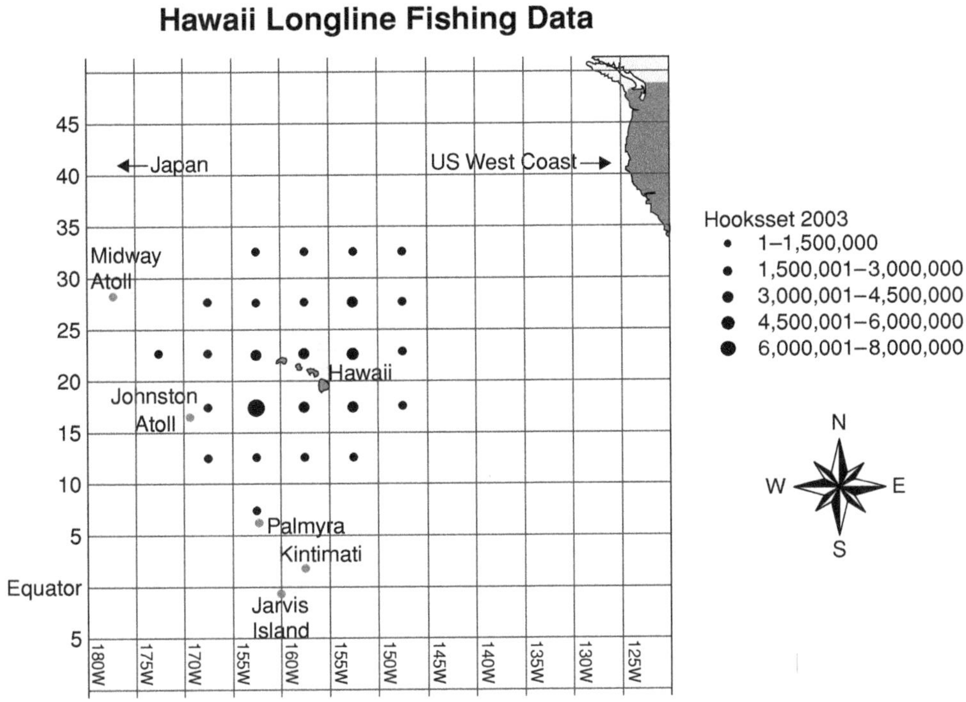

Figure 2.5. Example of the spatial distribution of effort (2003 Hooksets) of the Hawaii-based Pelagic Longline Fishery (www.pifsc.noaa.gov/fmep/charts/index.html).

Fishery-Dependent Data for the Hawaii-Based Longline Fishery

Russell Ito, NMFS, Pacific Islands Fisheries Science Center

The Hawaii-based pelagic longline fishery is, in economic terms, the most valuable fishery in Hawaii. The longline fleet fishes in waters close to Hawaii, and also beyond the EEZ (Figure 2.5). Fishery-dependent data for the Hawaii-based pelagic longline fleet derive from both federal and state sources, the details of which are compared in Table 2.2.

Fishery Dependent Data, WPacFIN

Michael Quach, National Marine Fisheries Service, Pacific Islands Fisheries Science Center

The Western Pacific Fisheries Information Network (WPacFIN) was established by NOAA in 1981, with the mission of bringing the best possible information to bear in fishery management decisions across the US Pacific Islands. A Key objective is to assist agencies in establishing and maintaining monitoring programs by providing data collection design, data system design and development, data analysis, training in all aspects of fisheries monitoring, and offsite data storage and management.

Table 2.2. Comparison of Federal/State Fishery-dependent data collection mechanisms: Hawaii-based longline fishery.

		Federal			State	
	Logbook	Fish auction	Observer	Catch report (C-3)	Trip report (C-5)	Fish dealer report
Item						
Key taxa	Macro pelagic	Macro pelagic	Macro pelagic	Macro pelagic	Macro pelagic	Macro pelagic
Geographic region	Pacific	Pacific	Pacific	Pacific	Pacific	Pacific
Sampling interval	Daily	Random day	Daily	Monthly	Trip	Date of sale
Duration	1990–present	1987–2001[a]	1990–1994 (voluntary), 1994–present (mandatory)	1948–present	1948–2001[b]	1948–present
Parameters reported						
Date	Date of operation	Date of landing	Date of operation	—	—	—
Location	Latitude and longitude	—	Latitude and longitude	State statistics areas	Latitude and longitude	—
Effort	Hooks set	—	Hooks set	—	—	—
Biological data	Catch number	Number and weight	Length	Catch number and weight	Catch number and weight	Catch number and weight
Species						
	Protected spp. interactions	—	Protected spp. interactions	—	—	—
	—	Price and buyer of individual fish	—	Value and buyer	Value and buyer	Value, buyer, and fisher

[a]Superseded by state dealer data.
[b]Replaced by Federal data.

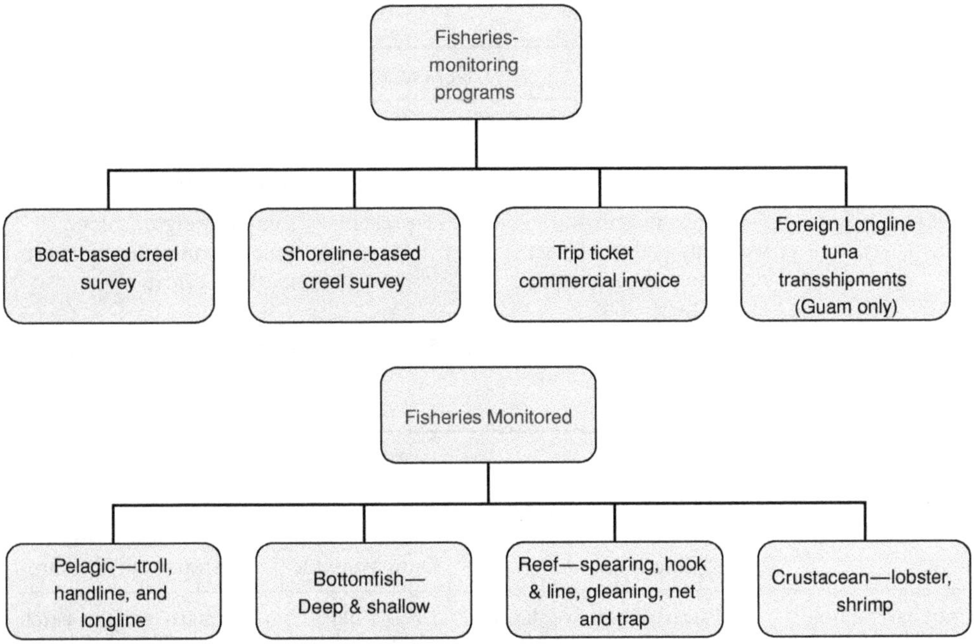

Figure 2.6. General components of the US Pacific Islands Long-Term Fisheries Monitoring Program developed and supported through WPacFIN.

Four agencies are partners in WPacFIN: (1) the Territory of American Samoa, Department of Marine and Wildlife Resources; (2) the Commonwealth of the Northern Mariana Islands (CNMI), Division of Fish and Wildlife; (3) the Territory of Guam, Division of Aquatic and Wildlife Resources & Bureau of Statistics and Plans; and (4) the State of Hawaii, Division of Aquatic Resources.

The overall approach of WPacFIN is to utilize the various data collection programs to capture best available fishery-dependent data needed for purposes of effective management (Figure 2.6). General information and non-confidential data can be accessed freely through the WPacFIN website at www.pifsc.noaa.gov/wpacfin.

Coverage, duration, and frequency of data programs differ among agencies, with little or no data available for earlier decades. Because Hawaii was covered in previous presentations, this presentation focused on American Samoa, CNMI, and Guam.

Boat-based creel surveys are conducted at ports, harbors, and access ramps, and data collection consists of two main components: (1) participation counts as measured by the number of fishing trips and (2) interviews with fishermen. Survey days are randomly selected and range from 3 to 8 days per month. Surveys are stratified by weekdays, weekend days, and day- and nighttime. Data expansion algorithms are based on port, type of day, and fishing methods used. Summary details are presented in Table 2.3.

The shoreline-based surveys focus on accessible shoreline areas, and data collection methods also involve participation counts and interviews. On Guam, aerial surveys are used to access the extent of fishing activity along less accessible shorelines. Survey days are randomly selected, and range from 2 to 4 times per week. Data expansion

Table 2.3. Boat-based creel survey data.

Island	Sampling interval	Data available	Parameters measured
American Samoa	3–4 times/week, weekday and weekend, day and night	1985–present	Length, weight, catch, species composition, gear, method, effort
CNMI	6–8 times/month, weekday and weekend, daytime only, nighttime as of March 2005	2000–present	Length, weight, catch, species composition, gear, method, effort
Guam	2–4 times/week, weekday and weekend, day and night	1985–present	Length, weight, catch, species composition, effort

Table 2.4. Shoreline-based creel survey data.

Island	Sampling interval	Data available	Parameters measured
American Samoa	3–4 times/week, weekday and weekend, day and night	1991–1994, 2002–present	Length, weight, catch, species composition, effort
Commonwealth of the Northern Mariana Islands	2–4 times/week, weekday and weekend, day and night	Begin 2005	Length, weight, catch, species composition, effort
Guam	2–4 times/week, weekday and weekend, day and night	1985–present	Length, weight, catch, species composition, effort

algorithms are based on island region, type of day, and fishing method. Summary details are presented in Table 2.4.

The Trip Ticket Commercial Invoice program is used to monitor sale of seafood products. Data derive from invoices submitted by fish dealers and by hotels and restaurants that purchase fish directly from the fishermen. Each invoice indicates daily catches. This program is mandatory only in American Samoa; it is voluntary on Guam and in the CNMI. Summary details are presented in Table 2.5.

On Guam, information about transshipment of tuna is documented via a census of offloading foreign longline vessels administered by the Guam Customs Authority. Transshipment forms document species, number of pieces, and weight of fish transshipped.

Group Discussion

This series of presentations described fishery-dependent datasets from Hawaii and other areas of the Western Pacific, including logbooks, creel surveys, and commercial data. Presenters described: how data are collected and maintained, how new datasets could be integrated to produce compatible historic datasets to support time-series

Table 2.5. Trip Ticket Commercial Invoice program details.

Island	Sampling interval	Data available	Parameters measured
American Samoa	Ongoing daily sale trip tickets	1990–present	Fisher, buyers, weight, price, value by species or species group
Commonwealth of the Northern Mariana Islands	Ongoing daily sale trip tickets	1983–present	Fisher, buyers, weight, price, value by species or species group
Guam	Ongoing daily sale trip tickets	1981–present	Fisher, buyers, weight, price, value by species or species group

analysis, and the importance of working with local data managers and biologists to ensure that data are consistent across the archipelago.

Discussion focused on the time depth of some of the data, cross-validation of contemporary data (e.g., logbook, observer, and dealer reports), and the overall quality of the data. Kurt Kawamoto and Russell Ito were asked if they could discuss any efforts that had been undertaken to validate certain data, to identify biases, or to examine completeness or accuracy. The presenters responded that there was cross-referencing across datasets at the federal level, that is, between logbook, observer, and market data. These sources were triangulated to verify whether fishermen were accurately reporting their catch. With regard to data quality, it was suggested that observer data were typically of the highest quality in terms of interactions with protected species. Problems associated with the identification of tuna species were discussed. Sam Pooley noted that the Hawaii auction data were relatively accurate in terms of species identification.

Carl Walters asked how comfortable PIFSC staff are with data regarding mean body weight, since these could help indicate mortality rates and whether size at capture has declined significantly since the 1950s. Fishing areas and technologies have changed significantly during recent decades. It was felt that such factors could confound analysis of mortality.

Use of VMS to monitor and record vessel location was discussed in terms of its potential for assessing spatial aspects of fishing effort in the region. At present, VMS is used for enforcement rather than research.

An observation was made that while there are data gaps, there are surplus data for others. Some of the older data may be at risk of being lost, and in some cases, it was suggested that establishment of a school of fisheries at the University of Hawaii could improve the pool of resources needed to collect, maintain, and analyze data needed for an EAFM in the region.

Resource and Habitat Data

This session focused on historic and recent efforts to characterize important physical and biological resources associated with marine ecosystems in the Western Pacific.

Multi-Disciplinary Monitoring, Mapping and Research Data for Conservation and Management of the Coral Reef Ecosystems of the US-Affiliated Pacific Islands

Russell Brainard, National Marine Fisheries Service, Pacific Islands Fisheries Science Center

Much of the work to map and characterize shallow benthic habitats and associated fish and invertebrate communities around the US Pacific Islands is funded by NOAA's Coral Reef Conservation Program and its numerous program partners. The program uses a multidisciplinary ecosystem approach to undertake the following: (a) mapping of benthic habitats and habitat utilization patterns; (b) long-term ecological monitoring of fish, coral, invertebrates, and algae; (c) long-term oceanographic monitoring, including hydrodynamic modeling and processes influencing ecosystems; and (d) study and modeling of ecological interactions, including biophysical and trophic interrelativity.

The program's benthic habitat mapping work involves work in deep and shallow water areas. A variety of tools are used, ranging from IKONOS satellite imagery to boat-based multibeam surveys. Ground-truthing involves use of remotely operated vehicles (ROVs) and standard towed-diver surveys. The approach is thematic, combining bathymetry, imagery, rugosity measures, and habitat classification. Mapping is integrated with ecological assessment via habitat utilization patterns and provides boundary conditions for coupled hydrodynamic–biological models. Mapping data are available via a geographic information system (GIS) web interface similar to NOAA's Ocean Atlas.

The program's ecological assessments are designed to meet seven principal objectives:

1. Implement complementary assessment methodologies (belt transects, stationary point counts, towed-diver/video surveys, rapid ecological assessments) for fish, corals, other invertebrates, and algae.
2. Systematically monitor reef resources to characterize natural spatial variation and temporal change.
3. Conduct inventories of key reef species as indicators of overall patterns of biodiversity.
4. Apply innovative ecological research techniques related to ecosystem-based management of reef-related resources.
5. Statistically interpret density and abundance estimates with reference to pertinent ecosystem parameters, including relevant natural and anthropogenic factors and processes.
6. Improve understanding of ecosystem linkages among organisms and across trophic levels.
7. Provide the scientific basis for (a) estimating locally sustainable harvest levels of reef resources, (b) placing and evaluating MPAs or other ecological reserves, (c) identifying ecosystem indicators, and (d) contributing to adaptive management.

Ecological assessment and monitoring utilize a variety of methods. These include: (a) *in situ* observation using traditional fish belt transects, roving diver surveys, photo quadrates, video transects, and towed-diver surveys; (b) bioacoustic surveys such as ship-based echosounder surveys and passive acoustic moorings; (c) bottomfish bait stations for fishery-independent bottomfish assessments; and (d) ocean observations via shipboard surveys, oceanographic moorings, and satellite-tracked drifters.

Prior to the present program, relatively little data was available to characterize the shallow water areas of the Western Pacific region, except at occasional intervals and localized scales. The overall objective of the NOAA Coral Reef Restoration and affiliated programs is to provide baseline data with which to develop models with the potential to improve understanding and management of reef fisheries across the region.

Ecological and Functional Studies: Snapshots in Time

Frank Parrish, NMFS, Pacific Islands Fisheries Science Center

Knowledge regarding functional aspects of groups or complexes of organisms such as corals and reef fish tends to have accrued in relation to small-scale, shallow water areas. Research has relied on cruises, field camps, remote sensing, collaboration with other projects and scientists, and use of archival data. Important data sources include the following:

- The NWHI Resource Assessment Survey of bottomfish, shrimp, lobster, and kona crab, conducted between 1975 and 1982.
- Shrimp and bottomfish assessment work, conducted in the Mariana Archipelago between 1982 and 1984.
- Renewed groundfish assessment work conducted around the Southern Emperor-Northern Hawaiian Ridge Seamounts between 1985 and 1988.

More recently, data were generated during the 1986–2004 NWHI Lobster Surveys, and the 1996–2000 subtropical Frontal Zone Annual Oceanographic Series. The NWHI Habitat Community Assessments undertaken between 1990 and 2004 examined reef, bank, slope, and subphotic communities. Associated field camps undertook studies of Monk seal foraging, seabird reproductive success, and trophic aspects of local reef fish populations.

An overall qualitative assessment of existing knowledge about ecosystem processes and interactions across the Hawaiian archipelago is summarized in Table 2.6. Ecosystem groups were ranked as good, moderate, or poor based on data available via the Pacific Islands Fisheries Science Center or the general scientific literature. In some cases, an abundance of historical data is available to support development and implementation of an EAFM in this region.

NWHI Monk Seal and Green Sea Turtle Demographics

Bud Antonelis, NMFS, Pacific Islands Fisheries Science Center

A number of monk seal studies have been conducted annually since 1984. Labor-intensive fieldwork has focused on populations at Kure, Laysan, Lisianski, Pearl, and Hermes Islands; French Frigate Shoals; and Midway Atoll. Investigations were undertaken at both subpopulation and archipelagic levels of analysis. Various research approaches have been used, including visual surveys, tagging, individual identification, and scat and spew analyses. Targeted parameters have included abundance,

Table 2.6. Qualitative knowledge bases by data type and species group for Hawaiian archipelagic ecosystem.

	Knowledge bases by data type						
Group	Trophic	Movement	Life history	Abundance	Spatial	Temporal	Energy flow
Nutrient flux	—	P	P	P	M	P	P
Phytoplankton	G	G	M	G	G	G	M
Zooplankton	M	M	P	M	P	M	P
Pelagic micronekton	M	P	M	M	M	P	P
Benthic algae	M	M	M	M	M	M	P
Corals	M	P	G	M	M	P	P
Infauna	P	P	P	P	P	P	P
Ocotpods	M	P	M	M	M	M	P
Macro inverts	M	M	M	M	G	M	P
Lobster	P	M	M	M	M	G	P
Shallow reef fish	G	G	M	G	G	G	M
Reef slope fish	M	P	M	M	M	M	P
Comm. bottomfish	M	P	M	M	M	P	P
Subphotic fish	P	P	P	M	M	P	P
Jacks	G	M	M	M	P	P	P
Small coastal pelagics	G	P	G	P	P	P	P
Sea birds	G	M	G	G	M	M	P
Turtles	G	G	G	G	G	G	M
Reef sharks	G	M	G	P	M	M	M
Monk seals	M	G	G	G	G	G	M
Tiger sharks	M	P	G	P	P	P	M
Small cetaceans	P	P	P	P	P	P	P
Large cetaceans	P	P	P	P	P	P	P

G, good; M, moderate; P, poor.
Knowledge base scale: good—multiple regional studies, plus suitable body of related literature from other locations; moderate—limited regional studies or literature from other locations as proxy; poor—no regional studies.

survival, fecundity, migration, body condition, causes of mortality, and composition of diet.

While monk seals have a protracted breeding season, much of the reproductive activity is concentrated in spring and summer. Fieldwork has been successful in large part due to the collaborative and cooperative efforts of many partners, including the US Fish and Wildlife Service and the State of Hawaii. While these investigations have required substantial effort over a long time period and have led to a better understanding of monk seals, in reality they have yielded relatively little information about certain important aspects of monk seal populations.

Green sea turtles are found throughout the Hawaiian archipelago. However, some 90 percent of nesting typically occurs at French Frigate Shoals in the NWHI, where data have been collected since 1973. Green turtles also appear to use regular migration patterns throughout the NWHI. Annual sampling has been undertaken at French Frigate Shoals and elsewhere in the NWHI, and periodic sampling has been undertaken at selected foraging sites in the Main Hawaiian Islands (MHI). Additionally, turtle basking data, collected incidental to monk seal research, have been collected since 1982. Methods used for sea turtle research have included nest surveys, tagging, individual identification, stranding investigations, foraging site surveys, and fishery by-catch data collection. The main parameters investigated as part of the sea turtle program include an index of population abundance, patterns of migration, somatic growth rates, causes of mortality, and composition of diet.

Group Discussion

Several questions were posed regarding the availability of recently collected data that could be used to characterize data. The group determined that new data are now often available on the internet, although some biological data take time to process and post. All but the most sensitive data, such as the location of endangered species, will eventually be posted for public access.

Scientists are tending to rely on historic data to characterize trophic interactions among many key species, but that new data could help unravel some of the compelling questions about ecosystem linkages. It was noted that examples include 1986–2004 data on spiny lobsters and available data regarding shark predation on monk seals. It was stressed that such data need to be "rescued" before they are lost because of decay of antiquated storage media such as old computer disks and paper. It was also noted that substantial data may be hidden in research papers at the University of Hawaii, and that there is an urgent need for such data to be sought out and reviewed for its potential contributions to an EAFM.

Oceanographic Data

A single presentation described efforts to integrate oceanographic observations into a web-enabled format for inventory and analysis. Both *in situ* measurements and remote sensing data were discussed.

Oceanography for Ecosystem-Based Research and Management

Russell Moffitt, NMFS, Pacific Islands Fisheries Science Center

The purpose of collecting oceanographic data is to improve understanding of the horizontal and vertical structure of the ocean. Although extensive data regarding major physical ocean phenomena and climate relationships are available on broad, ocean-basin scales (Figure 2.7), data regarding smaller scale, near-shore environments

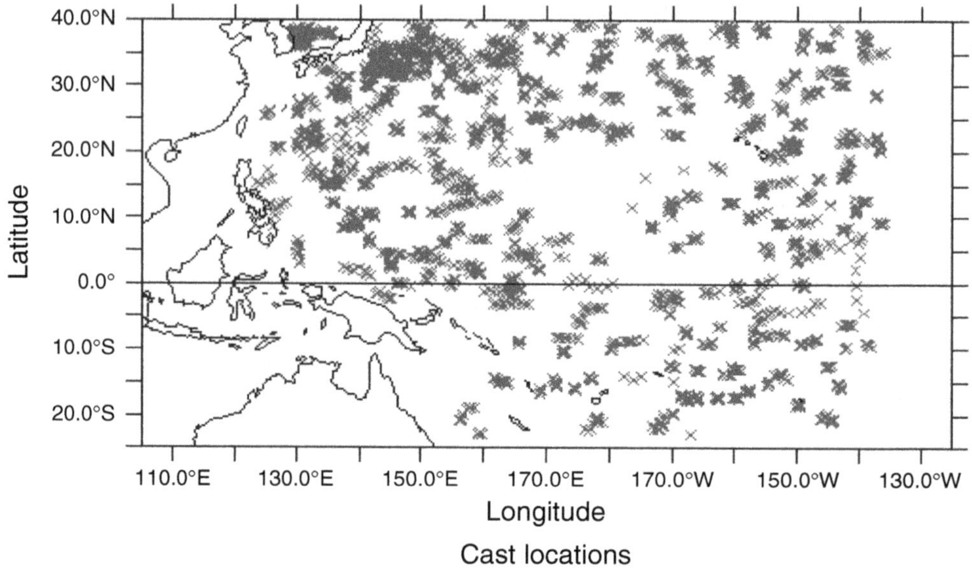

Figure 2.7. Example of spatial distribution of large-scale oceanographic data sampling throughout the Western Pacific Ocean.

is limited by a lack of localized measurement and monitoring. Because such data will be needed for an (Figure 2.8) archipelagic EAFM, an important next step will be to determine what specific types of oceanographic data are needed and at what level of resolution.

The principal oceanographic factors and processes currently being observed in the Western Pacific include the following: (a) temperature; (b) salinity; (c) dissolved oxygen and other gases; (d) concentrations of other chemicals and nutrients; (e) concentrations of organisms and organic material such as chlorophyll pigment; (f) flow dynamics associated with currents, tides, swells, and waves; (g) sea-surface height anomalies; (h) air–sea flux in relation to global climate; and (i) patterns of change in horizontal and vertical structures.

Three basic methods of modes are currently being employed to understand these factors and processes. These are: (1) *in situ* observations that are precise and available at a high level of resolution for vertical phenomena at specific sampling sites, but sparse across ocean basins and irregular over time; (2) remotely sensed observations with synoptic coverage and high temporal resolution, but which are restricted to surface layers and which involve measurement limitations for certain parameters; and (3) modeled parameters.

In sum, large-scale oceanographic data are available to indicate spatial and temporal variability of the physical oceanographic and climatic environment at a basin-wide scale of resolution. Data of a finer resolution will need to be developed for an effective archipelagic EAFM, but the appropriate level of resolution and the most sritical types of data need to be determined.

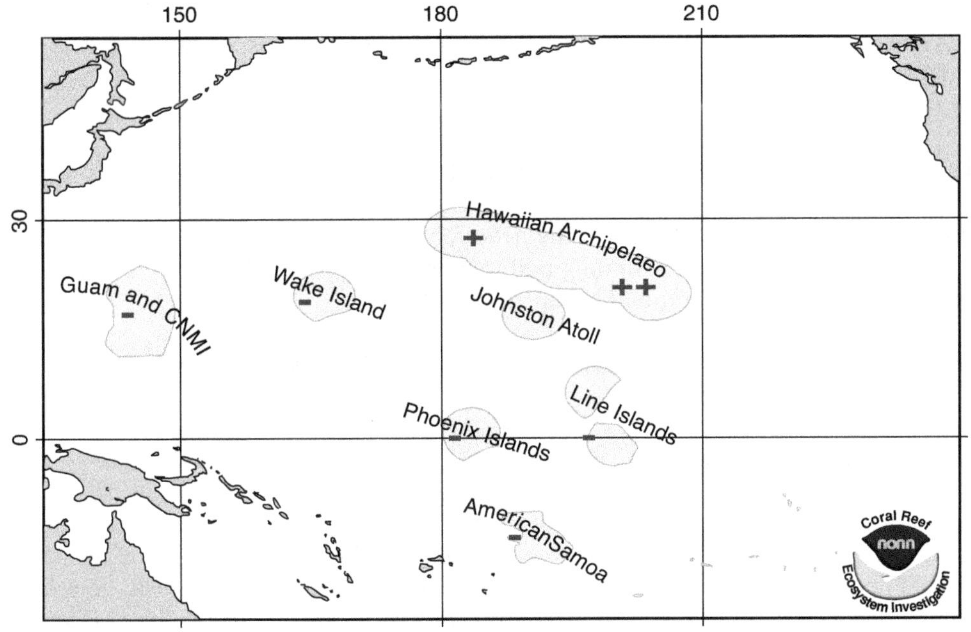

Figure 2.8. Examples of sampling locations for higher resolution data collection associated with NOAA's Coral Reef Ecosystem Investigation Initiative.

Group Discussion

Ensuing discussion focused the potential for using existing oceanographic models to generate a better understanding of coastal and shallow water processes around the region. The paucity of data needed to characterize near shore areas or boundary conditions incurred discussion of ongoing oceanographic research activities that might bear some utility for EAFM in the Western Pacific.

For example, the array data collected through the Tropical Ocean Global Atmosphere (TOGA) program in the Central Pacific was discussed as an example of high-resolution information that may prove useful, although the time-depth of the information is limited. It was noted that some nutrient–phytoplankton–zooplankton models go back many years, but that only recent data are available for ground-truthing. A model of ocean circulation developed by the US Navy was reviewed, but the group determined that its coastal boundary conditions are poorly defined. A high-resolution Hawaii circulation model is being developed, as is a near-shore thermodynamic model of reef conditions, but discussants noted that both are in early stages. It was noted that the OSCARS model being developed by NOAA's Pacific Marine Environmental Laboratory in Seattle to examine surface layer velocity may be helpful with respect to connectivity, and some large-scale pelagic fish models are currently being refined to smaller scales and may ultimately prove valuable.

Finally, it was noted that the University of Hawaii (UH) has developed useful oceanographic models, and the Pelagic Fisheries Research Program (PFRP) at UH has been working to create an atlas of oceanographic information that may be applicable to EAFM. Discussion about using models to fill in some of the apparent data gaps was conditioned by warnings that such approaches must be carefully defined and tested.

Reconstructing Time-Series Catch Data

This session extended discussion of the need for time-series data in the development of EAFM. The challenges of finding and reconstructing such data were reviewed in some depth by the presenter and group as a whole.

Reconstruction of Coral Reef Fisheries Catches for US Associated Islands in the Western Pacific Region: 1950–2002

Dirk Zeller, Fisheries Centre, University of British Columbia

A lack of sufficient historic data has constrained fisheries analysis in many regions of the world. Zeller et al. (2005a) have recommended using a combination of available historic and contemporary information and clearly defined assumptions to generate conservative estimates of historic catch levels for various species of particular interest to fishery managers. This could help fill gaps in understanding about the historic status of fisheries and fish stocks in the Western Pacific.

Near shore fisheries have long supported subsistence, cultural, and food security functions in the Western Pacific. However, data associated with near-shore fisheries are limited when compared with that of the higher profile pelagic fisheries of major international commercial importance, such as tuna and billfish. This is partially due to the spatially scattered nature of landing sites for nonpelagic fisheries, which renders catch reporting and estimation of catch relatively difficult. The problem is also often driven by the misperception that noncommercial and nonpelagic fisheries are somehow less "important" than commercial pelagic operations (Zeller et al. 2005b).

Estimation of historic catch rates in the absence of time-series data requires that analysts make distinct assumptions and interpolations about often widely spaced data "anchor points." As defined here, anchor points are based on data sources such as localized human health, and food consumption studies that are not specifically related to fishing. While such points and associated data tend to be scattered over time and space, such information does have value and thus it is counterproductive to continue the popular approach for dealing with "no time-series data" issues by reporting nothing. This will usually result in spurious interpretations of no historical catches or no accounting of historical catches in analytical and policy considerations.

Increasing demand for accountability and holistic considerations with respect to sustainability and EAFM require a more complete accounting of catches, even if catches from earlier periods are estimated at some level of uncertainty. Without such data we cannot begin to estimate the direct and indirect economic and cultural benefits of marine resources to society, or the costs of overfishing in communities around the Western Pacific.

The purpose of the project described here was to assemble available information on catches for nonpelagic fisheries of American Samoa, Guam, CNMI, Hawaii, and other isolated islands and atolls under US jurisdiction, for the period 1950 through

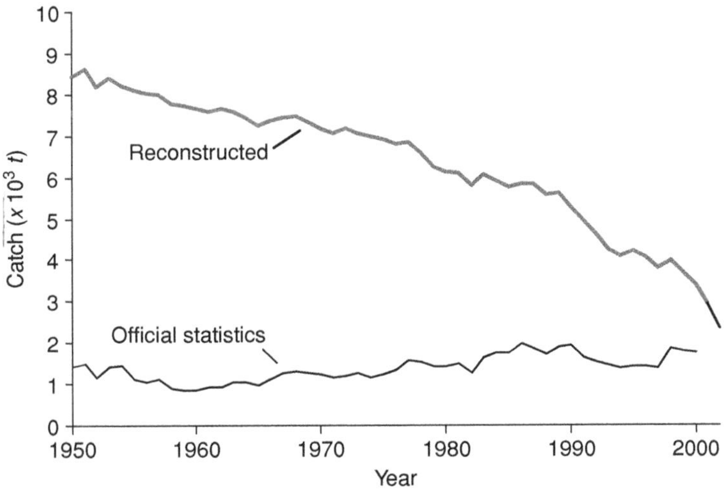

Figure 2.9. Total reconstructed catches of nonpelagic fisheries for all US-associated flag islands of the Western Pacific versus the officially reported statistics.

2002. The aim was to derive estimates of total harvest of key near shore resources for this period, excluding large pelagic species. In sum, the study indicated the following:

- Reestimated catches for all islands combined suggest a substantial decline of about 72 percent in total catches between 1950 and 2002. This was in contrast with the pattern observed from the officially reported data alone, which suggested an increase of about 19 percent (Figure 2.9).
- The official reported data underrepresented the reestimated total catch for this time period by a factor of about 5 (Figure 2.9). This implies a substantial inadvertent underrepresentation of the status of local, nonpelagic fisheries.

Using American Samoa as an example, the reestimated catch suggests a 79 percent decline in landings of non-pelagic fish species between 1950 and 2002 (Figure 2.10). A 17-fold difference was identified between the reestimated catches and the official statistics (Zeller et al. 2006).

Similar patterns were observed for Guam, where an 86 percent decline and a 2.5-fold difference was identified between reestimated and official statistics were estimated. In the CNMI, a 54 percent decline and 2.2-fold difference were estimated.

Historic catch reestimations such as undertaken in the study presented here will be useful for developing or complementing baselines of historic resource fish populations and use patterns as needed for an effective EAFM. The distinct differences between existing catch statistics and estimates based on systematic reconstruction of catch data shed new light on issues and concerns for fisheries sustainability and conservation. It is suggested that management considerations regarding use and conservation of near-shore coral reefs, particularly around the main inhabited islands, are no longer about sustainability, but rather about rebuilding depleted stocks. Moreover, the economic, social, and food security costs of not rebuilding such populations will have to be considered when evaluating and implementing ecosystem-based management strategies.

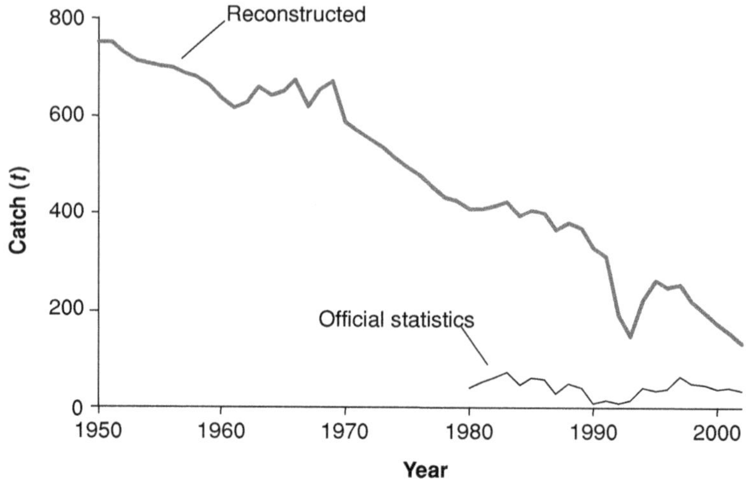

Figure 2.10. Reconstructed nonpelagic fisheries catches for American Samoa.

Group Discussion

This presentation stimulated a number of comments and questions that led to extended discussion. Workshop participants acknowledged the need to creatively employ all sources of knowledge in cases where existing data were insufficient. It was suggested that reef fisheries were already at multiple sustainable yield (MSY) by 1950 or shortly thereafter, at least around the populated main islands. It was suggested that potentially significant differences in factors associated with local rates of productivity between Pacific islands, such as upwelling and nutrient circulation, should be considered in future analyses. It was also pointed out that research at Wake Island and Midway and Johnston Atolls indicates that even small human populations can significantly reduce local stocks.

Participants also mentioned the need for clearly stated confidence intervals and/or approximate value ranges to better illustrate uncertainty in the reconstructed estimates. Participants asserted that some of the consumption rates for fishes in the Mariana Islands seemed relatively high; yet, they were within the range of published fish consumption rates in Micronesia. Dirk Zeller noted that the annual per capita consumption rate of 165 kg of seafood is too high, and that he had adjusted it downwards in keeping with a conservative approach to the analysis.

Because near shore systems in the NWHI are considered relatively pristine compared to those in the MHI, the NWHI was suggested as a good region to control for factors not associated with fishing. Discussants reviewed the utility of experimental design as an important element of an EAFM.

Dirk Zeller was asked if any of the bait catches from the pole-and-line fleets that historically operated in the Western Pacific might reveal anything about the productivity of lagoons in the region. This was deemed a logical question since pole-and-line fishing often occurs where bait can easily be captured. It was pointed out, however, that baitfish were generally part of the so-called "small forage fish" community, about which relatively little is known. Thus, it might be difficult to identify a clear association between the availability of bait and the status of near shore ecosystems. Given the often central role of forage fish in marine ecosystems, limited understanding of such organisms is one form of constraint on valid ecosystem modeling.

There was a question about how reestimation of data analysis tends to be viewed by the anthropological community. Notably, a similar approach is being undertaken by anthropologists in Palau, and has been well-received to date. Regarding historic sources for Hawaii, early reports of fishing were used as an anchor point for the Hawaii analysis.

Biological measures for calibrating catch estimates were also discussed, as were biomass estimates and their potential for analyzing the structure of historic harvests. However, discussants agreed that biomass estimates for early time periods were even more difficult to obtain than catch information, although spatially segregated data sources may be used to generate such estimates. It was suggested that such modeling efforts should be undertaken with caution and with due attention to factors that are known to affect the productivity of coral reef ecosystems.

Discussants agreed that part of the challenge of reconstructing historic data is the difficulty of assigning harvests to allocating catches to individual taxa since historic information about taxonomic composition is usually limited. As a consequence, in locations where historic data are limited, species composition has been projected backward with the analytical caveat that changes in exploitation rates and areas will have varied over the course of time.

References

Zeller, D., Froese, R., Pauly, D. (2005a) On losing and recovering fisheries and marine science data. *Marine Policy Issue*, 29, 69–73.

Zeller, D., Booth, S., Pauly, D. (2005b) Underestimating small-scale fisheries: Contributions to GDP. Fisheries Centre Working Paper 2005–02. The University of British Columbia, Vancouver, Canada. Available online at: www.fisheries.ubc.ca/publications /working/2005/series5.pdf.

Zeller, D., Booth, S., Craig, P., et al. (2006) Reconstruction of coral reef fisheries catches in American Samoa, 1950–2002. *Coral Reefs*, 25, 144–152.

Chapter 3
Ecosystem Models and Modeling

Lessons from Other Regions

Ecosystem modeling incorporates a variety of assumptions, explanations, and confounding factors. This session of the biophysical workshop addressed some of the modeling challenges being encountered by those seeking to develop ecosystem management approach in island and coastal settings around the world.

Ecosystem-Based Management of Australia's Great Barrier Reef Fisheries: Modeling and Policy Initiatives

Neil Gribble, Queensland Department of Primary Industries, Cairns, Australia

An ecosystem model has been developed to provide insight into management problems associated with the Great Barrier Reef (GBR) prawn trawl fishery (Figure 3.1). The prawn fishery is one of the principal fisheries in the GBR region, providing jobs, seafood products, and other benefits across a large region.

Other important fisheries are conducted in the region. These involve the participation of numerous fleets and individuals, including the following:

- Commercial trawl fleet (~500 endorsements).
- Commercial hook and line fleet (~1,300 endorsements).
- Commercial gillnet fleet (~1,400 endorsements).
- Recreational fisheries, significant for some gear types, especially hook and line.
- Indigenous fisheries, significant for some protected species, such as the dugong.

The GBR was designated as a Marine Park in 1975 and a World Heritage Area in 1981. Management efforts aim to strike a balance between fisheries and conservation interests. The ecosystem-based management policy developed for the GBR is based on a suite of state and federal legal instruments and shared state-federal jurisdiction. With specific regard to fisheries, the major federal legislation is the Ecological Protection & Biodiversity Conservation Act of 1999, which requires every federally managed fishery, and every fishery involving an export license, to justify ecological sustainability by the year 2005. Given that most of Australia's seafood is exported, this is a powerful federal conservation tool.

Participants in GBR fisheries are required to demonstrate that pursuit of target and bycatch species is undertaken in a sustainable manner. For example, the trawl fishery conducted in the World Heritage Area is limited by the quantity of syngnathids (seahorses) caught as bycatch. Thus, this fishery is to some extent managed for the

Ecosystem-Based Fisheries Management in the Western Pacific, First Edition. Edward Glazier.

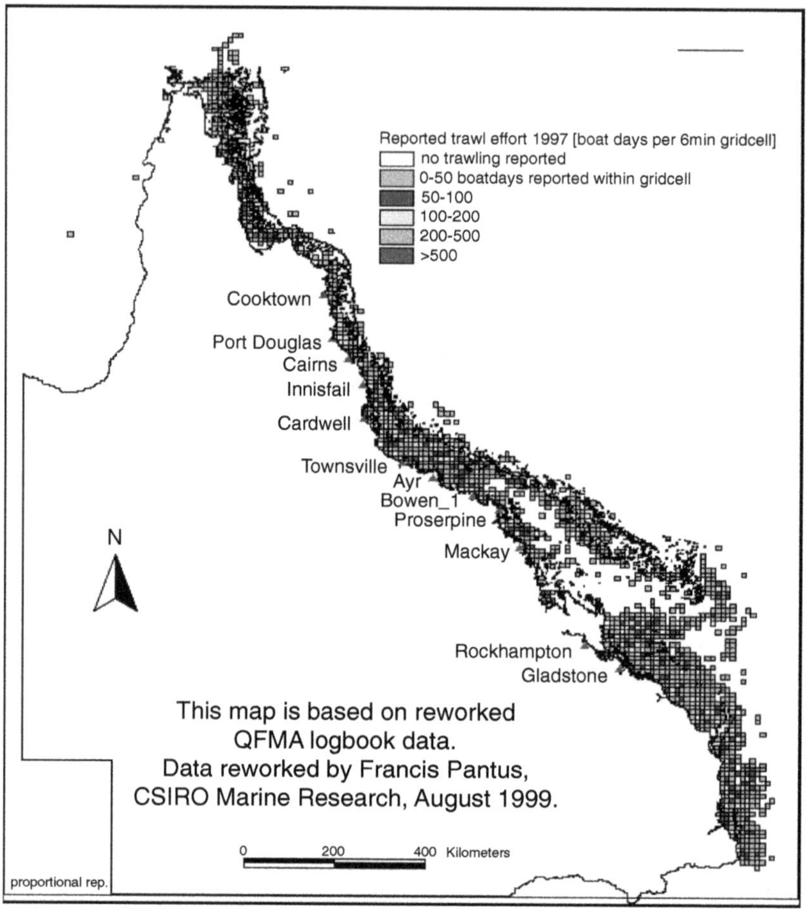

Figure 3.1. Estimated spatial distribution of effort for the prawn trawl fishery on Australia's Great Barrier Reef. (For a color version of this figure, see Plate 2.)

sustainability of the most vulnerable species rather than the sustainability of the target species. This may be an important characteristic of an EAFM in the future.

Legal instruments used by state agencies to manage marine fisheries in Australia include the 1994 Queensland Fisheries Act, legislated FMPS, 80 Fish Habitat Areas that prohibit development around ecologically sensitive areas, and several dugong sanctuary areas.

Ecosystem-based fisheries management in Australia's GBR thus relies on two fundamental approaches: substantial spatial closures (~30 percent of all habitat types are closed to all extractive uses), and requirements that the region's fisheries are sustainable across all species captured.

The ecosystem model for the Northern GBR prawn trawl fishery (Figure 3.2) was developed by the author while at the University of British Columbia, with assistance from Carl Walters and Villy Christensen. The aim was to develop a modeling tool to assist evaluation of management plans for the GBR World Heritage Area.

Data sources for the model were numerous and relied heavily on collaboration among several institutions, including Queensland Department of Primary Industries,

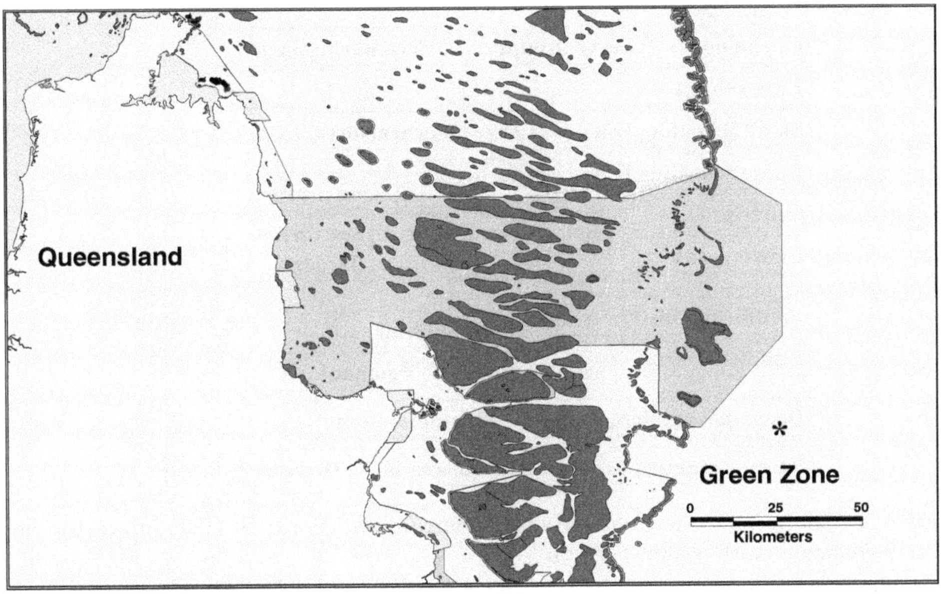

Figure 3.2. Great Barrier Reef ecosystem model, depicting regional prawn trawl closure. (For a color version of this figure, see Plate 3.)

Commonwealth Scientific and Industrial Research Organization, Australian Institute of Marine Science (AIMS), and the Reef Cooperative Research Centre at James Cook University.

The model structure is based on the trophic guild modeling work undertaken by Opitz (1996) in the Caribbean, but was substantially respecified for the GBR (Figure 3.3).

The model involves use of five parameters for each of 1,000 species in 25 functional groups. Commercially important species and bycatch species are included. Validation was carried out by comparing simulations to logbook data and interviews with fishermen. The goal was to capture major biomass dynamics and basic trends in the much more complex real system. Spatial simulations were eventually conducted to examine the effectiveness of spatial closures. Data obtained through the compulsory vessel monitoring system were crucial in assessing compliance with closure regulations and to compare logbook-based information about spatial distribution of effort with actual distribution of effort (Figure 3.4).

In summary, the cross shelf closure simulations suggested the following:

- Fishing effort is likely to concentrate on the borders of closed areas; this relates to the size of the closed area and the historic level of effort in the area.
- Vulnerable species (e.g., long-lived, slow growing, or rare) will likely benefit from MPAs;
- Opportunistic species may not benefit from reserve protection to the same degree as many species.

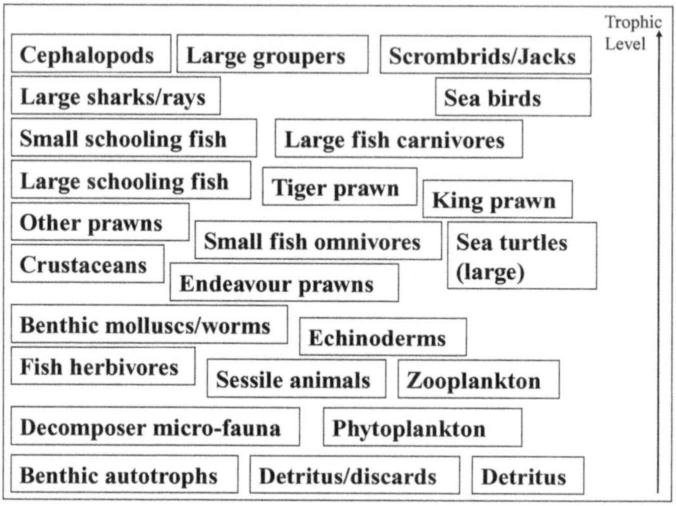

Figure 3.3. Species and functional groups comprising the Northern GBR prawn trawling model, including relative trophic position.

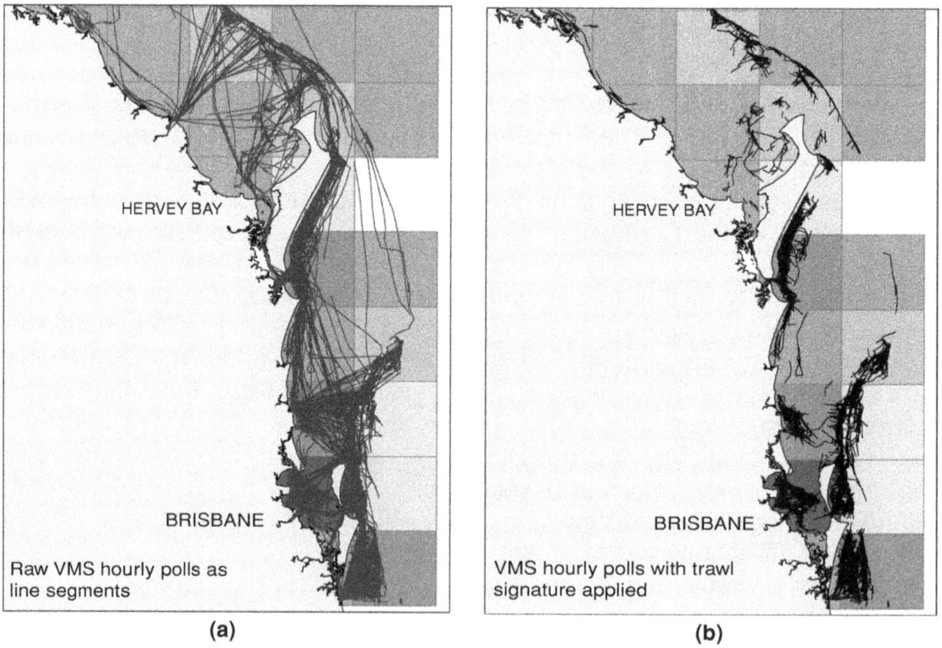

Figure 3.4. Map depicting VMS signatures of prawn trawlers overlaid over compulsory daily logbook 30 minute effort grids. (a) Raw VMS hourly polls as line segments. (b) Trawl signature filter applied (i.e., speed-dependent net deployment). (For a color version of this figure, see Plate 4.)

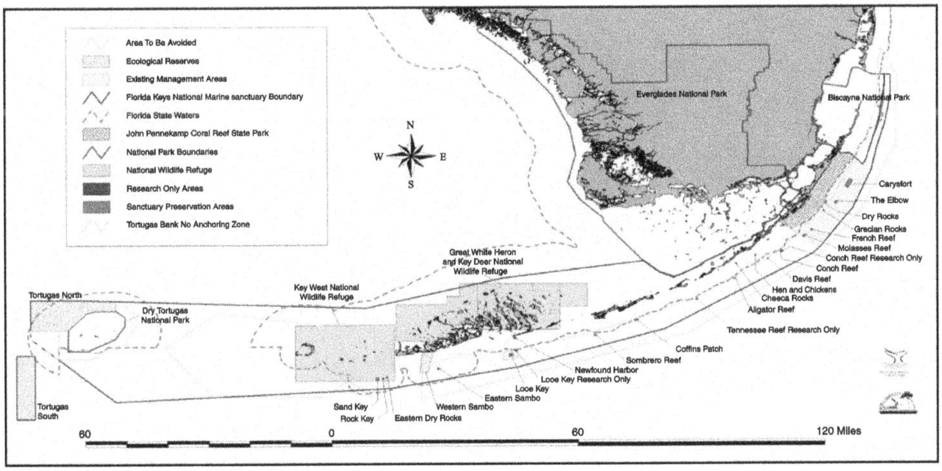

Figure 3.5. Multiple management jurisdictions in the Florida Keys and South Florida. (For a color version of this figure, see Plate 5.)

- MPAs are only as effective to the extent that fishermen comply with the closures, emphasizing the importance of both stakeholder buy-in as well as rigorous enforcement.

Sustaining Florida's Multispecies Coral Reef Fisheries

Jerald Ault, University of Miami

Florida was home to 16 million people in the year 2000. In 2001, reef ecosystems system along the coast of South Florida generated 71,000 jobs and $6 billion in economic activity. The Florida legislature designated the state as the "fishing capital of the world," and in 2004, recreational fisheries generated $10 billion in economic activity. Ecosystem goods and services are thought to be threatened by increased exploitation and environmental changes.

Management of Florida's coral reef systems is complex in jurisdictional terms (Figure 3.5). However, there is growing public awareness that the citizenry is responsible for maintaining critical ecosystem connections and functions.

University of Miami research program examining an EAFM for Florida, uses, fishery-dependent data and fishery-independent data to understand various ecosystem attributes. Assessments undertaken between 1979 and 2005, included over 300 species and covered over 12,500 km^2 of ocean. The approach allows analysis to link measures of abundance to specific benthic habitats, which in turn allows simple preliminary evaluations to be undertaken at an ecosystem-scale, such as reef fish diversity by benthic habitat type. Examples include: trends in fisheries components of the ecosystem, management benchmarks for the South Florida multispecies reef fish communities (Figure 3.6, Ault et al. 2006), and effective design of marine reserves (e.g., Figure 3.8; Meester et al. 2001, 2004, Ault et al. 2005b).

Community control rules centered on biomass (B/BMSY) and fishing mortality ratios (F/FMSY) suggest that most species are being overfished (Figure 3.7,

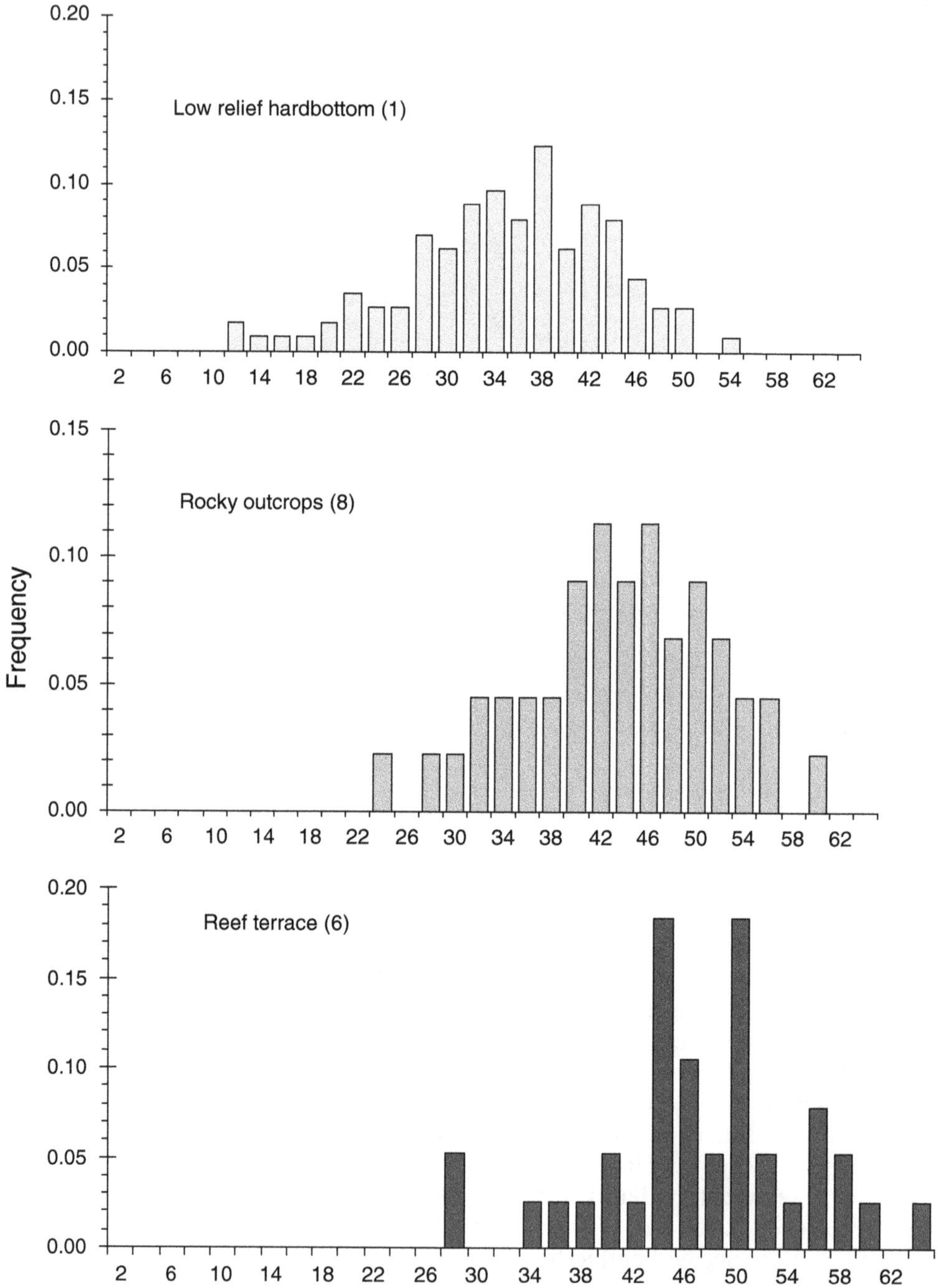

Figure 3.6. Example of coral reef fish diversity patterns by Habitat Type: Southern Florida reef ecosystem.

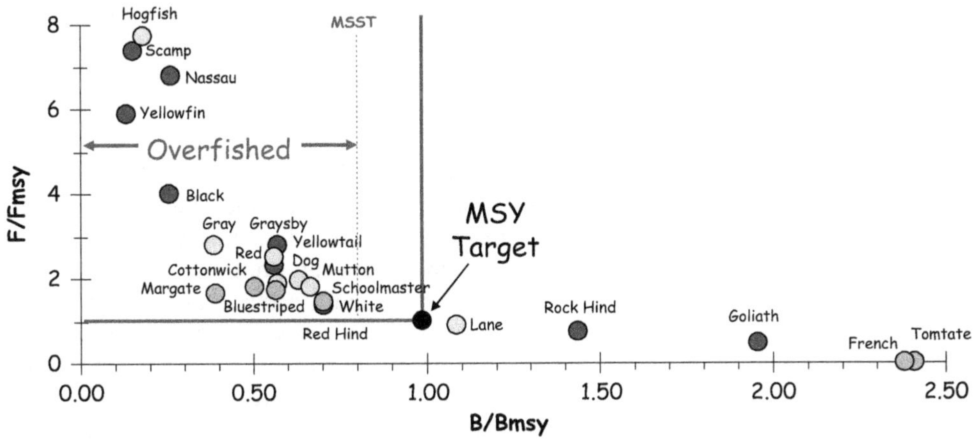

Figure 3.7. Management benchmarks for coral reef ecosystems in the Florida Keys and South Florida. (For a color version of this figure, see Plate 6.)

Ault et al. 2005a). In order to improve our understanding of the ecosystems in questions, however, we must do a better job of identifying and using scales that reflect actual biological processes. Moreover, we must develop a better understanding of habitat foraging dynamics and interactions rates between predator and prey.

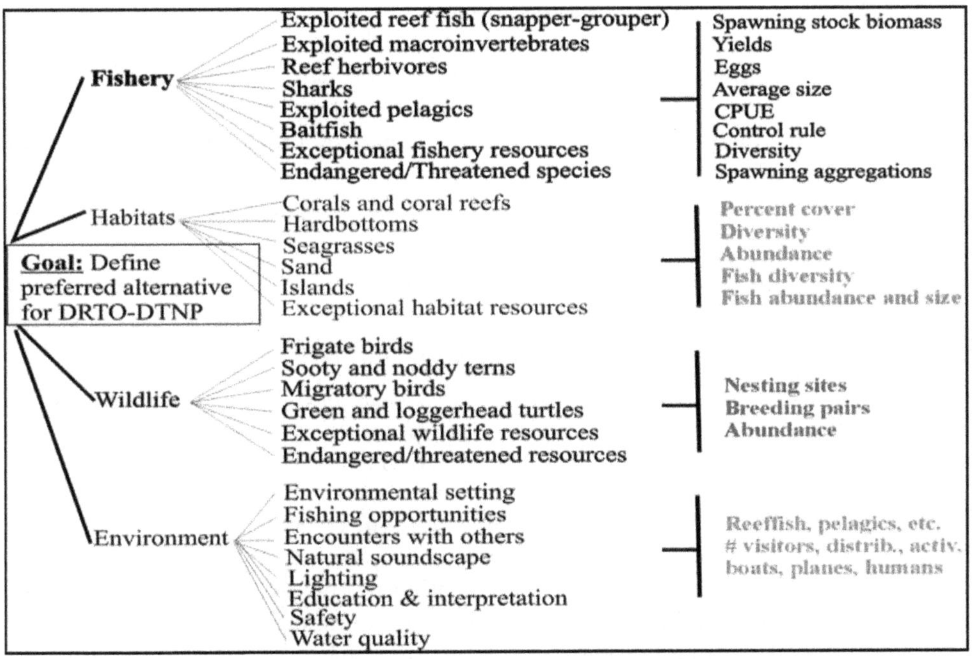

Figure 3.8. Factors and measures being considered in the design of marine reserves as part of the ecosystem-based management approach to fisheries in the Southern Florida coral reef ecosystem.

Applications of Ecosystem Modeling in Canadian Fisheries: Have They Been of Any Use?

Carl Walters, Fisheries Centre, University of British Columbia, Canada

Modelers clearly have not yet enabled autopilot-type fishery management. Models cannot remove responsibility for subjective judgments about uncertainties. If managers or scientists are looking for a model to provide value-laden answers, then they are going to be disappointed by anything that modelers can do. Earlier in the workshop, Mike Orbach made a key point that scientists can't provide answers to value judgments. One of the difficult things happening with ecosystem modeling, as has happened previously with single-species assessment, has been that managers try to pass the responsibility for value judgments to scientists. Managers ask the scientists to model the decision choice problem in terms of tradeoff relationships, but scientists can't do that. A model can generate alternative predictions about what might happen under different circumstances; it might expose uncertainties that ought to cause a responsible manager to think carefully about experimental and adaptive strategies. However, a model cannot definitely relate the appropriate level of risk to take in a given decision-making process, or dictate the appropriate precautionary approach to undertake. When model simulations are run, it is best that the subjectively determined information needs or values of managers are somehow incorporated in the model's parameters. Scientists can develop models that lay out various design alternatives that address certain objectives or generate weightings for different performance measures. However, modelers cannot say which objectives are the "right" ones to pursue. It is not the role of modelers to make prescriptions.

In short, modelers should maintain the objective principles of science in their work. In truth, the efficacy of many ecosystem models would be enhanced through recognition that we are not yet at a highly advanced level of understanding. One modeling strategy that maintains objectivity and simplicity in design involves a three-step process, as follows:

(1) Use broadly conceived models to expose tradeoffs and risks, without a pretense that the predictions are all precise.
(2) Build simple models or calculations to explain, justify, and check major findings from broadly conceived models.
(3) Build detailed spatial and other models to examine specific regulatory options.

This approach is capable of enhancing basic analysis of management tradeoffs. Canada's Atlantic cod fishing is a good case in point. Shrimp populations have expanded dramatically following the crash of the cod fishery, and shrimp have become more valuable in the marketplace than cod ever were. Any prospective attempts to revive the cod fishery would involve some tradeoff of costs and benefit between the fisheries. Based on available data, the cod-shrimp tradeoff is apparently concave (Figure 3.9). Thus, "balanced" policies attempting to enhance both fisheries would fail, ultimately producing low levels of biomass and diminished catch rates for both species. This illustrates that in some cases, modelers can expose the potential severity of tradeoffs among management objectives. Some dislike of modeling is occurring in the Canadian fisheries arena, not because the models were right or wrong, or the underlying science was problematic, but because they have exposed the severity of certain

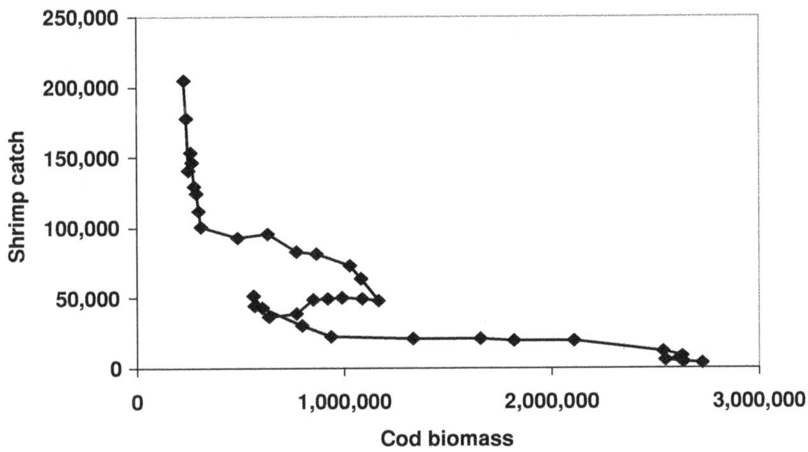

Figure 3.9. The potential tradeoff between Atlantic cod biomass and shrimp catches on Canada's east coast, illustrating that "balanced" policies that attempt to maximize biomass and landings for both stocks have a high likelihood of failing.

management tradeoffs. There often is no easy compromise or a simple prescription for the "right" course of action in fisheries management.

Models have also been used to help structure "who-done-it" debates that surround many major fisheries regulatory changes (that is, who caused the basic problem). This often relates to arguments about the effects of fishing versus the effects of environmental changes. What has been found is that models have generally supported "interactive effects" hypotheses, that is, the amplification of "natural" changes associated with fisheries exploitation, such as may occur when predator species are heavily fished in a given area. Such models have also exposed uncertainty about the efficacy of culling programs (e.g., seals, cormorants) because of possible "third party" responses (e.g., unexpected increases in hake rather than salmon along Canada's Pacific coast).

Time should be spent discussing models based on empirical reference points in areas where fishing pressure has been minimal or non-existent. If an unfished reference point area does exist (e.g., parts of the Northwestern Hawaiian Islands (NWHI) or remote Pacific islands), then the models may not need to back-calculate historic ecosystem attributes. Reducing at least these uncertainties while retaining the appropriate functions of and approaches to ecosystem modeling would likely enhance EAFM in the Western Pacific.

Using Ecosystem Modeling for Fisheries Management: Where Are We?

Villy Christensen, Fisheries Centre, University of British Columbia, Canada

Ecosystem modeling work has forced an understanding of basic ecosystem attributes to summarize what is known about ecosystems, such as what the resources are, how they interact, and how they are exploited. Some attributes have been shown to be pivotal to ecosystem modeling. Concepts and processes such as "foraging arena" (Figure 3.10), predator–prey relationships, carrying capacity, total catch history, climate change, nutrient loading, and habitat considerations are increasingly important in conceptual

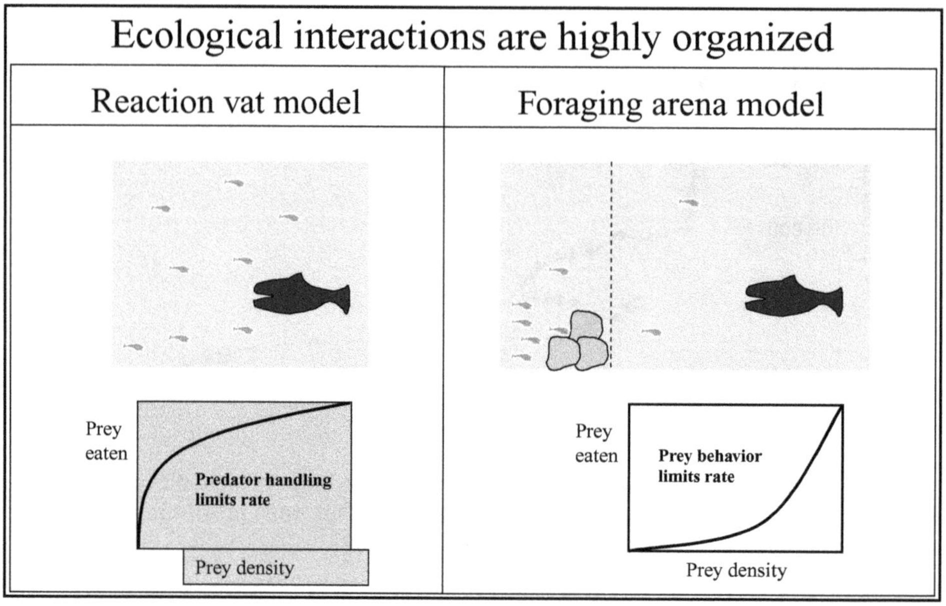

Figure 3.10. Simplified visual representation of the "foraging arena" concept underlying much of the recently improved simulations of trophic relationships in many ecosystem models, such as ecosim.

terms. The importance of evaluating models by replicating historic trends (Figure 3.11) and making extrapolations to new situations is also critically important.

In general, the ecosystem modeling for purposes of fishery management has led to some basic findings:

- It is possible to simulate development of ecosystems by incorporating environmental and fisheries parameters, as these manifest over time.
- More data are needed, particularly regarding "ecosystem history."
- Ecosystem models should supplement single species assessment – not replace it.

Notably, even though ecosystem models are being used to address some management issues, they are rarely used in actual management *decisions*. We can begin to describe, with some credibility, the agents of mortality and trophic interdependence and mortality among ecosystem components, and we can begin to evaluate the relative impacts of fisheries and environmental factors (at least at the "looking for correlation" stage). But, at this point in its development, ecosystem modeling is not readily used for prediction or planning, especially in the single species tactical management response regime that characterizes many fisheries settings around the world today.

In summary, EAFM is currently based on less than optimal set of decision tools. Development of strategic and predictive approaches may help to move fisheries management out of current approaches that are (focused on short-term horizons. This raises questions about whether scientists, managers, and stakeholders can or are ready to use models for strategic planning and decision-making. While ecosystem modeling cannot yet replace single-species assessments and decision-making, an adaptive approach that recognizes the potential strategic value of ecosystem models is clearly called for.

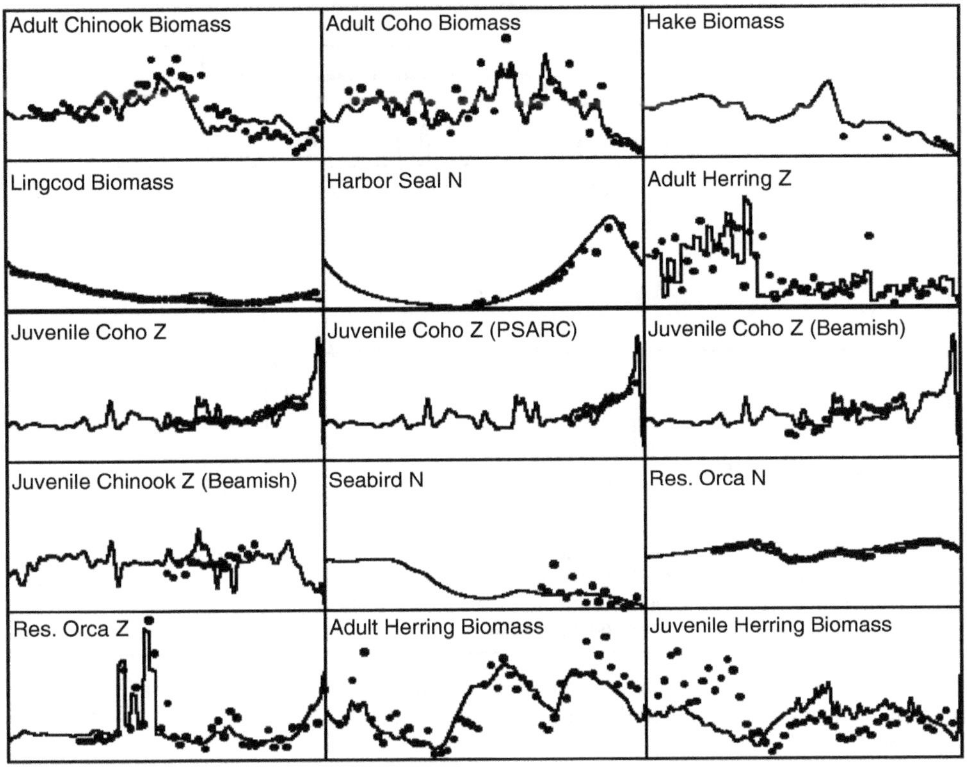

Figure 3.11. Strait of Georgia (Canada) time series from assessments or surveys (dots) compared to ecosim simulation runs (lines), illustrating the utility from evaluating models by fitting to observed time-series data.

A Spatial Ecosystem and Population Dynamic Model

Patrick Lehodey, Secretariat of the Pacific Community, New Caledonia

When developing ecosystem models, much emphasis tends to be placed on trophic relationships and information regarding the physics, primary food production, and commercial fish species in question. A simple definition has been to describe forage groups in three vertically stratified layers in the pelagic system: the epi-, meso-, or bathy-pelagic layers. Some species remain in their layer while others migrate vertically at night (Figure 3.12). We have identified six components, which correspond to what is generally observed in the vertical movement behavior of predators. We use historical data for recorded species, isotope analyses, and indirect validation. Our work make clear that for large-scale modeling of pelagic systems, such as those associated with tuna, many of the required basin- or region-scale data layers do exist and can be used in a beneficial way.

A number of projects are currently underway to improve understanding of pelagic ecosystems in the region. These include the following:

- The "Mixed-resolution models for investigating individual to population spatial dynamics of large pelagics" project, undertaken in collaboration with Senina,

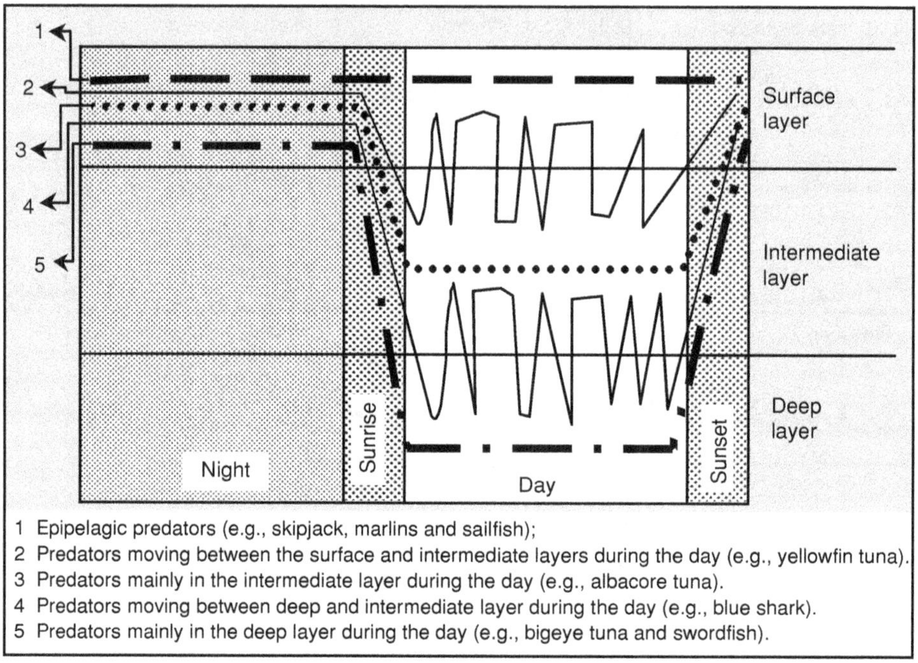

Figure 3.12. Five typical vertical movement behaviors simulated using a three-layer and two-prey-type pelagic system. (Adapted from Dagorn et al. 2000.)

Allain, Sibert, Murtuggude, Kirby, Ancheta, and Polovina and funded by the Pelagic Fisheries Research Program (PFRP) at UH.

- Implementation of an optimization model function, also in collaboration with Sibert, Senina, and Ancheta from the PFRP.
- Simulations in the Atlantic and Indian Oceans via Global Ocean Ecosystem Dynamics (GLOBEC), CLimate Impacts on Oceanic TOp Predators (CLIOTOP) and WG4).
- Comparative analysis of modeling approaches: IBMs, trophic models (e.g., Ecopath with Ecosim), statistical models (e.g., multifan-cl).
- Development of a regional network of modeling collaborators and end users.

Status of Hawaii Ecosystem Models

Jeff Polovina, National Marine Fishery Service, Pacific Fishery Science Center

Hawaii's ecosystem models include trophic models (Ecopath with Ecosim) for the NWHI and the Central North Pacific, and spatial models, such as a time–area–gear model, a passive movement model, and an active movement model. One of the reasons for reevaluating the old NWHI Ecopath model (first published in 1984 in *Coral Reefs* 3, pp. 1–11) is to take advantage of work undertaken by Frank Parrish. The original Ecopath model was parameterized with help from Frank Parrish's father. Twenty years later, a multiyear survey of reef fish in the NWHI being conducted by Frank Parrish, Ed DeMartini, and others has generated updated parameters, and the need to revisit the earlier model. The model in question is for French Frigate Shoals. It

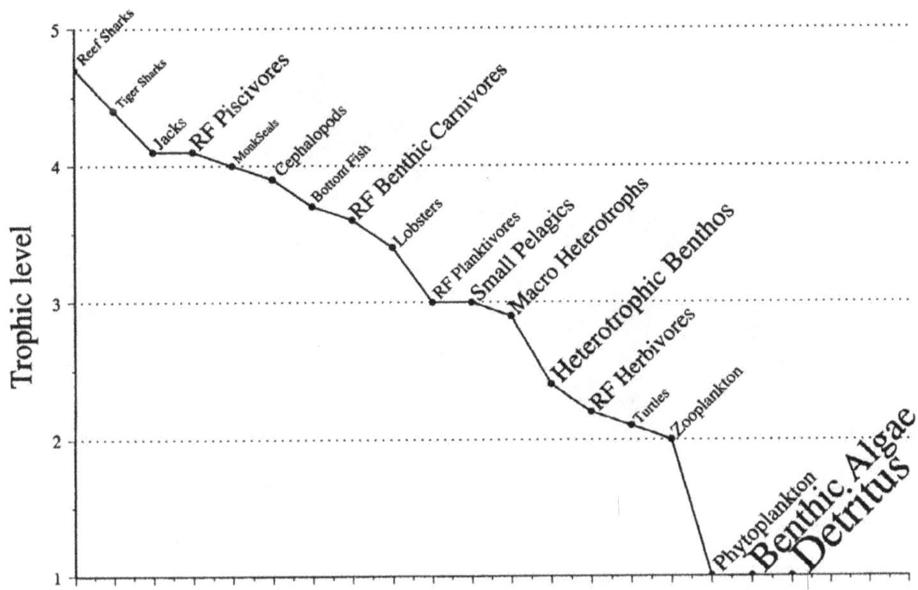

Figure 3.13. Trophic levels and biomass scales (font size) for the updated ecosim French Frigate Shoals model.

incorporates 20 functional groups. But whereas the original model contained only a single group for reef fish, the updated model is now separated into piscivores, benthic carnivores, planktivores, and herbivores (Figure 3.13). The new model also makes use of some IKONOS habitat characterization data (Figure 3.14).

Models such as that being used for French Frigate Shoals can help:

- determine the structure and dynamics of the ecosystem energy budget;
- determine biomass of a species of interest may respond to various forcing functions;
- identify species that drive ecosystem dynamics;
- identify ecosystem indicator species; and
- explore the carrying capacities of one or more species.

One potential use of such a model, possibly with more detailed specification of key target and not-target bycatch species for which catch history data exist, may be as a reformulated early time period model for the MHI (i.e., under conditions of low exploitation). This would permit the opportunity to examine the potential response of the modeled ecosystem to archived fisheries catch and effort data.

Three spatial models are either being used or in development in the Western Pacific. One examines impact of the longline fishing fleet on the pelagic ecosystem. The model was developed to examine how the spatial and temporal distribution of fishing effort could be changed to reduce the incidental take of sea turtles, and how related management measures and their effects could best be evaluated. The historical spatial distribution of effort by the Hawaii-based longline fleet is based on logbook and observer data (Figure 3.15).

Figure 3.16 depicts simulation of 350,000 possible time/area closures against inadvertent take of endangered leatherback turtles. For example, a time/area closure

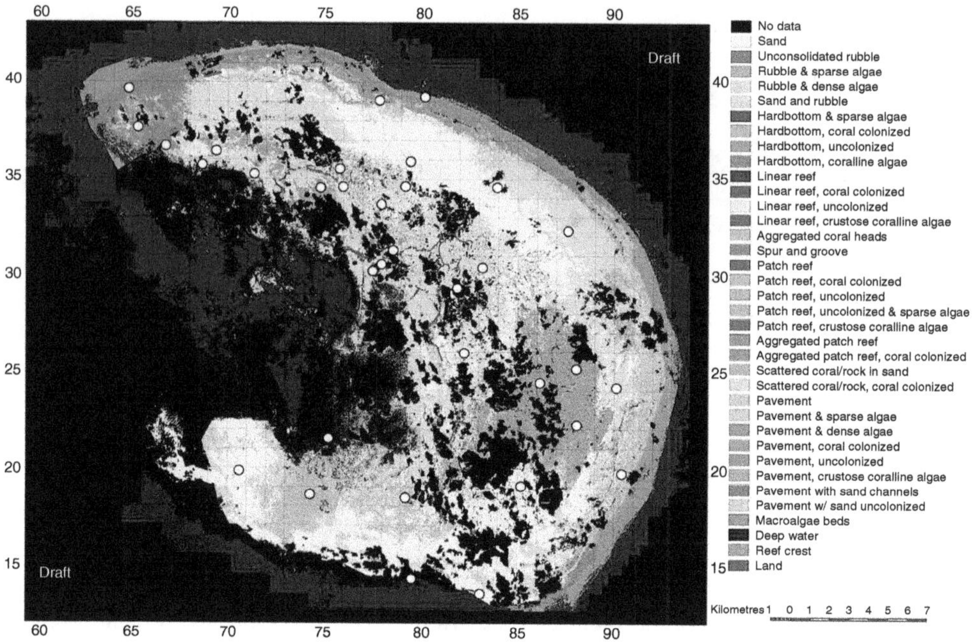

Figure 3.14. Habitat information for French Frigate Shoals, as derived from IKONOS imagery. (For a color version of this figure, see Plate 7.)

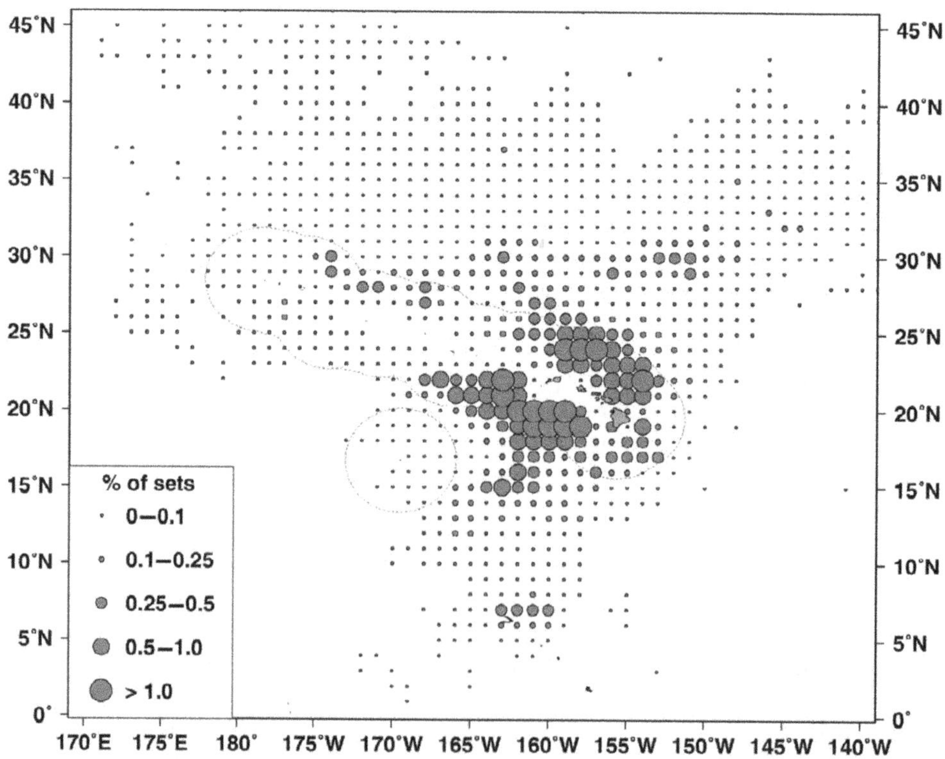

Figure 3.15. Distribution of Hawaii-based longline fishing effort (1994–1998).

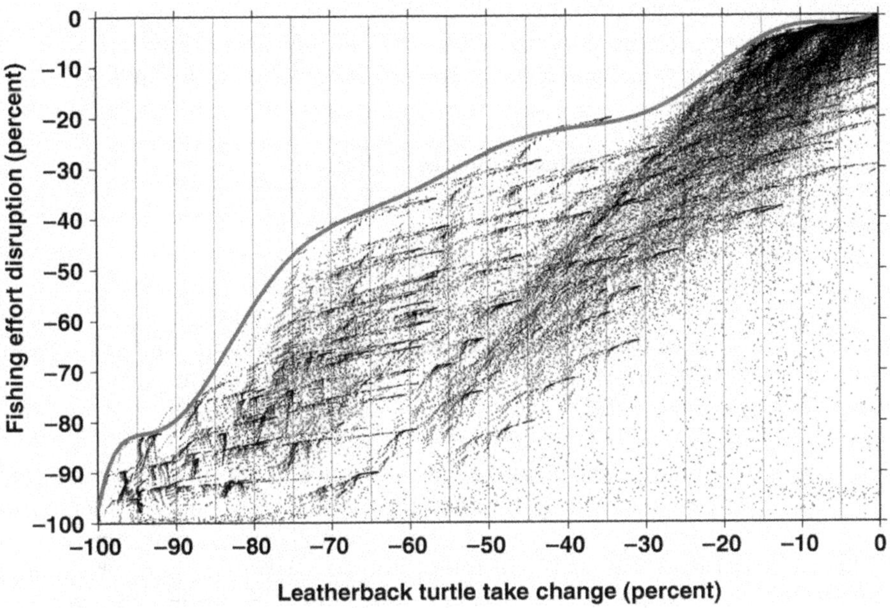

Figure 3.16. "Efficient frontier" identification in closure scenarios, based on 350,000 possible time/area closure simulations.

that reduces the leatherback take by 70 percent may require that managers reduce fishing effort by approximately 40 percent. The model can be used to estimate the best time/area closure needed to most effectively reduce leatherback takes as per conservation targets founded on sound population dynamics data. Fisheries effects can then also be assessed.

Potential applications of this model include the following:

- Describe the impacts of changes in spatial distribution of effort.
- Assess which time, area, and gear configurations achieve a desired level of incidental catch or bycatch while minimizing fishery impacts.
- Test whether observed decline in landings result from changes in fishing grounds or fishing methods.

A second spatial model, originated in work done with Pierre Kleiber and Don Kobayashi at PIFSC. This simulates the transport of larvae in the NWHI. The model incorporates surface and subsurface water movement based on satellite wind and altimetry observations. Simulated larvae were released into the environment, some from O'ahu, some from Necker and Maro Islands, and some from Midway Atoll. The simulation released "larvae" every day, and removed them after 6 months, thereby representing settlement out of the pelagic environment.

Satellite-based observations of temperature and chlorophyll encountered by the simulated larvae act as habitat indices for survival and growth. The model can assist in developing potential hypotheses for different metapopulation dynamics and the potential for dispersal and retention of larvae in the archipelago. Results thus far suggest that there is a certain amount of local retention and advection between banks in both parts of the Hawaiian archipelago. Work on this model is ongoing. New work is incorporating both Ekman and geostrophic transport to represent deeper flows. This is enabling simulation of the potential effects of diurnal vertical movement. Models such as the one described here may be helpful in developing testable hypotheses regarding:

- the spatial and temporal dynamics of larvas of interest;
- good and poor recruitment years;
- source and sink locations; and
- the source/sink debate for NWHI and MHI.

Further improvement to this model would require incorporation of key behavioral aspects of larvas such as the documented, substantial swimming ability of late-stage tropical fish larvas. Currently, model larvas are treated as passive particles and are thus misrepresenting their disbursal as well as capacity to self-disperse or relative position.

A third model examines the distribution and movements of loggerhead turtles in the North Pacific as per fishery bycatch and fishery independent tracking data (Figure 3.17). The observed pattern suggests potential connectivity between East Asia, Hawaii, and Baja California. An important question underlying the model is whether the existing tracking data can be used to infer population dynamics across space. The results suggest a potential continuum of loggerhead movements across the entire North Pacific Ocean. These data and models would benefit from further investigation.

In sum, data, models, and management are part of a circular relationship (Figure 3.18), in which management questions determine the data needed to model the process of interest. The resultant outcomes can assist in visualizing the question posed and may

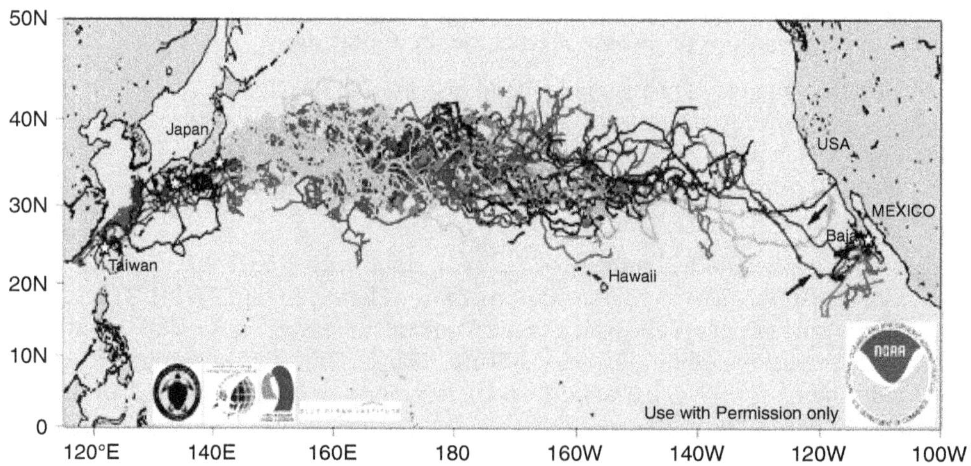

Figure 3.17. Distribution and movement of loggerhead turtles in the North Pacific Ocean. (data consist of both loggerheads found in the 1990–1992 high-seas driftnet fishery, and track lines indicating movement of free roaming turtles fitted with satellite tracking transmitters between 1997 and 2005. Prepared by D.M. Parker in collaboration with J.J. Polovina, G.H. Balazs, I. Cheng, P.H. Dutton, N. Kamezacki, W.J. Nichols, and I. Uchida). (For a color version of this figure, see Plate 8.)

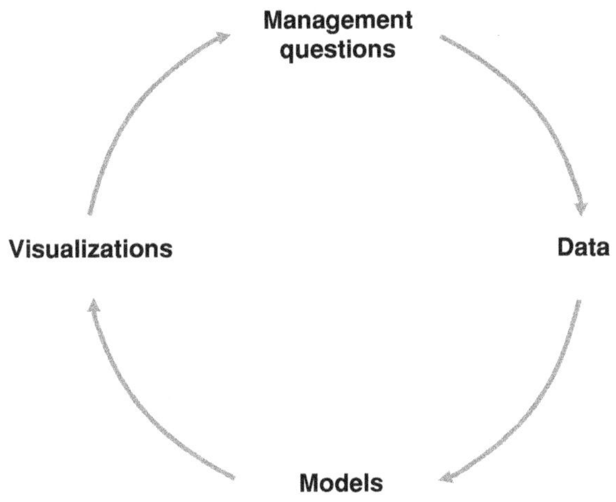

Figure 3.18. Conceptualization of the relationship between management questions, data, and models.

help managers in clarifying the questions. This process may lead to more questions and serve as an important feedback loop in the spirit of experimental and adaptive approaches to ecosystem-based management.

Group Discussion

This session elicited many questions and extensive discussion. Villy Christensen asked if the simulation of fishery effects on Australia's GBR had been used for purposes of management. Neil Gribble noted that it was used along with other information and analysis, such as examination of limiting prawn fishing pressure on the reefs.

Dirk Zeller asked if satellite VMS was planned for use to monitor fleets besides the prawn trawl fleet. It was determined that plans to monitor the hook and line fishery with VMS are complicated by the need to track both mother ships and many individual dories. David Kirby discussed the potential for VMS tampering and suggested that additional surveillance methods are required to protect against illegal fishing and to validate the VMS information.

Charles Alexander asked if the no-take zones in Dr Gribble's model were set up with consideration for the edge effects of illegal fishing. This variable reportedly was considered wherever possible, but boundaries were based primarily on biodiversity considerations.

Michael Orbach asked if there had been any attempt to model the relative costs and benefits of the management options suggested by the odel. Dr Gribble noted that the next step in the modeling process would be to incorporate three fisheries into the model, and then to explore how and to what extent the management plans for each fishery generated ecosystem responses.

He added that the modeling approach used included a module for addressing cost and benefits, but that such analysis needed to be undertaken by the management agencies themselves. It was pointed out that a number of indicators could be extracted directly from the models, such as value of catch and profit generated by the fisheries.

Carl Walters commented on the presentation by Dr Jerald Ault, noting that much effort was being applied to further our understanding of how marine animals respond to environmental gradients in the marine ecosystems. He noted that Steve Martel has undertaken similar work. Reported that a considerable amount of time had been spent thinking through spatial aspects of his South Florida coral reef ecosystem data. The dataset in question is rather large and articulated, and includes a variety of assemblage constructs and variables including abundance, frequency of occurrence, and spatial distribution of reef fish species. The performance of the models is being scrutinized closely in South Florida, where public and political interest in fisheries management is considerable.

Carl Walters added that reviewers had been over the Florida reef modeling work with a fine-toothed comb because related policy options included fishery closures. But reviews did not identify factors likely to invalidate the model with the exception of potential movement of large fish to waters too deep to be picked up by visual surveys work.

Edward DeMartini initiated discussion Walters about the shape of the tradeoff relationships, and whether there might be any fundamental difference in the way lower trophic levels should be considered. The idea that tradeoffs ought to be linear or concave was considered a question of thermodynamics, and that if potential energy was modeled to support a fishery, then the system, it could not support as many species as it would sans thermodynamic considerations, regardless of trophic level.

John Sibert pointed out that the Commission for the Conservation of Antarctic Marine Living Resources (CCAMLR) managed prey species with direct regard for the predator species. In the earliest days of CCAMLR, an international convention took on questions related to ecosystem management. Australian scientists and others argued that the Antarctic varied in physical terms across various eco-regions, and that this was a perfect opportunity to undertake large-scale experiments and other adaptive management strategies. This could have helped examine the fate of the Antarctic food web, when species such as krill were exploited.

However, CCAMLR adopted a laissez-faire approach to management and such experiments were never realized.

Dr Sibert asked if CCAMLR had done any tradeoff analysis. It was determined that such analysis had been undertaken, but with little modeled consensus about tradeoffs. Carl Walters revisited the tradeoff issue, noting that as the cod collapsed in the late 1980s, shrimp surveys all showed an explosive increase in abundance and subsequent development of the shrimp fishery, clearly suggestive of an inverse relationship.

Chris Boggs asked whether adaptive management was allowed under the Sustainable Fisheries Act (SFA). David Fluharty commented that when an ESA issue is being addressed, adaptive management approaches could not be used. The SFA does not prohibit adaptive management, but when protected species issues are at hand, the range of management options are limited.

Carl Walters asked Patrick Lehodey if he had examined yellowfin tuna recruitment in the early 1980s to see if it could be explained with an ecosystem rationale. Dr Lehodey responded that he had not yet completed all the modeling simulations, and hence could not present the results. These are forthcoming.

Tim Adams pointed out that the work that Patrick Lehodey and colleagues were doing was actually influencing fisheries policy with substantial regional implications, as it demonstrates that ecosystems in much of the Pacific are affected by a single source of upwelling. This is important because it suggests that the Pacific is not in reality divided into discrete western and eastern regions.

Mike Fogarty asked Jeff Polovina, if it was common to find gyre systems associated with banks, which would contribute to larval retention. Dr Polovina answered in the affirmative.

Tom Hourigan asked whether the French Frigate Shoals reef fish model was applicable to management questions in the MHI. Given relatively good estimates of species composition and biomass, it was decided that it would be possible to assemble a first step model. Whether this would be applied on an archipelago-wide or bank-specific scale could be addressed in other fora.

John Sibert reminded workshop participants that ecosystems were never static and probably never had reached equilibrium. Instead, ecosystems appeared to be always responding to the last perturbation, and managers should not be given the impression that if humans would stop influencing ecosystems that everything would be constant from then on. It has to be made clear that ecosystems respond and adjust in due time, and that included response time to management that at least a 3-year time lag was often needed to reliably detect recruitment, or failed recruitment, as the case may be.

Rusty Brainard asserted that the likely impacts of fishery management decisions on other components of the ecosystem should always be borne in mind, and managers should consider issues of scale and scale of effects when considering the implications of new regulatory measures. Steve Murawski commented that one of the recurring themes in discussion of models relates to the effects of fishing overlaid onto many other drivers and that it might be worthwhile to simultaneously imbed anthropogenic and physical-environmental factors in the same model.

Rusty Brainard commented that much of the variability was related to environmental factors, with fishing an additional disturbance factor. Coupling hydrodynamic models and trophic interaction models was discussed as a potentially useful approach. Jerald Ault commented that a nesting of hydrodynamic and biological models was useful but challenging in his Florida coral reef modeling work.

At some stage, modelers must identify and explain processes underlying observed patterns and participants agreed that this must be kept in mind when developing complex models.

Carl Walters noted that large and highly complex models are not necessarily the answer to EAFM. They are also expensive and it is estimated that $200 million has been spent on marine ecosystem modeling in the US alone.

Workshop participants agreed that models are nothing more than intellectual devices to help scientists and managers think about problems and possible solutions. When models are too complex, scientists cannot easily interpret the findings. Complicated models also do not have the capability to address some of the more simple but compelling questions about marine ecosystems. Sophistry is to be avoided.

Thus, large models that incorporate too much complexity should not be pursued. Modelers should rather pursue answers to questions that are basic to understanding ecosystem relationships and to the management problems related to those linkages.

For example, high-resolution hydrodynamic nutrient cycling models have been used to examine nutrient loading and dispersion patterns along the Florida coastline. The findings have been used in basic models of productivity and trophic relationships. It was not reasonable to model the full complexity of the system all at once. Thus, a cumulative approach to modeling and knowledge may be the best way to proceed.

The depth of understanding that must be accumulated relates directly to the questions being asked and how deeply the problem need be looked into. Discussants decided there should be a marriage of clearly stated objectives, and models that address those objectives in the least complex manner possible.

The process would ideally progress in systematic fashion, wherein analytical needs for policy are clearly conceived and the appropriate models or modeling approach is developed. Models, it was agreed, are a good tool for analyzing ideas and policy options, and they are best used in conjunction with the collective wisdom of those in the fisheries science and management arena. Budget considerations are unavoidable.

Another Florida case suggests a need for budgetary planning. In 1991, an Everglades restoration ecosystem model was developed for $30,000. Since then, $6 million has been spent on increasingly complex modeling work. Notably, some critics argue that the money would have been better spent on collection of data needed to fill the gaps identified by simple modeling approaches. In response to discussion about the utility of dialog for addressing such problems, Carl Walters warned that it may not always be productive to sit down and talk about management objectives. Rather, it can be more productive to consider feasible management actions for which predictions could be made with some confidence. Dr Walters felt that this could lead to identification of relevant performance indicators and an overall modeling system that could respond to relevant scenarios and enable analysis of management tradeoffs. The willingness of the various stakeholders to compromise on such tradeoffs could be exposed as well, and win–win opportunities might start to become evident.

Mike Fogarty noted that modeling work must also relate to some very clearly defined guidelines and objectives for fishery management. These often relate to optimum yield and, in the United States, to maximum sustainable yield. Because such requirements may call for equivalent ecosystem reference points this can help structure basic elements of ecosystem models.

Further attention was given to the issue of funding for ecosystem modeling. A group, such as the one assembled for the present workshop, ought not allow funding problems to constrain the future of EAFM. Funds are increasingly available form non-governmental sources and federal funding levels may ultimately increase above current levels.

Discussants noted that funding is often tied to public interests, which tends to be cyclic in nature. Thus, the achievement of ultimate success may require a considerable amount of time and patience.

Carl Walters emphasized that even though funding and how a problem is approached and analyzed are related, nature may ultimately demand that we model ecosystems and related problems in a practical way. This way well require some refocusing on data about basic ecosystem processes. Here, expert input regarding data needed for effective, if basic, modeling is invaluable.

Steve Murawski suggested there may be a way to effectively use existing data from specific island ecosystems to confidently generalize to similar systems in the Western Pacific. This could potentially reduce some funding requirements.

Regarding such generalization, it may be useful to model replicates of the same system and to analyze observed versus predicted fishery exploitation outcomes. This could help determine whether an ecosystem model can approximate actual historic fishing scenarios regardless of inherent variability in the system. This approach has the potential to enhance understanding of interaction between fishing fleets and natural changes in many island settings.

John Sibert noted that the phrase ecosystem-based management was incomplete, because management implies the existence of an object on which action is taken. This raised questions about what exactly is being managed. For example, if one is truly concerned about coral reef ecosystems and overfished island ecosystems, then thought may have to be given to managing the problem of people, perhaps removing the people from the system until some level of recovery occurs. It may not be possible to take away people's right to fish, and thus, there is a basic conundrum involving human values, science, and the practical needs of fishery management.

Regarding the "what is to be managed" question, participants agreed that the implications of managing humans must be considered, and with some priority. Thus, if fishing capacity is to be reduced, thought must be given to what happens next to the fishermen living on the islands. Job alternatives may be required. This is not a trivial problem, and it is often at the core of ecosystem-based management.

Carl Walters added that a very good and experienced modeler, named Josh Korman had been involved in many ecosystem assessment projects. Some time ago, Dr Korman asked why so much effort was exerted in ecosystem modeling when managers had the option of simply recognizing that natural resource management problems are always people problems and that if anything should be modeled, it should be human behavior.

Participants agreed that, extensive or complicated modeling is not always needed; scientific questions can often be answered with relatively simple data, models, and tradeoff calculations. Difficulties emerge when modeling and science become confused with value-laden resource management decision-making.

Beth Flint noted that people who manage fisheries have been issued a new mandate; they now must consider the ecosystem, and not only fisheries effects on the target population but also on all components of the system. Attempting to incorporate the new mandate into management of target species now involves not only attention to maximum sustained yield or optimum sustained yield but also to effects on four or more trophic relationships to the target population. Such complexity requires sufficiently detailed modeling.

Svein Fougner countered, stating that it was a big step from saying that managers had to consider those impacts to saying managers had to base decisions on the status of other resources than the targeted fishery resources. This implies some kind of value-scaling in decisions about the ultimate use of ecosystem modeling. Because the Magnuson–Stevens Act and other applicable laws require managers to consider impacts on target species, bycatch species, endangered species, and fishing communities, these aspects of marine ecosystems *must* be considered. But ultimately, managers must also make subjective decisions regarding the full range of factors to model and the rationale used to select such factors.

Finally, Tom Hourigan commented that workshops were held in each of the US territories wherein fishermen were given the opportunity to discuss the status of local coral reef fisheries management. For the most part, members of the fishing communities were unhappy with the status of their fisheries and the way the resources are currently managed. This indicates a need for better communicative interaction between scientists, modelers, managers, and those who actually use marine ecosystems for purposes of recreation, commerce, food acquisition, and cultural fulfillment.

References

Ault, J. S., Smith, S. G., Bohnsack, J. A. (2005a) Evaluation of average length as an estimator of exploitation status for the Florida coral reef fish community. *ICES Journal of Marine Science*, 62, 417–423.

Ault, J. S., Bohnsack, J. A., Smith, S. G., et al. (2005b) Towards sustainable multispecies fisheries in the Florida, USA, coral reef ecosystem. *Bulletin of Marine Science*, 76 (2), 595–622.

Ault, J. S., Smith, S. G., Bohnsack, J. A., et al. (2006) Building sustainable fisheries in Florida's coral reef ecosystem: positive signs in the Dry Tortugas. *Bulletin of Marine Science*, 78 (3), 633–654.

Meester, G. A., Ault, J. S., Smith, S. G., et al. (2001) An integrated simulation modeling and operations research approach to spatial management decision making. *Sarsia*, 86 (6), 543–558.

Meester, G. A., Mehrotra, A., Ault, J. S., et al. (2004) Designing marine reserves for fishery management. *Management Science*, 50 (8), 1031–1043.

Chapter 4
Ecosystem Indicators

Indicators are valuable tools for scientists and managers who need to understand how an EAFM is performing in a given region. This chapter reviews some of the challenges and solutions associated with the development of ecosystem indicators. Experiences and lessons gleaned by regional experts were reviewed in depth and formed the basis for discussion by the full group of workshop participants.

What Do Scientists Need from Fishery and Conservation Managers in Order to Develop Models and Indicators Useful to Policy Development and Management?

David Kirby, Secretariat of the Pacific Community, New Caledonia

Many scientists charged with developing an EAFM are frustrated by a lack of clear goals and objectives for doing so. As noted in Table 4.1, the role of scientists, managers, and user groups in undertaking an EAFM do not always overlap. An EAFM is always complex, and its origins are based in multiple disciplines, as discussed below:

Ethics: Environmental and ecological ethics are branches of moral philosophy that were developed amidst a growing appreciation of the evolutionary origin of humans and our role in the global ecosystem. As a broadening of conventional fisheries management, EAFM is the pragmatic interpretation of ecological ethics applied to fisheries.

Ecology: The role of ecologists is to provide and analyze information, and correct misinformation about the structure and functioning of ecosystems. It is important for the credibility of the scientific process to communicate both knowledge and uncertainties. Under an EAFM, the scope of research needs to be broadened to address all ecosystem components.

Economics: Economic valuation attempts to assign quantitative values to goods and services provided by environmental resources. Indirect use values arise from supporting or protecting economic activities that have direct use values. Intrinsic values arise where individuals who do not currently make use of the goods and services of an ecosystem wish to see them preserved in their own right.

The practical legal foundation of EAFM is in various binding and nonbinding international agreements, such as the 1982 Law of the Sea Convention and the 1995 Code of Conduct for Responsible Fisheries. Notably, the 2002 World Summit on Sustainable Development agreed to "encourage the application by 2010 of the ecosystem approach" to fisheries management.

Ecosystem-Based Fisheries Management in the Western Pacific, First Edition. Edward Glazier.

Table 4.1. Suggested roles and responsibilities of different stakeholders under EAFM.

Issue	Scientist	Manager	User
Principles and concepts of EAFM			
Ecological relationships should be maintained			
Limit impact of fisheries on ecosystem			
Apply precautionary approach in the face of uncertainty	*Research & discuss*	*Discuss & decide*	*Discuss & implement*
Ensure both human and ecosystem well-being and equity			
Ensure management measures are compatible across entire distribution of resource (i.e., across jurisdictions and management plans)			
Operational implementation			
Identify specific issues	√	√	√
Develop collective values	√	√	√
Identify "ecological relationships"	√		
Determine how ecosystems function	√		
Improve knowledge of impact of fishing on system components	√		
Define ecosystem boundaries	√		
Define management areas		√	
Ensure stakeholder participation		√	√
Educate stakeholders about knowledge and uncertainties	√	√	
Develop appropriate bio-economic models	√	√	
Use them to explore management scenarios		√	√
Investigate the effectiveness of, e.g., MPAs*	√	√	
Decide where to put MPAs		√	√

*MPA, Marine Protected Area.

A high-level policy goal of EAFM is to balance diverse social objectives by *taking into account* knowledge and uncertainties about biotic, abiotic, and human aspects of ecosystems, and applying an *integrated approach* to fisheries management within ecologically meaningful boundaries. Useful implementation guidelines are presented in Garcia et al. (2003a,b). These suggest that for scientists to develop indicators and models useful for policy making, managers need to:

1. provide clear management objectives by heeding scientific advice and stakeholder values;
2. remove institutional barriers to effective collaboration in research and management;
3. rewrite existing legislation if not adequate; and

4. obtain/provide funding for the expanded research base needed to support an EAFM; this includes integration of empirical data and modeling, and documentation and development of the value systems that form the basis of prioritization in management decision-making processes.

Development of Marine Indicators for Assessment of the Biodiversity Loss of the Marine Environment

Robert Wakeford, Marine Resources Assessment Group—UK

A short-term project was recently undertaken on behalf of the European Environment Agency to assess the state of biodiversity in European marine ecosystems. The project focused on identification of important stressors and trends in various indicators, such as the status of sentinel species.

The indicator approach has facilitated the transformation of raw data into information that can be used for reporting complex trends at different levels of analysis (e.g., global, regional, and national). Indicators are used to help determine what is changing, to what extent (state); how (pressure); why (explanation); and how the system in question is changing or being made to change in response (response). This pressure—state–response (PSR) framework facilitates analysis of complex phenomena in the marine environment, and simplifies reporting and communication of results to a wide audience.

The PSR framework requires suitable indicators. A number of criteria are useful for choosing indicators, but their final selection depends on the intended user group. Realistically, most indicators cannot meet all the criteria. Rather, optimal characteristics may be used to select indicators most suitable for the questions and audience at hand.

In Europe, several marine management regions have been classified for assessing the ecological state of the environment. These include the International Council for the Exploration of the Seas (ICES) statistical areas, European regional seas, ICES ecoregions, and large marine Ecosystems (LMEs). Each classified region has a number of merits and complications for assessing the status of marine biodiversity, including data availability and ongoing changes to political and physical boundaries.

A number of pressures on the marine environment have been identified through use of indicators. Increases in fishing pressure and measured fishing capacity (i.e., engine power), have shown a general decline since the 1990s. However, it is notable that fleet capacity reductions are likely to have been offset by increases in overall fleet efficiency and effort. Other pressures include increases in sea temperature and levels of marine pollution.

In total, five categories of species were determined to have the potential to serve as sentinel species in the oceans and seas around Europe. These include: habitat creators, planktonic predators, top predators, water body indicators, and feed indicators. Case studies were developed for each indicator species in the Black Sea, Baltic Sea, North Sea, Mediterranean Sea, and Celtic-Biscay shelf.

In conclusion, the results showed that the major anthropogenic pressures on the European marine environment include excessive fishing pressure and pollutive contamination and eutrophication. Sentinel species were used to identify apparent relationships between such pressure and levels of biodiversity. Additional information about the full range of interactions within such systems is needed to validate such findings and the overall utility of sentinel species in an EAFM.

Indicators of Change for the Georges Bank Ecosystem

Mike Fogarty, National Marine Fishery Service, Woods Hole

Indicators can help improve understanding of ecosystem change but may not necessarily evolve into established reference points. The interpretation of indicators is often quite difficult and without a reasonably sufficient mechanistic understanding of how a system is operating, use of indicators can be spurious.

Georges Bank and the Continental Shelf along the Northeast coastline of the United States has long been recognized as a highly productive region for nearshore and offshore fisheries. On Georges Bank, an anticyclonic gyre is associated with strong rotary tidal currents. This creates relatively high probabilities of retention of organisms on the Bank and promotes nutrient mixing.

Retention and focused mixing exert important influences on the distribution of fish and shellfish in the region. Production in the central part of the bank involves much recycling and may not articulate with higher trophic levels. The outer parts of the bank, particularly around the 60 m isobath, are where new phytoplankton production often occurs, driven by upwelling from deeper waters. Based on the overall biomass spectrum for the area, however, zooplankton production is lower than expected on Georges Bank. Consequently, energy is tightly constrained, which in turn affects population and trophic dynamics across the system. On the decadal scale, the North Atlantic Oscillation (NAO) appears to affect production in the region. When the NAO is negative, less Labrador Slope water penetrates to the Mid-Atlantic Bight and primary production in the system drops substantially.

Time-series plankton surveys have shown that the dominant copepods, *Calanus finmarchicis* and *Pseudocalanus* spp. have been relatively stable over time but that some of the smaller bodied forms associated with warmer waters, such as *Centropages*, *Oithona*, and *Metridia*, have increased in number at Georges Bank.

There has also been a large increase in scallop biomass in the system likely due in part to imposition of area closures. This may have important implications for energy flow in the system.

Analysis of fish biomass on Georges Bank indicates dramatic changes over time. Most notably there has been a major increase of pelagic fishes, and a decline of gadoid fishes. Over the period of observations (1963–2002), there was a change in the relative importance of planktivorous, benthivorous, and piscivorous fish assemblages in the system. Planktivorous fish have increased substantially over time, which suggests lower levels of fishing mortality and possibly lower predation mortality on this component of the system. The large increase in productivity in the planktivores suggests there has been a fundamental difference in the way energy is redistributed on the Banks, with much more energy going into the pelagic food web than was the case in the early 1960s and 1970s.

Interestingly, the Shannon Index of diversity does not indicate major changes despite shifts in species composition over time. The results suggest that the Index may not be a particularly sensitive indicator of change – in this system.

Further, there was no apparent shift in species richness on Georges Bank. There has been a slight decline in the mean trophic level of fishes on the Bank, due in part to the loss of some important piscivores such as cod and an increase in biomass of smaller pelagic fishes. The size spectra for benthivores showed a decline during the period of intensive exploitation in the 1970s and 1980s.

An overall recovery was indicated by the 1990s, when more effective management measures were emplaced. The piscivores show little overall change nor do the planktivores, although in the late 1960s, early 1970s, the planktivore community became dominated by sand lance and other smaller bodied creatures.

In summary, the various fishery ecosystem indicators for the Georges Bank could be grouped in three broad matrices, representing the physical, human, and biotic elements of the system (Figure 4.1) the following conclusions emerge from this analysis of biological diversity on Georges Bank:

- Increased biomass and production is detectable at the lower trophic levels.
- There has been a basic change in energy pathways to macrobenthic and pelagic food webs.
- Clear changes in the composition of fish communities as evident.
- Changes in fish community dynamics appear to be reversible.

Finally, it should be noted that Georges Bank ecosystems have historically been dominated by fisheries exploitation effects, with environmental influences largely subordinate, or at least less evident, the system provides instruction about ecosystem dynamics under conditions of prolonged fishing pressure.

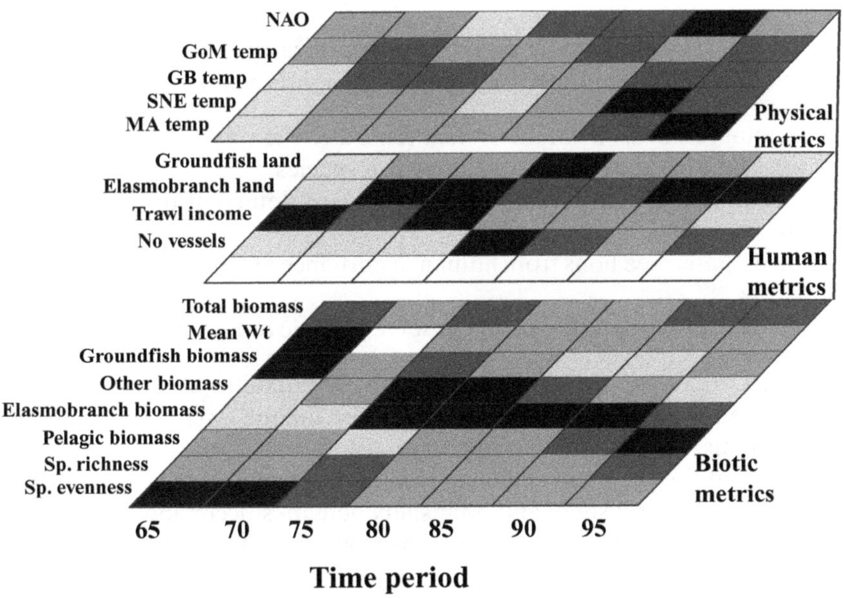

Figure 4.1. Ternary matrix of ecosystem indicators for US fisheries on Georges Bank. (For a color version of this figure, see Plate 9.)

Ecosystem Approach and Indicators for the North Pacific

David Witherell, North Pacific Fishery Management Council

The ecosystem approach being undertaken by the North Pacific Fishery Management Council is built upon four basic goals:

1. Maintain biodiversity consistent with natural evolutionary and ecological processes, including dynamic change and variability.
2. Maintain and restore habitats essential for fish and prey.
3. Maintain system sustainability and sustainable yields for human consumption and nonextractive uses.
4. Maintain the concept that humans are part of the ecosystem.

The existing ecosystem-based management objectives used to achieve these goals include the following:

- Strong science and research programs.
- Effective reporting and in-season management.
- Comprehensive observer programs.
- Precautionary and conservative catch limits.
- Limits on bycatch and discards.
- Limited entry programs.
- Habitat protection (e.g., year-round closures, Figure 4.2).
- Ecosystem considerations.
- Stakeholder involvement.
- Interagency coordination.

With regard to special ecosystem considerations, measures have been taken to minimize potential impacts to marine mammals and seabirds, via the following measures:

- Limits on total removals from the system (e.g., 2×106 ton annual limit).
- Prohibitions on directed fishing for forage fish species.
- Requirements that longline vessels deploy seabird deterrent devices to minimize incidental bycatch.
- Protection of Stellar sea lions from human disturbance and human take of important prey.
- Establishment of marine reserves to conserve benthic biodiversity.

The NPFMC Ecosystem Committee is actively pursuing additional avenues to further implement an EAFM in Alaska. Given the unique environment of Aleutian Islands ecosystems, the Council is planning to use this area to develop distinct FEP, and to test the utility of a sub-regional ecosystem council that would discuss and exchange information on fishery and nonfishery activities occurring in the Aleutians. Details of the FEP, including possible designation of an Aleutian Island Ecosystem Plan Team, are being developed.

Additional considerations are being given to various EAFM governance options (Figure 4.3). A likely scenario would involve an independent ecosystem council

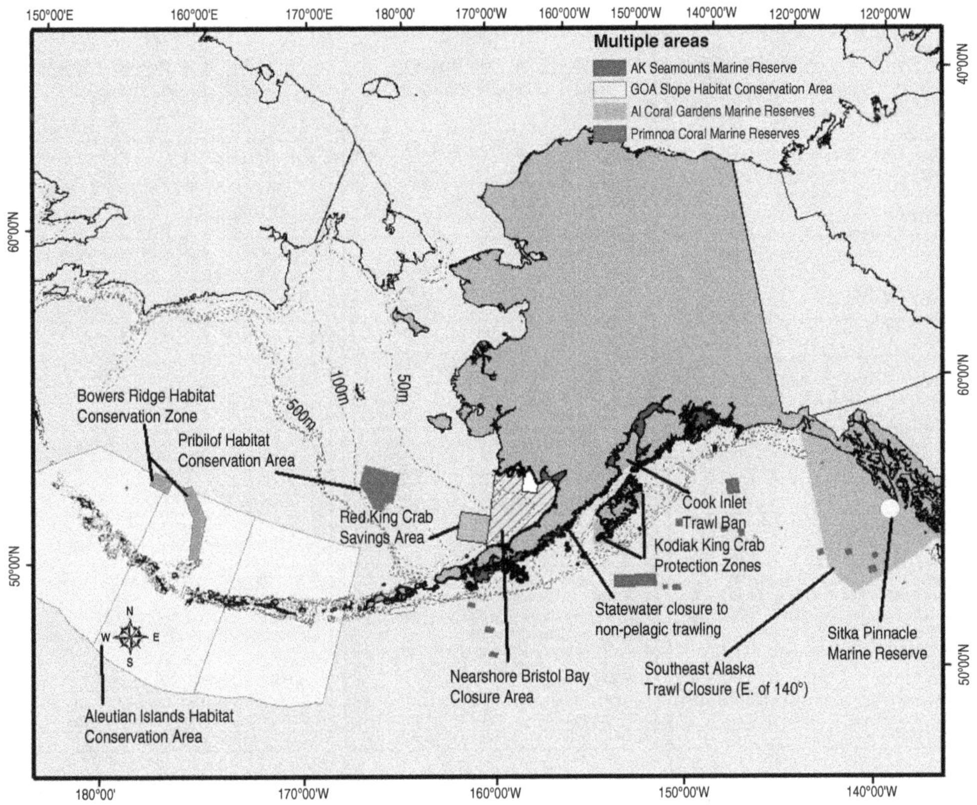

Figure 4.2. Conservation zones and area closures affecting Alaska commercial fisheries. (For a color version of this figure, see Plate 10.)

comprised of representatives of agencies with jurisdiction over the area in question (Figure 4.4).

Since 1995, the North Pacific Council has produced an ecosystem chapter in its annual groundfish SAFE report. The chapter has been developed through voluntary contributions from scientists in several agencies, including the National Marine Fisheries Service (NMFS), Alaska Department of Fish and Game (ADF&G), the North Pacific Council, and the United States Fish and Wildlife Service (USFWS). The work was initiated to document marine mammal and seabird trends but has evolved to employ a variety of comprehensive ecosystem indicators (Figure 4.5).

Environmental indicators include the status of the Pacific Decadal Oscillation, sea surface and bottom temperature, and extent of ice cover, among others. Biological indicators are built around three primary ecosystem-based objectives that call for maintaining: (1) predator–prey relationships, (2) biodiversity, and (3) energy flow and balance.

Indicators of predator–prey relationships include pelagic forage availability from survey data, concentration of fisheries on forage species, removals of top predators, and introduction of nonnative species. Diversity (species, functional, genetic) is

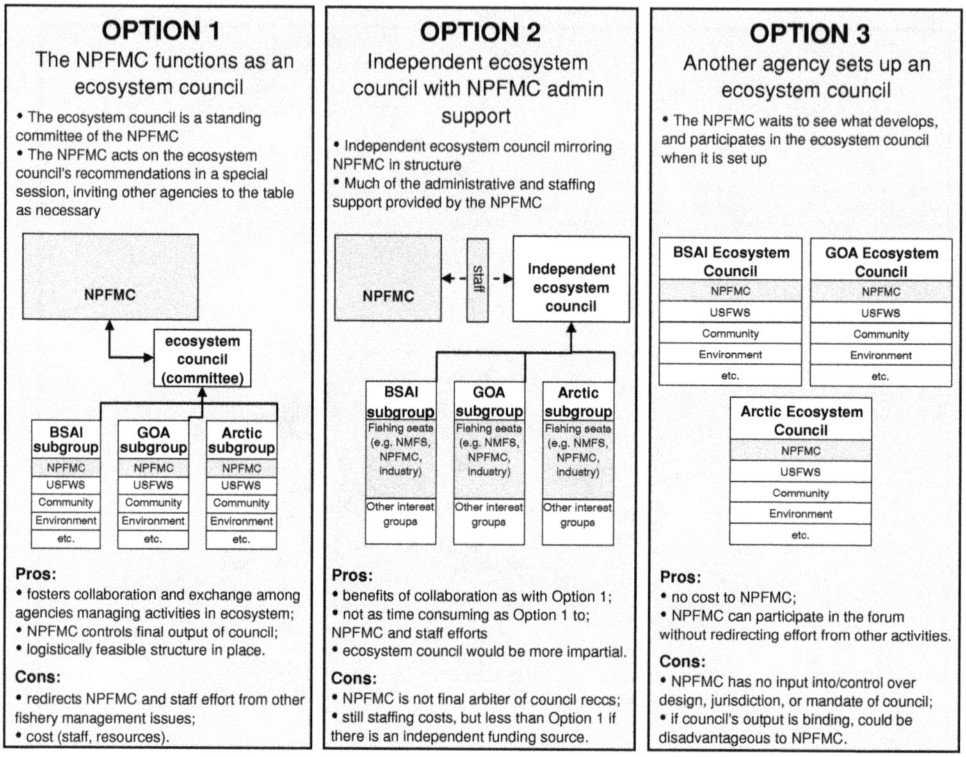

Figure 4.3. Options being considered by the North Pacific Fishery Management Council for implementing an ecosystem council initiative.

monitored through trawl surveys (e.g., richness, evenness, trophic levels, size). Indicators of energy flow include overall removals and species composition, trophic level of catch, fishing mortality rates, and discard amounts. Indicators are summarized in text tables as well as "traffic light" figures. The next step in the process is to evaluate how these indicators can be better incorporated into ecosystem decision making.

The presentation concluded with some final thoughts on the need to:

- develop regional objectives and goals for EAFM;
- develop ecosystem reference points for assessing regulatory changes;
- packaging scientific data for ease of use by managers; and
- begin preparing the FEPs across Alaska so that information gaps can be addressed as soon as possible.

Group Discussion

Many resource managers see themselves as being "downstream" from legislation that ultimately affects their ability to manage. Indeed, it is sometimes the case that managers are actively excluded from the policy and legislative process. Unfortunately,

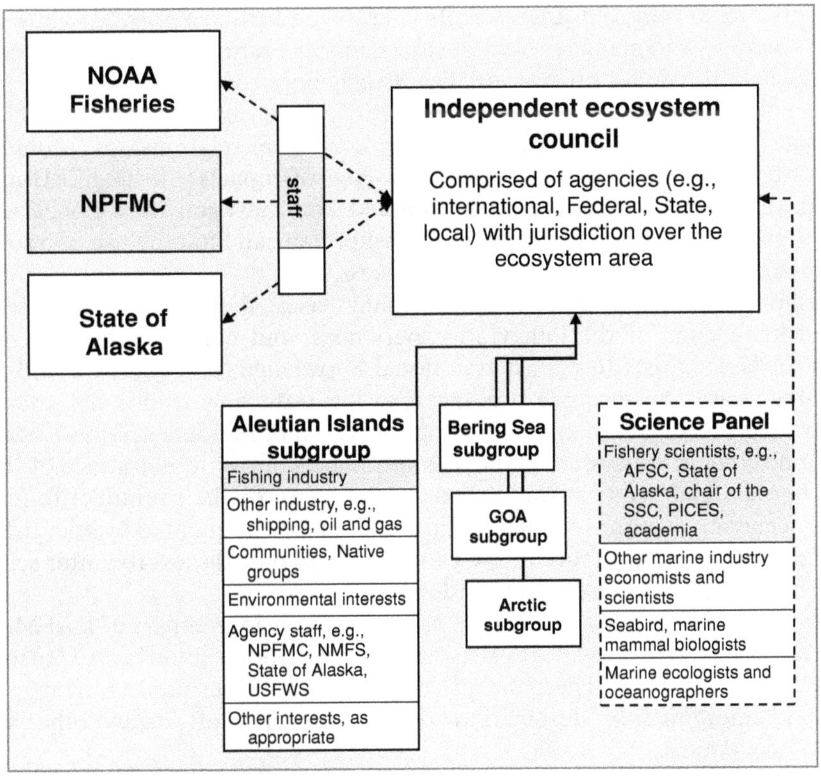

Figure 4.4. Potential EAM management option for implementing an independent ecosystem council.

Indicator	Observation	Interpretation
Oceanography		
PDO	PDO neg. w/ cool waters 98-02, PDO positive > Aug 02	higher production with positive PDO
etc....		
etc....		
Habitat		
Target fish		
Forage fish		
Misc. species		
Mammals		
Seabirds		
Aggregate indicators		
trophic level of catch	constant high level since 1960s	not fishing down food web

Figure 4.5. Examples of comprehensive ecosystem indicators developed in the ecosystem chapter of the annual NPFMC groundfish SAFE report.

if managers are so removed, then scientists are even further removed. Thus, dialog between scientists and managers and the mechanisms for articulating their needs and knowledge with the policy process are increasingly important.

It was pointed out in Table 4.1 of David Kirby's presentation that only scientists were associated with the categories "identify ecological relationships," "determine how ecosystems function," and "improve knowledge of impacts to fishing." Meanwhile, managers and resource users are often excluded from such activities. Concerns were raised about a lack of direct participation of fishermen can limit the use of traditional ecological knowledge in the management process.

Workshop participants agreed that, in many cases, there is a general need for improved knowledge of the behavior, experiences, and beliefs of fishermen. This requires social scientists to collect traditional knowledge, and compare and merge it with other data. An example was given to illustrate how traditional knowledge has contributed to scientific understanding. In Lapland, reindeer scientists had been working on reindeer ecosystem management policies but were not aware of the fact that the herders had been shaking lichens out of trees for their reindeer to feed on. This was occurring at a time of year and in locations not frequented by scientists. The cause of change in the structure of the ecosystem was only discovered after scientists engaged in dialog with the reindeer herders.

Marine Protected Areas (MPAs) are sometimes considered part of EAFM. However, it was suggested that a wide range of management tools should be used to manage fisheries under an EAFM. These include the range of options used to manage single species, and emerging tools designed to address interaction effects and other aspects of marine ecosystems.

Several workshop participants disagreed about jurisdictional responsibilities in the NWHI. Participants recognized that there will likely always be some institutional tension in fisheries management, particularly if one agency emphasized extractive values while others emphasized conservation. Such tension is natural and effort is required on all sides to ensure that ideological differences do not obstruct cooperative approaches to EAFM.

There was discussion about incorporating experimentation into ecosystem-based management. For example, fisheries data can be used to provide a sense of how the ecosystem responds to anthropogenic change, or how it behaves in the absence of human impacts. Clearly, the problem here is the uncontrolled nature of the fishing "experiment." That is, fisheries effects can rarely be adequately understood without controlling for intervening variables such as those associated with the natural environment.

One of the principles of EAFM is the incorporation of public input into management decisions. In the case of the NWHI, a number of institutions represented diverse stakeholder views. As an institution, NOAA has developed principles for ecosystem management that must at once apply across a broad set of constituents and take into account diverse stakeholder perspectives.

NOAA has many divisions, including National Ocean Service (NOS) and National Marine Fisheries Service (NMFS), each of which have unique objectives. The National Park Service (NPS) focuses on generating a relatively pristine environment for maximum human enjoyment. In the case of the NWHI, management was challenging in part because of the vast and remote nature of the area and related communication problems between agencies. Solving such problems under an EAFM will require some degree of institutional analysis and collaborative planning on the part of the agencies involved.

Another institutional problem relates to funding and a generally limited pool of monies for implementing ecosystem-based approaches to management. In reality, substantial increases in funding are needed across the board. This problem cannot be taken up by a single source. However, if all agencies and institutions associated with the ecosystem-based approach aligned their stated needs, a strong centralized voice could serve to bring the desired effect.

A point was raised with regard to mechanisms for enhancing dialog among the marine resource user and stakeholder groups. While the current workshop is a good example of scientists and managers interacting, other mechanisms are needed to elicit additional perspectives in a nonthreatening and open manner. It was noted that when fishery managers consult user groups, the discussions often relate to new regulatory changes. This can be a very difficult environment in which to conduct open dialog. Consequently, serious thought should be given to the structuring of multiparty interaction in a way that would facilitate cooperative and productive input from the interested public.

It was noted that the cases presented by Robert Wakeford were good examples of how science often identifies correlations but is (inherently) eluded by firm causality. For example, it is not clear whether the collapse of tuna stocks in the Black Sea that started the sequence of biomass changes had anything to do with fishing or whether it was related to loss of cold deep water habitat. This is a classic example of potentially confounding effects. Similarly, most ecosystem models suggest jellyfish blooms are a symptom and not a cause of change in marine ecosystems, and generally result from depletion of planktivores. In both instances, counterfactuals make it difficult for managers to understand ecosystem dynamics. Those types of issues needed to be discussed, and policy makers need to consider precautionary responses and management options for such scenarios.

This provoked further discussion about ecosystem change and how to determine underlying causes of change. Could adaptive experiments be devised to investigate key ecosystem processes? It was felt that this could require long periods of experimentation (e.g., the Effects of Line Fishing experiment on Australia's Great Barrier Reef) with the associated need for long-term management resolve, sufficient resources, and scientists who can undertake experiments and monitoring work.

The issue of ecosystem indicators was raised at the 2004 Paris Symposium on Ecosystem Management. It was noted of the conference that indicators tend to be identified and used in a variety of ways. Some modelers have developed systematic means for choosing ecosystem indicators, and principles have been formulated to guide the process. Nevertheless, choice of indicators can be problematic. Dr Wakeford indicated that his team did look at a range of potential indicators, but that the determinant force was the availability of time-series data for an indicator deemed likely to be useful. Thus, data availability is a key consideration.

This generated further review of indicator selection, particularly as regards non-target species. It was deemed unlikely that there would ever be a fixed set of indicators useful for understanding all scenarios in an ecosystem. It was noted that formal decision analysis has been undertaken to determine whether to conduct large policy experiments in marine fisheries. The analysis suggested that experimental initiatives should seek to avoid detrimental socioeconomic effects (e.g., rapid closures of fisheries in the Baltic Sea). In the case of the Western Pacific, a suite of potential areas is available to facilitate experimental approaches without potential social or economic impacts.

Participants reviewed the question of whether indicator selection could lead to something comparable to biological reference points, but on an ecosystem level. This would require a mechanistic understanding of relationships between ecosystem components and system drivers. But a fundamental problem here is a lack of understanding of many of the basic control and response relationships. With respect to biodiversity, a useful indicator could address ecological properties of a system and the way these change over time. There would, in this case, be a continuous need to monitor species composition. This led to discussion about whether a model could generate a useful type of "multivariate" reference point. This was thought to be possible in principle, but that such an approach may not provide the guidance needed to take prescriptive action in response to ecosystem change. It was agreed that reference points are one of many tools available to scientists and managers.

Discussion regarding David Witherell's presentation focused on differences and similarities between the North Pacific archipelagic approach and that of the Western Pacific. It was noted that the systems both contain a number of islands and areas that are remote and little used. While some fishing has occurred even in the most remote locations in both areas, certain areas could be used to represent a relatively undisturbed baseline of biological conditions.

Another similarity between regions is the level of oceanographic investigation of currents and gyres around the islands. However, it was noted that land use has been relatively less extensive in Alaska. Moreover, jurisdictional issues are generally less complex in Alaska than in the Western Pacific. It was thought that the general approach the North Pacific Council is taking in terms of an overall cap on harvest and is a good example of an interim approach to EAFM.

Dr Witherell noted that the North Pacific Council has reanalyzed the optimum yield cap principle, as there were some concerns that relative catch allocation under that limit might depend on which species commands the highest price; thus, conservation benefits were not equally distributed among all the target species. This generated further discussions about tradeoffs. It was noted that not all components in an ecosystem could remain at optimum levels of exploitation, as there needed to be constraints to protect both target and nontarget species and other elements of the system. It was decided among participating scientists and managers that in an EAFM, some target stocks might have to be exploited at lower rates than those considered from the single-species perspective.

Further discussion of caps on total commercial take related the biomass ratio to the fishing mortality ratio for the North Pacific. Caps were set conservatively and this was deemed a useful precaution, as the stocks that were closest to the trigger points are the ones with the highest economic value and thus are sensitive to "last minute" fishing pressure.

The Alaska example was deemed a good lesson on the importance of spatial fisheries data. Discussants placed much emphasis on using GIS to understand the spatial dynamics of fishing effort in relationship to spatial distribution of the resources. There was extensive discussion about reference levels that could trigger management actions, particularly with regard to nontarget species. Nontarget species generated considerable discussion, and workshop participants were interested in the North

Pacific Council's perspective that harvesting activities should not drive any nontarget stock into critically low levels of abundance. North Pacific stock assessment scientists have long been investigating nontarget species and were seeking to develop indicators or reference points, and control rules.

It was noted that substantial declines in productivity of currently dominant commercial species such as cod, pollock, and flounder could lead North Pacific ecosystems to be dominated by pelagic species. The issue was raised whether the North Pacific Council was planning a response to such an eventuality. Dr Witherell related that this and other changes were being addressed in a general way through development of a management program capable of responding quickly to any such changes, even if specific responses were not preprogramed. Ideally, this program will be capable of effective interaction with those fishermen who are most directly affected by new conditions in the ocean environment upon which they and other participants in the various fishery support sectors depend for a living.

References

Garcia, S. M., Zerbi, A., Aliaume, C., et al. (2003a) The ecosystem approach to fisheries. FAO Fisheries Technical Paper (Suppl. 443).

Garcia, S. M., Zerbi, A., Aliaume, C., et al. (2003b) Technical guidelines for responsible fisheries 4. FAO Fisheries Technical Paper (Suppl. 2).

Chapter 5
Working Groups

Breakout discussions were held during the workshop to facilitate review of ecosystem data sources, models, and indicators. Summaries of the discussions and key points made in each session are provided in the following pages.

Prelude to Breakout Sessions

In preparation for the breakout group discussion, Steve Murawski reviewed a variety of important considerations for developing an effective EAFM for the Western Pacific. These considerations included use of reference points, valid indicators, and the appropriate data.

The concept and use of reference points in fisheries management is fairly straightforward. Model-based fishing mortality and biomass reference points are directly related to the market value of target stocks and catches. The problem in ecosystem-based management is that some management objectives relate obviously to market based goods, while others relate to less tangible ecosystem services, such as aesthetic enjoyment. The problem with reference points in this context is the difficulty of properly evaluating ecosystem services, and using these values in development of reference points.

Environmental status indicators are common in many different fields. For example, salinity, turbidity and temperature are important indicators for oceanographers. Similarly, pounds landed, ex-vessel value, and net revenue are important indicators for fishery economists. However, setting the reference levels for such indicators is a subjective decision, heavily influenced by social issues. Many fisheries management decisions and responses under single-species management involve value-based judgments.

Science should support natural resource management decisions, in general. But the ultimate decision itself is inevitably rooted in a societal value system or systems. Making these decisions is outside the sphere of science; science can only inform the decision makers (managers) by providing insights into the potential tradeoffs associated with each choice. Furthermore, quantifying the costs and benefits of monitoring and using an indicator can be a difficult choice of itself. The US Forest Service has struggled with this issue, as has the Environmental Protection Agency (EPA). Choosing indicators for management of marine ecosystems could benefit from a review of approaches taken and decisions made by agencies in both the marine and terrestrial spheres of natural resource management.

An analogy that may be instructive is the use of leading economic indicators to monitor the economy. The Dow Jones Index is routinely used to monitor the

Ecosystem-Based Fisheries Management in the Western Pacific, First Edition. Edward Glazier.

performance of the nation's economy. However, if the Federal Reserve Bank raises the interest rate, or if any regional economic conditions improve or worsen, more than one indicator may be needed to understand how the overall economy may respond. Indicators are useful when one os seeking to understand the relationship between the action occurring (the driver) and the resulting effect on the dependent variable, in this case, the economy. Notable, certain indicators or a suite of indicators may be sensitive to differing or overlapping drivers.

In environmental settings, selection of suitable indicators may require not only a thorough evaluation of available data and past trends and patterns but also targeted experimental management approaches (adaptive management). By examining the effects of controlled manipulation of certain conditions, it may be possible to identify which variables best indicate resulting changes. At the same time, sufficient understanding of the whole system is needed to detect and adjust unexpected or poorly understood vectors of change, such as oceanic regime shifts, international economic crises, or improvements in fishing technology.

Thus far, at the present workshop, the phrase "reference point" has been used in three ways: as a benchmark, as an objective, and as an indicator. Reference points can also be thought of as a threshold for risk, wherein a given fishery should not move to the wrong side of that point, because this may set in motion undesirable changes in population and ecosystem function. Indicators enable quick assessment of the performance of the system and reference points serve as benchmarks for those indicators. Thus, we have an indicator such as abundance of a given species and a point of reference for understanding functional or desirable levels of abundance under baseline or managed conditions.

The value of an indicator is that it may suggest some action to be taken or to be avoided, depending on the "closeness" of the indicator value to the reference point. There are two types of indicators. One is a status indicator for example, the status of protected resources, fishery resources or trophic dynamics of the system. The other type is the pressure indicator; for example, how much fishing effort is affecting the system, the direction and magnitude of the Pacific Decadal Oscillation Index (PDOI), and so on.

Indicators always have some level of associated error or uncertainty. Scientists and managers need to consider uncertainty when using ecosystem indicators.

Caution is urged in investing large amounts of time and resources in the computation of control rules and reference points. Eventually, there will be a larger than anticipated change in the control rule design. It may be more productive for assessing large, complex systems such as the Western Pacific to envision and troubleshoot a range of scenarios, including both expected and unexpected situations. Indeed, it is important to examine unexpected scenarios and whether or not managers possess the strategic and tactical capability to respond to such scenarios. One of the key lessons from experience in managing large systems is that nature is not readily predictable, and thus an EAFM and the science and policy that supports it must address system uncertainties and the potential for unexpected conditions. For example, the Pacific salmon fishery is a case in point. Original fisheries policy design was ultimately challenged by unexpectedly poor returns in recent years. The original control rules were necessarily abandoned, as they did not anticipate the extreme changes that have occurred.

There may well be a dichotomy among the scientists working with indicators. One group prefers to quantify time series data, and then establish criteria to set an objective level, for example, an avoidance target or a management goal. The other group suggests using a traffic light approach to indicate changes (green, yellow, and red). This also requires quantification, but it reduces the false impression of a fixed objective level of change. For example, simply being able to detect that a mean trophic level is changing may be more important than knowing the absolute value of that trophic level.

There has not yet been extensive discussion about objectives in an EAFM. Literature on multiobjective decision making suggests that it is hard to achieve an optimum result once many objectives are involved. This is something about which economists may offer useful advice. The economist and political scientist Herbert Simons stated that sometimes phenomenon are too complex to think through in their entirety, because the elements of that phenomenon are not comparable. At times, basic rules of thumb and simple logic must be used in the absence of complete understanding. Given the complexity of marine ecosystems, this may be how EAFM may be implemented. For example, the 2 million ton overall harvest cap used for an important Bering Sea fishery was, in manner of speaking, a "rule of thumb" developed from information that was available at the time.

One problem relates to the possibility that some indicators will not be well understood by management and policy personnel. The challenge will be to make ecosystem indicators readily understandable by scientists and nonscientists, alike.

A meeting sponsored by the Global Coral Reef Monitoring Network concluded that marine scientists around the world need and use need a wide range of indicators. But that policy-makers generally do not fully understand indicators of their utility.

It was decided that for sake of clarity and simplification scientists should choose a small number of particularly useful indicators and render their mechanics apparent to the non-scientific community.

Scientists present fishery managers with a range of different ecological scenarios. These are often expressed as a decision matrix for each policy choice, with likely outcomes and tradeoffs provided each option. Unfortunately, when such matrices been used to inform managers about risk, they have often been either misinterpreted, or have led to policies that merely obviate the least desirable outcome, rather than enable the potentially most beneficial decision. Part of the problem is that the tables are often hard to understand. This problem suggests that natural resource managers must acquire some basic degree of understanding of the complexities of the science and modeling work that feeds into the fishery management process.

In the Western Pacific, nearshore fisheries are generally conducted on coral reefs and in reef-associated lagoons. The nearshore fisheries are so complex in terms of species composition, catch rates, effort distribution in space and time, that it actually limits the available management options. Management controls such as catch quotas, total allowable catch, and effort limits may be meaningless since most of the information required to make grounded decisions will be impossible to obtain, and it would be highly challenging to regulate effort or catch. For example, in Hawaii alone, over 200 nearshore coral reef finfish species are harvested, excluding aquarium fish, invertebrates, and algae.

Caution should be expressed about the notion of "ecosystem health." It is noted that ecosystems are not organisms, and that it can be misleading to think of them as such. The concern about loss of individual parts and services is at the heart of many ecosystem management issues, but ecosystems cannot be measured in the same way an organism is measured. Nevertheless, the legal mandate for sanctuaries and refuges is to maintain "healthy ecosystems."

It may be difficult, however, to find a suitable substitute for the notion of "health," and one that better captures the imagination and understanding of the general public. A potential substitute may be "ecosystem status."

Regarding "health" or "status," of marine ecosystems, Pearl Harbor is often thought to be a dead coral reef ecosystem, while the NWHI are often considered to be relatively "healthy" coral reef ecosystems for several reasons. First, the reefs off the NWHI exhibit very low rates of disease, and corals are a defining component of this ecosystem. In fact, the NWHI exhibit the lowest prevalence of coral disease of any major reef ecosystem. This is partly the result of the location of the NWHI near the thermal biogeographic limit for coral growth and the positive effects of relatively low sea surface temperatures. Second, reefs in the region exhibit a high degree of endemism, and to our knowledge, endemic species have not been displaced by introduced species. Third, with respect to fish endemism, species such as hāpu'upu'u (*Epinephelus quernus*) appear to have maintained high levels of abundance in the NWHI, although their status in the MHI is less certain given ongoing pressure on MHI bottomfish stocks.

Thus, the NWHI appear to present a baseline of conditions that have not been measurably changed as a result of fishing pressure. Such relatively undisturbed ecosystems as the NWHI may provide clues to the processors that maintain high levels of abundance and large mean sizes of various long-lived species.

Finally, it should be noted that there is no reason to "reinvent the wheel" when it comes to selecting indicators. In the case of coral reef ecosystems in the Western Pacific a variety of relatively easily measured indicators are available to aid in effective management.

Data Needs Working Group Report

The Data Needs Working Group began with the basic question: "What are data good for?" Three major categories were delineated—the human, the environmental, and the ecological. The working group noted that such data fill dual central roles, as indicators of ecosystem performance, and as a source of information for modeling. Indicators and models are then used to evaluate tradeoffs associated with alternative management options. Requirements vary over time and space.

Generally speaking, valid data are needed to describe current ecosystem conditions and the history that led to those conditions. Many types of sources of data are needed to develop such understanding and the data itself must be made readily available for modelers and other analysts. The incorporation of key forms of information into relational databases is essential. Emphasis should be placed on minimally restricted global use and availability of data. Newly collected data could be restricted for an initial period (e.g., 1–2 years) to allow any confidentiality or proprietary information issues to be addressed and to enable publication of analyses.

Researchers should work and liaise with long-time users of the ecosystem such as fishermen, ecotourism operators, and native elders to help fill in much needed human dimensions information and to aid in validating information gathered by biophysical scientists.

An immediate data-related step to establishing an EAFM in the Western Pacific is determination of ecosystem data clearinghouse for the region. The clearinghouse could be developed gradually, ultimately becoming a comprehensive centralized repository of data and relational databases. The importance of a centralized archive of knowledge and the funding needed to develop and fund it cannot be overstated in this context.

The Data Needs Working Group identified several issues to be addressed with respect to EAFM data needs in the Western Pacific. These include:

- The status of protected, endangered, and threatened species
- The status of habitats
- Biodiversity
- Physical/natural variability
- Feedback effects
- Human pressure(s)
- Definition of boundaries

The Working Group framed and discussed those data needs as relevant to human needs and interests, and environmental and ecological considerations.

Human Dimensions Data

With respect to data needs to understand the role of humans in marine ecosystems and their effective management, a variety of data needs were identified. It was determined that data regarding historic utilization patterns and the history of management measures and their effects of fishing fleets and marine ecosystems were of tantamount importance. Integration of traditional ecological knowledge into historic and contemporary habitat utilization could be very useful to managers. While there may be extensive current and historic data for some areas, it may be limited elsewhere, requiring novel approaches to derive baselines of historic use patterns. Data mining will be needed, across the relevant natural and social sciences. Overall, more and better use of sociological information is needed to improve ecosystem science and management.

It also was reiterated that the effects of current and historic management measures must be documented. Analysis should address near- and long-term outcomes. It was noted that there is good information about the nature of past management measures, but little information regarding the effects, successes or failures of the measures in different locations and time periods. Fortunately, for most Council actions, there are established processes for assessing fishery and conservation management measures, but this is not universally the case among governing bodies in the region.

A point was raised that there is a need for better characterization of minimally impacted ecosystems and adjacent human communities. Analysis of geographically remote fishing communities and how they are or are not changing could contribute significantly to understanding of human-ecosystem dynamics and thus to improved EAFM in the region. Periodic socioeconomic surveys are also needed. The Working Group also recognized the need to seek out and systematically use not only commercial fishermen data, but also data regarding recreational use of marine ecosystems. Subregional data shortages were noted in this regard.

Environmental Data

The need for data with which to evaluate temporal variability in marine ecosystems around the Western Pacific was deemed particularly important given current concerns about climate change. High-resolution mapping data was clearly prioritized by the Working Group. With regard to mapping and habitat characterization, the group determined that emphasis be placed on the following:

- Baseline maps of select historic conditions and human use patterns.
- High-resolution spatial data collection technologies (e.g., high-resolution multi-beam).
- Representative, if limited ground truthing of spatial data.
- Need for clear habitat-resource linkages (species habitat preferences).
- Improved mapping linkages between shallow and deeper waters.
- Other new and developing technologies (semiautonomous vehicles, etc.) that could facilitate collection of valid spatial data.

It was recognized that the nature of connectivity between resources and particular ecosystem zones and processes needs to be better understood, including terrestrial watershed influences and the effects of urbanization in island settings. The issue of connectivity needs to incorporate passive and active movements of animals at various life stages at archipelagic scales. This may require large-scale and long-term tagging studies combining traditional methods with advanced technologies such as acoustic telemetry, satellite technology, pop-up and genetic tagging work, and so on. Studies of larval behavior and various localized studies will be needed to build sufficient understanding of ecosystem processes at the archipelagic scale of analysis.

Among the cost-effective types of habitat and environmental survey approaches suggested was the use of passive acoustics and bio-optical sensors for recording physical variability, and oceanographic methods such as drifters to track ocean circulation changes and deduce the potential for passive movement.

The Working Group issues such as marine debris monitoring, and the need for data programs to ensure adequate documentation of events or activities that may somehow exert exogenous influence on trends in the data, such as ENSO events or large-scale economic change. It was noted that there is some history of marine debris tracking, but it would be useful to have an ongoing at-sea debris data inventory.

Ecological Data

Participants in the Data Needs Working Group related the need for good temporary and spatial resolution of fisheries catch and effort data. It was thought this could enhance understanding of the effects of fishing on ecosystems in diverse areas across the region. Given the importance of noncommercial fishing in the Western Pacific region, reliable estimates of noncommercial catch and effort are urgently required. Useful indications of likely sources of discrepancy between reported commercial data and potential total catch are presented in a study of unreported catch conducted by Zeller et al. (2005a).

Participants also prioritized the need for reliable data on bycatch, thought best obtainable through properly designed and executed observer programs (which would also serve other data needs). There was also agreement about the urgent need for fishery-independent surveys to provide reliable estimates of abundance, biomass, and size spectra. A habitat-dependent stratified sampling scheme was discussed.

With respect to survey of insular area bottom fish stocks, it was suggested that, in addition to using standardized commercial-type gears, there could be a potential for focusing camera stations to monitor and analyze resource diversity. This will require work to enhance reliable estimation of abundance and biomass. Use of four-camera 360 degree stations may permit time limited stationary point-census approaches not possible with single camera systems. Stationary point-census counts are one method used for visual census of coral reef and reef-type habitats amenable to visual abundance estimates.

It was suggested during the data needs breakout session that spatial distribution of fishing effort and potentially site-specific fishing effects could be determined using VMS records. These, when used for research and monitoring, in combination with observer data and VMS data, have been shown to provide highly reliable catch and effort data estimates. Spatial stock structure analysis may require carefully designed genetic studies, and the recently developed tagging approaches that use "genetic fingerprinting" may enhance measurement of exploitation rates for certain species.

The need for valid trophic relationship data is critical for modeling. Acquiring such data is conditioned by extensive variability between regions and sub-regions. Methods such as fatty acid and isotope analyses, and diet studies were discussed in some depth.

The Working Group applied specific focus on the need for data regarding threatened and endangered species, especially if these are affected by fishing activities. Needed data include species distribution, genetic signatures, nature/extent of fisheries interactions, role in food webs, abundance, and local movement and migration. The work would be enhanced in the near-term through study of the spatiotemporal proximity of endangered populations to fishing fleets and derelict gear.

Finally, working Group participants discussed the need for data regarding the effects of introduced species on natural communities. Factors such as predation, competition, and human vectors of introduction were reviewed. Participants agreed about the need for study of fouling organisms on vessels, and abundance of invasive species in the benthic zone.

Recommendations and Priorities

The Data Needs Working Group recommended establishment of a formalized *Data Needs Working Group* to guide research design and data collection in support of an EAFM in the Western Pacific. As a priority, this working group should be guided by and focused on meeting data requirements for effective modeling, which in turn is driven by clear management questions and priorities. Identified data needs include, but are not limited to the following kinds of information:

- Catch and effort, including noncommercial data.
- Bycatch/by-product/fishery interactions and tradeoffs.
- Trophic interactions and diet.
- Habitat-species associations and habitat-fishery interactions.
- Stock identification and spatial distribution.
- Environmental variability.
- Traditional ecological knowledge.
- Ontogeny/life history.

- Spatially explicit processes.
- Consequences of and responses to climate change and oceanic regime shifts.
- Eutrophication/habitat alterations/inherent ecosystems productivity.
- Social/economic dimensions.
- Carrying capacity/lower trophic level and forage base interactions.
- Spatial contrasts that reveal processes under differing use impacts.

Ecosystem Models Working Group Report

The Ecosystems Models Breakout Group focused on a key question: "Can we develop a flexible, quantitative framework to address management issues and the range of policy and regulatory options required to sustain resources under an EAFM?"

Placed in the context of the Western Pacific region, this led to extensive discussion about the many and varied management situations that confront fisheries managers throughout the region, ranging from localized NWHI bottomfish fishery management issues to regional and international pelagic fisheries management issues. There was discussion about the extent of transferability of data, indicators, and models across the archipelagos. Questions were asked about whether it would be possible to draw inferences about trophic relationships in the MHI from what is known about the NWHI, and whether such information would be useful for understanding ecosystems in the Northern Mariana Islands and American Samoa. It was noted that trophic relationships may vary too extensively to allow reasonable inferences or useful comparative analysis.

The group recognized challenges associated with setting for modeling priorities for the region as a whole, particularly given the variable nature of fisheries, fisheries data, and nearshore ecosystems among the archipelagos. The variable needs and interests of humans were recognized as particularly important considerations for developing modeling priorities and associated data and fishing needs. This raised the awareness of the likely need to employ risk-based procedures in ecosystem modeling and application of modeling results.

There was consensus that the modeling endeavor must include the following processes:

- Identification of resource and management issues.
- Identification of potential management policies and options.
- Matching of the model to management policies and options.
- Identifying data needs for the model(s) chosen.
- Inventorying and obtain the data available.
- Identifying biophysical processes that could confound ecosystem analysis.

The Working Group noted that there are differences between generalized models that forecast changes in oceans and climate over time, and predictive models that are pointed to the specific needs of ecosystem modeling for purposes of fisheries management. The group emphasized the critical importance of clearly specifying the objective of any modeling exercise before it is undertaken.

For fishery managers, a model can function as a tool to help gain a sense of the likely impacts and implications of specific management actions. Hence, fisheries-oriented models must inform managers about some pertinent biological, ecological, social, and/or institutional aspects of specific fisheries. Models cannot be expected to solve problems, but rather they must inform decision-makers about the potential implications and risks of alternative management measures.

It was noted that, if tightly structured, the modeling exercise itself can also help *define* the nature of the management problem or issue being *addressed*, the components of the ecosystem that need to be encompassed by the model, the necessary data, and the tradeoffs involved in the policy environment, including the potential risks and impacts of what could turn out to be inappropriate decisions.

Participants recognized that it is difficult to model likely fishery and ecosystem responses to regulatory changes. This calls for adaptability among managers and modelers alike, and it was suggested that one approach would involve "preliminary" models that can be field-tested and refined over time.

It was noted, however, that the fishery councils may find it difficult to be able to "test" preliminary models, since the potential costs to marine resources and/or to fishermen may be high. On the other hand, the potential risk and costs of not undertaking adaptive approaches to modeling and management may could be even higher in the long term. As is the case with all management strategies, an adaptive approach to ecosystem modeling requires understanding, participation, and acceptance from the persons and institutions affected by the process.

With respect to likely models that are most likely to be developed in future years, Working Group participants agreed that these will incorporate, data layers addressing hydrodynamics, biological community dynamics, species-habitat associations, and the behavior of fishermen.

Recommendations and Priorities

The group agreed to the following priorities:

- Clearly identify management issues by area, and potential policy options available to the Council and affiliated decision-makers.
- Identify data needs and obtain available and/or new data suitable for the models chosen.
- Refine model capabilities, accommodate new data and modeling techniques, and adapt to changing ecosystem and fishing dynamics.

Indicators Working Group Report

The Indicators Working Group discussed the range of biophysical indicators that would likely prove useful in the Western Pacific region. A classification scheme was developed to organize indicators for: the state of the ecosystem, ecosystem stressors, and ecosystemic responses to pressures and stresses.

It was noted that while some indicators might be used to assess the general state of an ecosystem or ecosystem component, others were needed to address specific management needs, such as those related to protected species interactions and overfishing among commercial and recreational fishing fleets.

Session participants felt it important to recognize that most indicators cannot function as holistic ecosystem indicators *per se*, but could rather capture selected properties of the ecosystem in question. These include primary production, which affects all levels of the ecosystem, and the assemblage properties of an ecosystem.

It was suggested the North Pacific Council's approach to ecosystem-based fishery management may provide instruction on the use of indicators in a real-time management setting. This approach incorporates the concept that humans are important elements of marine ecosystems. The North Pacific Council's use of indicators relates to its overall objectives for EAFM, which are to:

- Maintain biodiversity consistent with evolutionary and ecological processes, including dynamic change and variability.
- Maintain and restore habitats essential for fish and prey.
- Maintain sustainable yields for human consumption and nonextractive uses.

The North Pacific Council's ecosystem planning process has evolved since 1995, when it initially focused on marine mammals. It later broadened to include other nontarget species, and now incorporates environmental and ecological indicators relevant to an EAFM. The goal is to monitor the efficacy of ecosystem-based management efforts and to track changes in ecosystems not easily addressed by single-species assessment. Specific objectives now include maintenance of species diversity, trophic diversity, structural habitat variability, genetic diversity, and energy flow and balance.

As a cautionary note, the temperate and subarctic ecosystems of the North Pacific region differ in various ways from the tropical ecosystems of the Western Pacific region. This confounds direct comparison of EAFM between the regions.

For the present workshop, the Indicators Working Group developed indicators for use in a pressure–state–response framework. It was noted that the final choice of indicators must be linked to specific management objectives, and the distinction between pressure and state indicators can be blurred in some instances. It was considered likely that for some indicators, no data would be available to indicate the status of or pressure on certain ecosystems. The group stressed the need for awareness about the origins of a given indicator; that is, whether it was modeled or developed from empirical data. A key if obvious consideration in developing indicators is availability of relevant data. A data availability study or review was considered to be of paramount importance and funding sources should be identified to undertake such work.

The Working Group discussed indicators for coral reef ecosystems and the coastal margins of the Western Pacific region. Viable indicators are thought to include coral cover; species abundance; population structure; and distribution of target, bycatch, and introduced species. Other potentially useful indicators include morphological diversity within a given area. Numerous studies have suggested that the link between rugosity and species diversity and/or abundance is often difficult to discern, is often species- and size-dependent, and is typically influenced by the measure of rugosity employed. The fractal nature of habitat rugosity and the vast size spectrum of marine life on coral reefs can render measures of rugosity useful for narrowly focused and targeted inquiries only.

Mapping was identified as a useful tool for documenting habitat distribution. Tools such as side scan sonar, light detection and ranging technology (LIDAR), and hyperspectral imagery were seen as increasingly important. The value of spatial indicators was thought to be enhanced when the data are appropriately grounded.

Participants listed various indicators that could be used as measures of the status of and/or pressure on marine ecosystems. Exemplary indicators and accompanying ranking criteria are depicted in Tables 5.1 and 5.2.

Table 5.1. Potential indicators relating to pressures on ecosystems.

		Ranking criteria					
Metrics	Sources/contacts	Available	Feasible	Cost	Sensitivity	Uncertainty	Mandate
Physical metrics							
Cyclones							
ENSO index							
PDO							
SST anomalies							
Satellite altimetry							
Thermocline							
Satellite wind data							
High rainfall							
Biological metrics							
Primary production							
Chlorophyll densities							
Chlorophyll transition zone							
Harmful algal bloom							
Coral bleaching							
Invasive species							
Introduced species							
Human metrics							
Commercial fishing							
Licenses							
Catch by gear							
Effort by gear							
Recreational fishing							
# Fishers per capita/area							
Catch rate							
Effort index							
Population density							
Total population							
Coastal population							
Linear coastline							
Per capita income							
Development							
Building permits							
Housing starts							
Area of natural to manmade							
Runoff							
Water quality							
Shipping traffic							
Anchorages							
Port of call manifests							
Bilge/ballast water handling							
Military							
Small boat harbors							
Track live-aboard vessels							
Boat-based visitor entry							
Marine debris							
Accumulation rate							
Ship groundings							
Eco-tourism							
# of divers/dives							
Military activities							
Bombing frequency							
Training exercises							
Port and airport security zones							

Note: Listed also are data sources or contacts if known, and suggested ranking criteria for use by expert panels in the region; List is likely incomplete.
Abbreviations: ENSO, El Niño Southern Oscillation; SST, social sciences team.

Table 5.2. Potential indicators relating to state of the ecosystems.

		Ranking criteria					
Metrics	Sources/contacts	Available	Feasible	Cost	Sensitivity	Uncertainty	Mandate
Habitat							
Live coral cover							
Incidence of disease							
Size structure							
Prop. alien/native							
Crown of thorns							
Diversity							
Rugosity(a)							
Habitat mapping							
Bottom complexity							
Apex predators/piscivores							
Diversity							
Abundance							
Biomass							
Mortality/productivity							
Herbivores/omnivores							
Diversity							
Abundance							
Biomass							
Mortality/productivity							
Interactions with habitat							
Sea urchins							
Holothurians							
Sea birds							
Population dynamics							
Migration/movements							
Sea turtles							
Population dynamics							
Migrations/movements							
Marine mammals							
Population dynamics							
Migrations							
Sentinel species							
Giant clams/oysters							
Napoleon wrasse							
Bumphead parrot fish							
Marlin							
Species diversity							
Genetic diversity							
Ecosystem/habitat diversity							
Endemism							
Exotics							
Protected species							
Ciguatera							
Fibropapiloma							
Viral epidemics							
Algal blooms							
Invasive species							

[a]But see note in text regarding usefulness.

Note: Listed also are data sources or contacts if known, and suggested ranking criteria for use by expert panels in the region; list is likely incomplete.

Recommendations and Priorities

The Indicator Working Group made the following general recommendations for developing an EAFM in the Western Pacific:

- Evaluate feasible, tractable, and potentially useful candidate indicators.
- Elicit expert recommendations for ranking prospective indicators.
- Select particularly high-ranking indicators for further consideration.
- Amend the list in keeping with management issues and needs for analysis of ecosystem-fisheries interaction.
- Develop approaches for using status indicators to help analyze the effects of ecosystem response to management of fishing pressure.

References

Zeller, D., Booth, S., Pauly, D. (2005a) Reconstruction of coral reef- and bottom-fisheries catches for U.S. flag island areas in the Western Pacific, 1950 to 2002. Report to the Western Pacific Regional Fishery Management Council, Honolulu.

Zeller, D., Booth, S., Craig, P. et al. (2006) Reconstruction of coral reef fisheries catches in American Samoa, 1950–2002. *Coral Reefs*, 25, 144–152.

Chapter 6
Synthesis

Dr Mike Orbach and Dr Dave Fluharty presented a synthesis of the workshop following the presentations by the breakout working groups and the final discussion of their conclusions and recommendations.

Summary Synthesis: Mike Orbach

Michael Orbach presented a synthesis of workshop findings and suggested steps for attaining an effective EAFM in the Western Pacific. In this context, "ecology" and "ecosystems" must be defined to include and indeed emphasize the effects of humans on the natural environment and the pivotal roles played by humans in the use and management of marine resources. The present meeting involved discussion primarily of biophysical aspects of EAFM. But, recurrent in the discussion were social, economic, and institutional impacts and actions. Management and policy are, in its final assessment, very much about social processes and value-driven decisions framed in an institutional "ecology." Understanding this aspect of ecosystem requires investigation of: legislative, judicial, and administrative aspects of governance in the context of use of the environment by humans. The various linkages, interactions, and issues between the human and biophysical dimensions of the world can be modeled and understood in a way that augments natural resource management.

Human constituents, often referred to as stakeholders, include fishery participants, consumers, interest groups, and members of the general public who interact with or care about fisheries. Ultimately, fishery management is about influencing and causing behavioral change in these groups of people. Rules are made by policy-makers and managers, and are ultimately legally endorsed by state and federal legislators. Thus, human actions and behavioral change are what are actually managed, not the fish, not the ocean, nor any other element of the environment.

Many participants in the current workshop conduct scientific studies. However, science does not tell people how to behave. It only analyzes and sometimes forecasts what might happen if humans choose different types of behavior. Governance, not science, is based on human values, advocacy, and subjective decision making.

An EAFM requires good scientific research and analysis. Thus, the issues presented and discussed here are crucial for the foundation of scientifically founded ecosystem-based management. However, the human component of ecosystem-based management needs facilitation, and in some cases, advocacy. Sometimes this is undertaken by a nongovernmental organization (NGO), sometimes by citizens. However, most often in human society, government agencies assume and support a specific stance or policy.

Ecosystem-Based Fisheries Management in the Western Pacific, First Edition. Edward Glazier.

All fisheries policy and management decisions have social, economic, and biological objectives and, subsequently, they generate social, economic, and biological effects. Scientists need to understand all three of these aspects of marine ecosystems in order to effectively analyze the tradeoffs of any given management action. All fishery policy and ecosystem management decisions result in tradeoffs. Hence, of fundamental importance in all likely modeling efforts, for example, is ensuring that the model helps understand tradeoffs for human societies and individuals.

Regarding Models

Models are heuristic devices. They are means for organizing information and enabling clear thought about a complex world. Clearly, models are only as good as the data that drive them, and they need to be designed with management objectives and salient issues clearly in mind.

Regarding Indicators

Both indicators and data collection processes need to be guided by principles. These principles should help guide which data are needed, which indicators are valid and useful, and which models are capable of generating insight into the management challenge at hand. The questions asked of data, indicators, and models should be formed carefully, and with full endorsement and involvement of resource managers and other decision-makers. In essence, managers need to be very clear about how to formulate salient questions for scientists and modeling.

Regarding Data

Dr Orbach suggested that data collection will need to better relate to the key questions and issues necessitated by development of an EAFM in the Western Pacific. Data collection is often based on a variety of motives, and not all scientific pursuits can or should serve ecosystem-based fishery management needs. Information needs for EAFM should relate directly to important management issues and to the models which are designed to address those issues and related questions.

Fishery Ecosystem versus Total Ecosystems

Fisheries are embedded in holistically conceived ecosystems. Ecosystem-based management must address all actions and processes that impinge on marine fisheries, including diminished water quality, loss of habitat, loss of diversity, and so on. The real challenge for managers and scientists is how to bound the system.

The ultimate ecosystem management question can be simply stated as follows: "*What would we do differently because of an increased understanding of the relationship of one species, habitat, or human behavior to one another?*" That is, what one does differently because of known linkages among numerous components, and how one judges the ecosystem tradeoffs that such understanding and related management decisions present. EAFM and all that is required to achieve effective EAFM must ultimately revolve around more effective analysis of tradeoffs in complex physical and human arenas.

Summary Synthesis: David Fluharty

David Fluharty also provided a general synthesis of the biophysical workshop. This is summarized as follows.

The accumulated body of scientific knowledge has been critical for advancing EAFM. Managers will often have to rely on this accumulated body of knowledge for decision making purposes, given that there will never be infinite resources available to address the necessary management measures in the required time-frame. In essence, this should serve as a reminder that we actually already know something about marine ecosystems and are not starting at ground zero.

A key point that became apparent during this workshop relates to contributions that social scientists can make to EAFM. This was emphasized independently in all three breakout sessions, and it became clear that virtually all participants perceive that EAFM should involve considerations about both biophysical and human aspects of the marine environment. Appropriately, the second workshop will address socioeconomic aspects of ecosystem-based management.

One way to develop a synthesis is to consider the criteria that others have put forward relative to ecosystem-based management and sustainable fisheries elsewhere. For example, as part of the National Research Council (1999) Sustainable Fisheries Report, several key recommendations were made for developing an EAFM:

- Adopt conservative harvest levels.
- Adopt a precautionary approach with respect to uncertainty.
- Reduce excess capacity and assign "rights" in fisheries.
- Establish Marine Protected Areas (MPAs) as a buffer against uncertainty and management failure.
- Include bycatch and discards in catch accounting for all sectors.
- Institute scientific and stakeholder reviews in transparent decision processes.
- Conduct targeted research on structure and function in ecosystems.
- Incorporate ecosystem-based goals in management decisions.

The 2004 report of the US Commission on Ocean Policy also made several recommendations that relate directly to ecosystem-based management efforts:

- Double the amount of ocean funding.
- Create regional ocean ecosystem councils.
- Refine the existing fishery management system . . . to strengthen the use of science and to move toward a more ecosystem-based management approach.

One may ask why it is prudent to embed marine fisheries in the ecosystem-based management approach. Given the large number of anthropogenic challenges to marine ecosystems, defining EAFM to directly include humans is critical. Fortunately, marine fisheries in the US is in some ways pre-adapted to an ecosystem-based regime. That is:

- Overall, fisheries are implicated in ecosystem change.
- A substantial body of scientific knowledge has been gained with regard to the nature and effect of marine fisheries.
- Generally, regular monitoring and assessment processes are established.
- Management systems are in place with broad regulatory authority.
- There are numerous opportunities for public participation in the management process.

It can be postulated that a "healthy" ecosystem (being aware of the anthropomorphic danger in using this word) is good for "healthy" fisheries. Hence, one could argue that implementing ecosystem-based fisheries management may lead to productive, and sustainable fisheries.

What are some of the main prerequisites for ecosystem-based fisheries management?

- Effective control over fisheries by the management system.
- Ability to enforce regulations.
- Ability to monitor all harvest, including bycatch.
- Ability to control fishing capacity and effort.
- Ability to establish incentives that match management goals.

Dr Fluharty asserted that fisheries stakeholders should recognize the inherent and often considerable uncertainty associated with natural systems. This suggests that management and exploitation of those systems be conservative and precautionary in nature. Stakeholders can expect that fisheries will change under an EAFM, specifically in the following ways:

- Fisheries will be managed for stock abundance rather than scarcity or productivity; that is, lower harvest rates from higher biomass can be expected.
- There may be less overall fishing capacity.
- Fishing incomes and use of technology.
- Fishing practices with high habitat impacts will be replaced with alternative techniques.
- Greater use of spatially explicit management measures will occur.
- Restrictions on fisheries will be implemented to accomplish other goals, for example, biodiversity protection, ecotourism, and recreational uses.

In summary, the policy advice one can give to fishery managers as they move toward an EAFM can be framed in six general recommendations:

1. Change the burden of proof.
2. Apply precautionary principle.
3. Purchase "insurance;" for example, spatial management options.
4. Learn from management experience in other areas and by applying "adaptive management" approaches.
5. Use incentives to achieve goals.
6. Promote fairness and equity within overall ecosystem-based management objectives.

Conclusions and Recommendations

EAFM requires application of conservative and precautionary approaches to managing major fisheries in a designated region. The impacts of fisheries on non-target species, the effects of fishing on habitats supporting ecosystem functions, predator–prey dynamics, and relationships between the biota and the environment must be considered and addressed. This comprehensive approach differentiates EAFM from traditional fisheries management approaches that generally focus on maximizing yield or value from targeted stocks on a single-species or multispecies group basis.

The complexity of marine ecosystems entails high levels of uncertainty in ecosystem science and management. This puts a premium on conservative adaptive approaches to management.

One problem managers have encountered in implementing an EAFM is the lack of experiential guidance on how to translate an ecosystem approach into a practical system of governance. In an integrated approach, the governance system examines a suite of information to develop management measures that achieve various strategic goals. This requires taking into account numerous perspectives and desired outcomes from a variety of stakeholders, including those representing nonextractive interests and ecosystem services.

An integrated approach must also rely on a source comprehensive data and a management decision support system to: synthesize the information and develop status indicators for individual ecosystem components, forecast status and trends, and evaluate the biological, social, and economic effects of policy choices.

The literature on EAFM suggests eight basic operational objectives that should be considered in developing FEPs:

1. Sustainable management
2. Minimizing bycatch
3. Managing tradeoffs
4. Accounting for feedback effects
5. Establishing appropriate ecosystem boundaries
6. Maintaining ecosystem productivity and balanced ecosystem structure
7. Accounting for climate variability
8. Using adaptive approaches to management

Key Points

During the discussions and plenary sessions several key points were repeatedly discussed, and are summarized as follows:

- Management/policy issues need to be clearly and precisely stated prior to data collection or modeling.
- Choice of indicators and models must be driven largely by management/policy issues and by available or readily obtainable data.
- Adaptive management experiments involving deliberate comparisons of policy options are of crucial importance for developing and implementing EAFM.
- Models cannot and should not determine the management decision. Models are conceptual devices intended to help scientists and managers think about problems and possible solutions.

- Some data collection efforts may not be appropriately scaled or may not target variables or parameters useful for EAFM. Research and monitoring efforts need to relate directly to key management/policy issues.
- New or different data may need to be collected to facilitate an EAFM. Data collection activities should include data "mining" and data recovery from old and/or unusual sources (e.g., research theses, unpublished grey literature, old print, and electronic media, etc.).
- Concerted effort is required to reduce or overcome agency-specific disagreements or miscommunication about the nature and objective of EAFM. It may be prudent to examine approaches taken and lessons learned elsewhere, such as experiences managing the Great Barrier Reef Marine Park and World Heritage Area.

Recommendations

Several recommendations can be extracted from the working group discussion and outcomes:

1. Clearly define and articulate management/policy issues and questions along lines of urgency and identified needs.
2. Assign a centralized resource entity with sufficient authority and resources to establish and maintain a centralized data clearinghouse.
3. Review and evaluate currently available data and data collection schemes (biological, social, economic, etc.), and initiate and maintain data "mining" and recovery activities.
4. Undertake review of available data pertinent to key management/policy issues identified by managers and stakeholders. The effort should be aimed at identifying strengths and weaknesses of current data and data collection programs, and pointing out obvious information gaps.
5. Identify and initiate appropriate adaptive management experiments.
6. Ensure that data collection and models/analyses used for ecosystem-based management are well-coordinated with managers.
7. Encourage simplistic models/analyses whenever possible, that is, avoid temptation to build large, exceedingly complex and expensive models.
8. Ensure adequate fiscal support for ecosystem-scale research and modeling.
9. Evaluate new and existing indicators and use in an adaptive fashion.

Overall, it was consistently emphasized that clear management objectives need to be identified before indicators and models can be proposed or developed. This is, in turn, influenced by availability of new and existing data.

A key recommendation is that a data availability inventory should be undertaken, addressing all potentially relevant quantitative and qualitative information and ideally combining physical environmental and socioeconomic data. This data inventory should be centralized, freely available, and comprehensive. An example is the Western Pacific Fishery Information Network (WPacFIN), and the University of Hawaii Pelagic Fisheries Research Program "atlas" of available data and models.

Data compilation and management should be a permanent feature of an EAFM in the region, and should be led by dedicated staff

Such an inventory should include all relevant types of data, including metadata and various forms of qualitative information. In the initial phase, this data inventory entity should facilitate the establishment of a *Data Needs Working Group* for research in support of an EAFM.

It should be noted that many of the currently available data were not initially collected for purposes of developing an EAFM, and hence the utility of the information for such an application has not been determined for all data. This should be considered under any data inventory initiative.

Much of the existing data were collected and utilized for narrow purposes (e.g., graduate research), and some may only exist in grey literature with limited print runs. It has been shown that recovering such "old" data can make significant contributions to science and contribute to historic baselines for current and future ecosystem-based science and management. This endeavor should be undertaken in close collaboration with experienced managers, and ideally with feedback from or coordination with a variety of scientists and knowledgeable fishermen.

Furthermore, ecosystem-based management will require more and better spatial data and information, with respect to both ecosystem components and functions and patterns of resource use. Use of a VMS may eventually become a requirement for extractive user groups in the context EAFM. Use of available VMS data may enhance research efforts. While this will require policy adjustments, such data can provide invaluable spatiotemporal information not otherwise attainable. VMS data may be useful, especially if combined with vessel specific catch and effort information. These data would aid understanding of spatial effort dynamics and why fishermen make the decisions they do. It should be kept in mind, however, that VMS programs will require extensive fiscal resources and the cooperation of the region's fishing fleet.

The use of MPAs as spatial fishing/exploitation experiments was identified as a key recommendation, lending itself to adaptive management within EAFM. The crucial importance of adaptive management experiments, involving deliberate large-scale and long-term spatial comparisons of policy options, was repeatedly emphasized as fundamental to an EAFM in the region.

It was felt that MPAs would bear utility in the Western Pacific only to the extent to which they incorporate ecological integrity at an archipelagic scale. The social and economic implications of MPAs were discussed in the context of the MHI.

It was deemed prudent that scientists and managers incorporate the long-term time horizons into planning, governance, monitoring, and enforcement aspects of EAFM, and to ensure stakeholder understanding of ecosystem timescales. It was felt that adaptive management experiments should be undertaken at appropriate spatial and temporal scales including both near-term and long-term scales.

In terms of ecosystem modeling, close interplay with policy and management options was identified as very important. Goals, objectives, and questions must be clearly expressed to modelers, and modelers must interact closely with those formulating the questions and processes to be modeled.

A model can generate a set of predictions for different circumstances. This can expose uncertainties that should cause a responsible manager to think carefully about the choices he/she has to make. In short, management actions and research efforts need to be carefully coordinated to enable better understanding of ecosystem dynamics, and the potential human impacts of management decisions.

Workshop participants also identified a need to foster participatory decision-making. This would serve to increase the use of traditional ecological knowledge and in-depth understanding of marine ecosystems as developed by knowledgeable fishermen.

The need to ensure adequate support for ecosystem monitoring, research, and modeling was deemed critical. Funding needs to extend beyond the focus on extractive resources, to include an emphasis on ecosystem goods and services, and appropriate metrics for accounting for nonconsumptive ecosystem services. Participants believe that nonextractive goods and services will become increasingly important considerations in EAFM.

With final regard to indicators, one of the biggest challenges in EAFM is how to link high-level principles such as maintaining healthy and productive ecosystems to informative performance indicators. Unfortunately, aside from basic fishery performance indicators, such as fishery-induced mortality rates and abundance, there are no established criteria for determining proper reference levels at the ecosystem level of analysis. Similarly, quantitative reference points have not yet been developed to gauge the potential or actual social benefits of EAFM.

It is important to recognize that most individual indicators are not holistic ecosystem indicators *per se* but rather capture elements or selected properties of the ecosystem. It may be necessary to prioritize indicators, which will be a subjective process based on perceived management issues needs.

There is no single suite of quantitative ecosystem indicators to support fishery management requirements in the Western Pacific Region. A critical task in the years to come will involve adaptive accommodation of new indicators based on trial and error.

It may be useful to develop an ecosystem indicator framework similar to leading economic indicators used to gauge the state of the nation's economy. It may be possible to select (or "evolve" or experimentally develop) a combination of indicators that, over time, would allow managers to consistently gauge the status of marine ecosystems prior to and after application of new management measures.

Workshop participants agreed upon a basic starting point for selecting and using indicators in an EAFM for the Western Pacific. These would include information about trends and conditions in the following:

- Habitat "quantity" and "quality"
- Keystone/functional species dominants
- Sentinel species
- Protected species
- Assemblage structure
- Biodiversity
- Pathogens
- Harmful events (e.g., severe pollution events)
- Fishery effects (catches, species, size, catch per effort, mortality)

Thus, a final recommendation generated during the workshop calls for fishery agencies in the region to evaluate a suite of such indicators for use in an evolving and adaptive process across the region.

Initially, this suite will be based on existing fishery, habitat, and protected species data, but should be reconsidered and amended, in line with management needs and related modeling requirements. The experiences of the North Pacific Regional Fishery

Management Council in this regard may offer insight into the likely challenges of this process for the Western Pacific.

Additional Comments

Several additional points were raised by workshop participants, and were marked for attention by participating management agencies.

The *Sustainable Marine Fisheries* report developed by the National Research Council (1999) put forward guidance for ecosystem-based fisheries management in the US, with several basic recommendations of relevance to the workshop. These include the following:

- Adopt conservative harvest levels.
- Adopt a precautionary approach with respect to uncertainty.
- Reduce excess capacity and assign "rights" in fisheries.
- Consider MPAs as a potential buffer against uncertainty.
- Institute scientific and stakeholder review as a part of a transparent decision process.
- Conduct targeted research on ecosystem structure and function.
- Incorporate ecosystem-based goals in management decisions.
- Ensure the capacity to enforce regulations.

Finally, as fishery managers in the Western Pacific region move toward an EAFM, six general points should be taken as clear policy advice consistent with consensus developed in other regions:

1. Industry and management should endeavor to be proactive in changing the burden of proof regarding the impacts of fishing by taking an active participatory role in research and monitoring, and resource conservation and sustainability.
2. Apply the precautionary principle as a default, but gauge the potential human impacts of doing so.
3. Purchase "insurance," for example, adequately sized spatial management options.
4. Learn from management experience in other areas and by applying "adaptive management" approaches.
5. Use incentives to achieve goals.
6. Promote fairness and equity among constituents when developing an EAFM in each of the archipelagos of the Western Pacific.

Reference

National Research Council. (1999) *Sustainable Marine Fisheries*. National Academy Press, Washington, DC.

Section 2
Ecosystem Social Science and Planning

Edward Glazier

Chapter 7
Background on Ecosystem Social Science and Planning

In 1998, the US Congress authorized NOAA Fisheries to establish an Ecosystem Principles Advisory Panel (EPAP) to examine ways in which ecosystem principles might be applied to the management of our domestic marine fisheries. The Panel subsequently determined that such principles would best be applied by gradually replacing existing Fishery Management Plans used by the nation's regional fishery management councils with plans that incorporate useful information about the ecosystems within which domestic fisheries occur. These would be called Fishery Ecosystem Plans and would involve a management approach that is "adaptive, specified geographically, takes into account ecosystem knowledge and uncertainties, considers multiple external influences, and strives to balance diverse social objectives" (EPAP 1999).

The Western Pacific Council subsequently incorporated ecosystem principles in the nation's first ever ecosystem-based fishery management plan—a plan for managing coral reef ecosystems, first implemented in 2001. The Council has since drafted place-based FEPs to further the ecosystem-based approach across the region (Western Pacific Fisheries Management Council 2005). A Draft Programmatic Environmental Impact Statement has also been completed (National Marine Fisheries Service 2005).

In keeping with EPAP recommendations, the Western Pacific Council has undertaken an incremental and collaborative approach to implementing FEPs across the region. One element of this approach, as described in this book, was the workshop series conducted to aid in the transition from FMPs to FEPs and to enhance application of ecosystem-based management principles over the long term in the Western Pacific. This chapter of the text reports on the Ecosystem Social Science Workshop held in January 2006.

Rationale and Questions for an EAFM in the Western Pacific

Evidence of decline in production associated with open access fisheries in the Northern Hemisphere during the late nineteenth century eventually led to development of international conventions for limiting fishing pressure in the Atlantic and North Sea. These were the first of a long series of strategies designed to improve the status of marine fisheries amidst growing pressures on natural resources in heavily populated areas of Europe and the United States. Management strategies and associated theory have since varied in nature and extent by region and over the course of time. Propagation theory, growth theory, biological productivity, equilibrium, logistic models, maximum sustainable yield, maximum economic yield, and a range of other conceptual

Ecosystem-Based Fisheries Management in the Western Pacific, First Edition. Edward Glazier.

approaches and applications have been tried. Modifications and paradigm shifts have emerged in response to ongoing challenges.

Most recently, fishery scientists, managers, and policy analysts in the United State and abroad have shifted attention to the principles and strategies of an EAFM. The definitions and parameters vary and continue to evolve, and there are similarities to previous approaches (Garcia et al. 2003), but there is general consensus that the ecosystem approach to fisheries management is novel in its attention to *whole marine systems and physical and biological relationships among the components that comprise those systems.* The Western Pacific Council defines ecosystems as "geographically specified systems of organisms, the environment, and the processes that control its dynamics." Significantly, it also considers humans and their societies to be an integral part of ecosystems (Western Pacific Fisheries Management Council 2005: 4) as per the EPAP 1999), which states that:

> Managers must also understand the complex linkages between natural ecosystems and the economic, social, and political dynamics of human systems. Humans are integral components of ecosystems and their interests, values and motivations must be understood and factored into resource management decisions. Information on human systems is as important as that from natural systems and must be included in any ecosystem research and management efforts. (EPAP 1999, p. 47)

The impetus for planning and implementing an ecosystem approach to fisheries management in the Western Pacific relates not only to this nascent paradigm shift, but also to the readiness of the Council to engage a strategy that is attentive to relationships within and between biophysical and human systems in island settings across the region. The ecosystem approach is seen as particularly amenable to the Pacific island context in that (a) historic management strategies undertaken here ultimately recognized human and biophysical relationships and interactions and, therefore, provide conceptual models for planning a new approach, (b) island settings foster common recognition of such relationships and interactions, and (c) an ecosystem strategy organized by archipelago may serve to improve focus on such interrelationships at local and archipelagic levels of analysis while reducing administrative burdens associated with management of single species pursued by multiple fleets across distant archipelagos.

The approach holds promise for enhancing existing fishery management efforts in the region. But its prospective development and application also lead to various questions and uncertainties for students and practitioners of contemporary fishery management. These include questions about human dimensions of ecosystems. One might justifiably ask, for example, whether establishment of the ecosystem approach will (a) necessitate collection and analysis of new forms of information about relationships within and between groups of fishery participants, governance entities, and the marine environment, (b) require development of behavioral modeling efforts to help predict its human or environmental outcomes, and/or (c) call for identification and development of indicators useful for assessing its economic or social impacts and/or ultimate effectiveness in simultaneously using and sustaining marine resources.

Purpose of the Social Science Workshop

Given the importance of such questions, the need to complement the initial biophysical workshop with analogous examination of social, economic, cultural, political, and demographic aspects of fishing, fisheries, and fisheries management was clearly recognized by the Council. Humans and human interests are pivotal aspects of fisheries management, and in the culturally diverse settings of the Western Pacific, such interests are highly complex. Thus, it is clear that social science has much to offer in terms of furthering understanding of marine ecosystems, associated resources, and their effective management under the "new" ecosystem paradigm in the Western Pacific.

The principal intent of the following chapters is to document the outcome of the social science workshop. This is, in itself, a relatively straightforward descriptive task. But because we wish to maximize the utility of the material for persons involved in fisheries management and ecosystem-related social science in this region and others, we provide additional context and draw on the workshop to move toward a general approach for incorporating the human dimension into ecosystem-based resource management in the region. Indeed, in reporting on the many human dimensions of EAFM and by describing key aspects of the regional context, we unavoidably and naturally arrive at basic conclusions and recommendations of potential value to resource managers and observers of ecosystem-based management in this region and elsewhere.

Organization of the Chapter

This introductory section and following sections build preliminary context. We begin by describing of the Western Pacific Council mission and revisit its rationale for moving toward an ecosystem-based management regime. This leads to discussion of unique aspects of Pacific islands and islanders and conditions that render the ecosystem approach particularly amenable in this setting. Some pertinent ecosystem models and lessons from the past are also reviewed. The subsequent section builds additional context with review of formalized ecosystem principles and Council actions on ecosystem issues to date, and brief discussion of the evolving role of social science in ecosystem-based fisheries management.

This is followed by summaries of speaker presentations and by a final section that elaborates on prospective social science approaches to ecosystem-based management in the Western Pacific.

Western Pacific Council Mission and Purview

As stipulated in the Magnuson–Stevens Fishery Conservation and Management Act (the Magnuson–Stevens Act), the Western Pacific Council was established as one of the nation's eight fishery management councils in 1976. It has thus been involved in the management of fisheries in the region for 30 years. The Council is the policy-making body for the management of fisheries in the US Exclusive Economic Zone (EEZ) of the Western Pacific. This includes fisheries conducted around the Hawaiian Islands, American Samoa, Guam, the Commonwealth of the Northern Mariana

Islands, various Pacific remote island areas, and in the vast open ocean areas of the region. The Council monitors fisheries and marine resources throughout the region with the cooperative interaction of NOAA Fisheries and develops and adjusts policies to ensure their sustainability over time.

As noted previously, that region is truly vast (see map, following). It is also complex in terms of national and international political–jurisdictional boundaries. It should be kept in mind that the region extends across the EEZ of Hawaii, American Samoa, Guam, the Commonwealth of the Northern Mariana Islands, and the Pacific Remote Island Area, which includes Palmyra Atoll, Baker Island, Howland Island, Johnston Atoll, and Jarvis Island.

Moreover, some of these areas share offshore jurisdictional boundaries with other nations. These include the following: (1) Palmyra Atoll and Jarvis Island, adjacent to the Republic of Kiribati-governed Northern and Southern Line Islands; (2) Howland and Baker Islands, adjacent to the Kiribati-governed Phoenix Islands, (3) American Samoa, adjacent to the Independent State of Samoa and Tonga, and to the Cook Islands, Niue, and Tokelau; (4) Wake Island, adjacent to possessions of the Republic of the Marshall Islands, (5) Guam, adjacent to possessions of the Federated State of Micronesia, and (6) the Northern Marianas, adjacent to various islands of Japan.

The Council is also responsible for managing migratory and highly migratory pelagic fishery resources across the region. This is increasingly complicated in that numerous groups and conventions now address management of those resources across international jurisdictional bounds, including those of the US EEZ. These include the InterAmerican Tropical Tuna Commission, the Interim Scientific Committee for Tunas and Tuna-like Species in the North Pacific, the Western and Central Pacific Fisheries Commission, the Secretariat of the Pacific Community, the Multilateral Treaty on Fisheries between the Government of certain Pacific Island States and the Government of the United States, and others (Figure 7.1).

Brief Overview of the Fisheries

Contemporary management of marine resources in the Western Pacific necessarily relates to the history and modern experience of indigenous peoples of the various island groups. The history of fishing and use of marine resources are truly ancient in this region. For instance, early voyagers settled in the Marianas by at least 3,500 years before present, in Samoa by at least 3,000 years before present, and in Hawaii by at least 1,700 years before present (Kirch 2000). Dates of earliest settlement of Hawaii are being disputed as a result of recent archaeological discoveries. Seafood was basic to the diets of the early voyagers and island residents; discarded fish bones, hooks, and other fishing-related items are often found at the earliest sites and throughout the archaeological record.

Pacific island societies and cultures have continually evolved over the millennia, of course, but modern indigenous groups certainly do retain knowledge, interest, and a political stake in traditional aspects of harvest and management of marine resources. Fish and fishing retain great social, cultural, and economic significance for Native Hawaiians, Samoans, Chamorros, Carolinians, and other indigenous persons residing in the region. Consideration of the historic and modern experience and perspectives

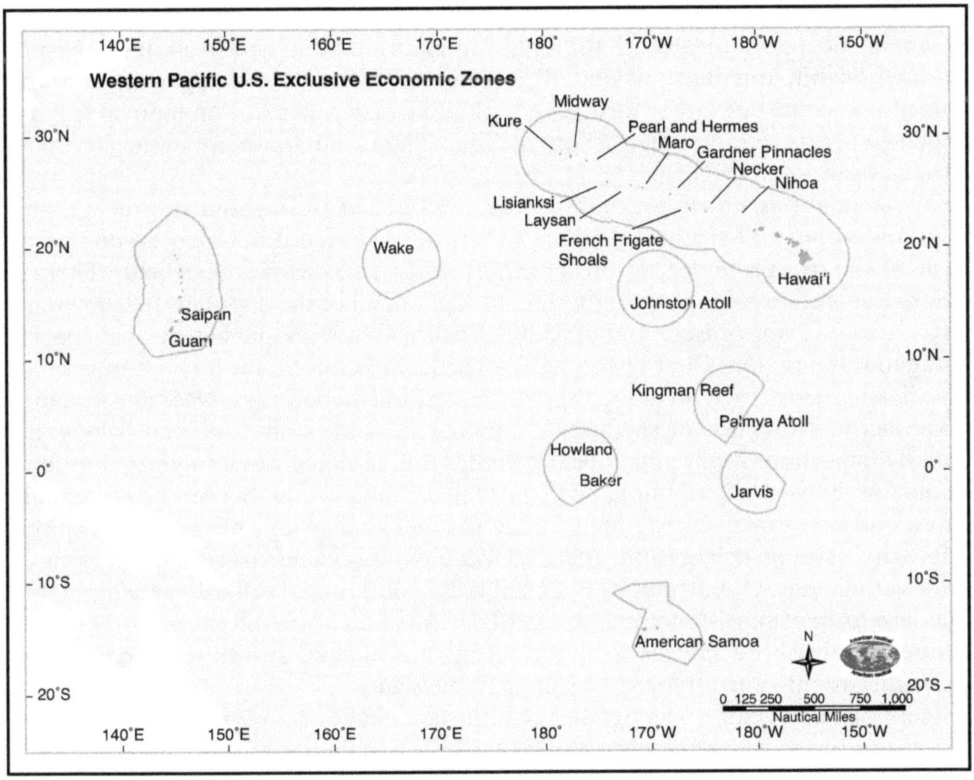

Figure 7.1. EEZs specific to US flag Islands of the Western Pacific. (For a color version of this figure, see Plate 11.)

of these groups remains an important dimension of contemporary marine resource management in the region. (Western Pacific Fisheries Management Council 2003).

Fishing and fisheries obviously are important to broader populations of island residents as well. Persons with fishing knowledge and skills have arrived from outside the region over the past century or more, gradually increasing the overall level of effort via new vessel and gear technology and subsequently the availability of seafood. Ongoing demand for seafood products in local and distant markets has led to extensive processing and distribution sectors in the larger island areas and the development of various commercial fleets (c.f., Pan and Pooley 2005).

Extensive nearshore and shoreline fishing occur throughout inhabited portions of the archipelagos, and because that deep water occurs in close proximity to these mid-ocean islands, commercial pursuit of pelagic fish is common. Participants in the Hawaii-based longline fleet pursue tunas and swordfish around the islands and in more distant locations throughout the region and beyond. Most participants in the Samoa-based longline fleet tend to pursue tunas close to the Samoa archipelago, but some fish distant waters as well, with permit arrangements to enter neighboring EEZs. Small-boat commercial trolling and handlining for tuna species occur in both Samoa and Hawaii, and commercial pole and line fishing for aku (skipjack tuna) occurs in Hawaii. Of note regarding the remote island areas, some Hawaii-based longline operators regularly fish around Palmyra Atoll, and operators of the US purse seine fleet periodically fish for tuna species around Palmyra Atoll, Jarvis Island, Howland Island, and Baker Island. The purse seine fleet is not regulated through Council decisions, but rather through a separate treaty process. Commercial trolling for pelagic species is common offshore of Guam and Commonwealth of the Northern Mariana Islands (CNMI).

Commercial pursuit of bottomfish is also important throughout the region, and deep, mid-slope, and shallow handline fisheries have been developed around all of the main islands. A commercial lobster fishery occurs in the Hawaiian Islands. There is extensive shoreline fishing and gathering throughout all of the populated island areas.

Recreational and mixed-commercial/recreational vessels pursue pelagic species throughout the region. Charter fishing is a particularly important form of tourism in Hawaii, and increasingly so in the other island groups, especially CNMI and Guam.

The act and practice of small-scale commercial and noncommercial fishing are similarly important. Many small local societies are, in many ways, organized around the pursuit, harvest, distribution, and consumption of seafood. Seafood is commonly shared and consumed in extended family settings and is an object of generalized reciprocity, sharing, celebration, and associated practices and customs for persons of various ethnicities. Opportunities for small-scale operators to sell fish actually enable a modern form of subsistence and associated ways of life. The full range of shoreline, inshore, and offshore species is important in this context, and thus, managers are necessarily attentive to all forms of fishing in the region.

Ecological knowledge is as significant in the context of inshore and offshore small-boat operations as it is for operators of larger vessels in the far offshore waters. As has long been the case for voyagers in Oceania (Gladwin 1970; Lewis 1972), accomplished fishery participants today often possess extensive and intricate knowledge of the ocean environment, the weather, swell and sea states, sea signs, bird activity, the habits of pelagic and other species, and the various bathymetric features, habitats, and ecosystems that surround the islands (Maly and Maly 2003; Glazier 2006).

Seafood is itself critically important in economic terms throughout the region. Quality seafood products are purveyed in mainland and overseas markets, and consumed by tourists visiting the islands. Indicative of the economic importance of the industry in the region. Honolulu was the 29th most productive US fishing port in 2008, with some 26 million pounds of commercial landings. Notably, the port ranked 5th in terms of value of landings, with total estimated ex-vessel value of some $73 million (National Marine Fisheries Service 2008). The commercial fishing industry is also linked in various ways to the larger economy of the region, and thus regulatory or other changes that affect commercial production are likely to affect the larger economy as well (Cai et al. 2001).

Although catch-and-release style of fishing is relatively rare, recreation-oriented fishing is also quite important in the region. This is especially so in MHI, where the rate of participation far exceeds that of other regions in the United States where the Marine Recreational Fisheries Statistics Survey (MRFSS) is conducted (National Marine Fisheries Service 2008:21–22). Per the National Marine Fisheries Service (2008:21–22), an estimated 407,000 Hawaii residents engaged in some form of marine recreational fishing in 2004. This was nearly 32 percent of the total population of 1,275,194 residents estimated for 2005 (US Census Bureau).

Research and management of marine resources involve numerous agencies, institutions, and groups across the vast area and complex sociopolitical configuration of the

Western Pacific. The principal marine fishery management entities at the level of the state, commonwealth, and territory include the State of Hawaii Division of Aquatic Resources (HDAR), the American Samoa Department of Marine and Wildlife Resources (DMWR), the Division of Fish and Wildlife (DFW) in the Commonwealth of the Northern Mariana Islands, and the Guam Division of Wildlife and Aquatic Resources (DWAR). At the federal level, the Council and the National Marine Fisheries Service (NMFS) are the principal entities involved in management of fishery resources in the EEZ. The US Department of the Interior Fish and Wildlife Service and the US Coast Guard are also involved in fishery issues in the region. Numerous nongovernment organizations and groups assert interests in the management of resources in the region.

A History of EAFM in the Pacific Islands

It is within this vast region and complex social and economic context that the Council has sought to achieve balance between the use and conservation of marine resources under its kuleana (purview). As for the other fishery councils around the nation, management efforts have, to date, assumed the structural form of fishery management planning and plans, wherein species, fisheries, and participants active in specific fisheries are considered in relatively distinct terms.

The now operational exception is the Western Pacific Council's Coral Reef Ecosystems Management Plan, the first ever ecosystem-based fishery management plan developed in the United States. As noted in Western Pacific Council (2003:8), the plan incorporates contemporary ecosystem principles in its establishment of a management regime for an extensive region and set of resources:

> The goal of the FMP is to establish a management regime for the entire Western Pacific Region that will maintain sustainable coral reef fisheries while preventing adverse impacts to stocks, habitat, protected species or the ecosystems. To achieve this goal, the FMP implements several management measures, including (1) the designation of zoned Marine Protected Areas (MPAs) for coral; (2) permit and reporting requirements to fish in designated low-use MPAs . . . , and if needed, a general permit program for all EEZ reef fisheries; and (3) a prohibition of nonselective/destructive fishing gears and conditions on the types and uses of allowable gears.

Council analysts note that the central feature of the Coral Reef Ecosystems FMP is adaptive management, "which recognizes the uncertainty, changing conditions and resilience associated with coral reef ecosystems" (ibid.). Significantly, the plan also recognizes the extensive and lengthy precedent of indigenous management of reefs and associated resources around the Pacific islands:

> Management systems for coral reef ecosystems have allowed Pacific islanders to survive for millennia . . . and are best viewed as adaptive responses over time. (Western Pacific Council 2003:8)

Clearly, both of these ecosystem-relevant concepts—adaptive management and indigenous management as adaptive process—can be applied to other marine resources and habitats in the region. Indeed, both may be particularly amenable in the context of the Pacific Islands.

Pacific Islands and Ecosystems

Several attributes render islands, and especially small Pacific islands, suitable for examining the roles of humans in ecological processes (Kirch 1997: 31) and, by extension, suitable environs for applying ecosystem-based approaches to management of natural resources. They are small relative to continents and oceans, and in the central and western Pacific they tend to be isolated. "Boundaries" between land and sea and their respective biophysical subsystems are readily envisioned (Berkes 1999:69). The sea itself is highly visible and its resources are important in the lives of many residents. Further, marine life congregates at islands (Sibert and Hampton 2003). At the same time, however, marine resources are perceived by islanders as finite and sometimes challenging to acquire (as they are everywhere), and increasingly, many goods and services are not available unless they are imported. Viewed in historical perspective, such limitations have clearly required islanders to develop extensive knowledge of marine resources and the factors that constrain or enable their availability, abundance, and acquisition (see Poepoe et al. 2003).

Of direct relevance on the biophysical front, as cited in Kirch (1997: 21), Vitousek (1995:11) asserts that islands afford the "opportunity [for scientists] to understand controls on ecosystem structure and function in a relatively simple, well defined set of ecosystems" and to develop models which "can then be applied as the basis for understanding more complex continental systems." Similarly, Kirch (1997) makes clear that understanding the long-term feedback effects of ecological change on Pacific islands may yield much insight into similar processes enacted in larger island and continental ecosystems around the world.

The latter discussion is significant in this context. Kirch (1997: 30–42) uses archaeological findings to compare the long-term responses of two divergent societies of islanders to ecological problems wrought by their ancestors. Early colonists of Tikopia in the Solomon Islands initially generated ecological challenges through deforestation and extirpation of various species. But subsequent generations overcame those challenges by developing cultural mechanisms to balance population size with highly effective means of resource production and conservation. The long-term response of colonists of Mangaia (Cook Islands) to similar problems caused by *their* ancestors was not so effective and ultimately led to further ecological damage, including damage to the marine ecosystems surrounding the island (p. 34). This, in turn, resulted in severe social problems and rapid decline of the human population. Of Mangaians, the author writes:

> . . . the social terror that pervaded late precontact Mangaia was inextricably linked to (I do not say "determined by") the sequence of ecosystem perturbations that had been precipitated [earlier] . . . . The Mangaians were in a very real sense the authors of their history, for in destabilizing and thus biotically impoverishing their island environment, they set up severe constraints that entailed severe cultural responses. (Kirch 1997: 37)

More highly adaptive responses prevailed on Tikopia. These involved certain sociocultural mechanisms for regulating population that may be considered draconian when viewed through our own ethnocentric filters. But when such controls were used in conjunction with effective strategies for managing resources, equilibrium was achieved between population density and food production:

> Protein is obtained almost exclusively from the reef and open sea through a sophisticated range of fishing and collecting strategies, the dangers of overexploitation held in check through the exercise of conservation strategies invoked by chiefly sanction (*tapu*) . . . Let it suffice to say that Tikopia is a model of the sustainable microcosm . . . (Kirch 1997: 35).

Interestingly, Kirch reports that hegemonic disruption of population control mechanisms by Christian missionaries during the period 1920–1950 preceded rapid population growth on the island. This peaked in 1952 and, in fact, exceeded the capacity of residents to produce sufficient food to respond to the effects of cyclones occurring that year and in 1953. Relief supplies arrived through the intercession of economic anthropologist Raymond Firth, who was still active in the area after his landmark work with the Tikopia in the 1930s (see Firth 1936, 1939, 1967). Kirch writes that the Tikopian council now closely monitors its population density and, in fact, some Tikopians have been forced to reside elsewhere in the Solomons. The chiefs reportedly are "acutely aware that their sustainable ecosystem depends upon a delicate balance between human numbers and productive resources" (Kirch 1997: 36).

The Antiquity of Ecosystem Concepts in the Region

History bears many lessons in the Pacific islands. It must be kept in mind that Polynesians, Micronesians, and Melanesians were developing detailed knowledge of and traditions regarding use of island ecosystems and resources long before the Viking expansion into continental Europe and longer still before arrival in the New World. Indigenous peoples in the Western Pacific gradually developed and modified settlement patterns and sociopolitical systems to utilize and/or produce marine and terrestrial resources in their respective island groups.

The early Chamorro peoples who first inhabited the Mariana Archipelago developed a "semi-mobile archipelago-wide settlement system adapted to irregular rainfall and frequent typhoons" (Amesbury and Hunter-Anderson 2003). The society gradually evolved in terms of cultural and political complexity, and the population expanded through subsequent millennia. Problematic contact with the Spanish preceded rapidly diminishing populations and changes in life ways during the late seventeenth century. While the Chamorros continued to inhabit the region, members of societies from around the Pacific islands and Pacific Rim gradually accompanied them. Carolinians practiced low impact subsistence fishing around Guam beginning in the early nineteenth century. Filipino immigrants followed, and subsequently Japanese, Americans, and others arrived—each with unique approaches and intensity of use of marine resources and ecosystems.

Indigenous residents of Samoa continue to dominate the political and cultural dimensions of life in that island group to the present day. A mixed horticultural-maritime economy characterized the region throughout much of its development, with fish and shellfish providing most dietary protein (Kirch 2000: 216). Missionaries influenced changes in the Samoan cosmology, and foreign governments occupied and eventually divided the islands into politically distinct America Samoa and Western Samoa (now independent Samoa). But the local political system of hereditary rank has remained intact. Indeed, local social institutions requiring allegiance to the *matai* or family chief are at the heart of *Fa'a Samoa*, or the Samoan way of life as enacted in both American and Western Samoa. Local village-level control over *fa'nua* (land) and nearshore marine resources is retained through the matai system, and fishing, seafood, regulation of fishing practices, and communal use and distribution of seafood remain critically important aspects of island life and organization of local society (Severance 2006; Severance and Franco 1989; O'Meara 1990; Tuilosega 2005).

Marine resources have long sustained Native Hawaiians. Seafood was originally consumed directly by nucleated groups residing on the lush windward sides of the islands (Kirch and Hunt 1985: 287–288). As new areas were explored and inhabited, society increased in complexity and seafood became a commodity for trade (see Sahlins 1992). Hawaiian society was increasingly disrupted through contact and interaction with Europeans, and so also were the social processes that sustained fishing, such as the expert crafting of hooks, line, and other gear. By the mid- and late-nineteenth century, Hawaiians were fishing primarily for purposes of consumption by the extended family ('ohana), or as a means for earning money in the context of an increasingly dominant cash economy. Methods developed during ancient times persisted in certain places through the Plantation era (see Maly and Maly 2003) and continue to be used around the Hawaiian Islands today.

The Ahupua'a and Other Forms of EAFM in the Pacific Islands

A particularly relevant lesson on ecosystems and ecosystem management derives from ancient Hawaii and the ahupua'a—the formalized system through which Hawaiians managed resources of land and sea. Ahupua'a were geographically influenced by political land divisions within which available resources from mountain to sea were produced, managed, and utilized, including pelagic resources from the deep sea (Kirch and Hunt 1985: 208; Goto 1986: 448).

The approach was particularly amenable to the geologic configuration of the Hawaiian Islands: characteristically steep mountains at center, uplands sloping downward to the coastline along ridgelines of sharp relief, resource-rich shoreline and nearshore areas, and deep water and pelagic resources occurring in close proximity to land. Ahupua'a bounds followed the topography, typically assuming a wedge shape, with the narrow point in the mountains broadening along the coast. People living or working in the forested upland areas would provide services and goods to people in the coastal portions, and vice versa. A hierarchy of persons and leaders held and transmitted knowledge and made decisions about the proper manner of pursuing, using, and managing resources in the respective zones. There was extensive interaction between commoners and leaders (ali'i) within and across each ahupua'a and island district (Sahlins 1992). The great Native Hawaiian historian Samuel Kamakau (1815–1876) provides a first-hand account of how this hierarchical political economy, still functioning during the early nineteenth century, had enabled management of natural resources and their efficient use across the Hawaiian Islands for hundreds of years (Kamakau 1992).

Marine resources associated with ko'a (areas of mounded reef) and other nearshore bathymetric features were pursued and used as food for residents of the proximate ahupua'a. Kuleana also extended to fishing locations and resources in the offshore waters (Kamakau 1992: 177–178). Even very specific grounds and resources in distant waters of the deep sea could be located by triangulating between landmarks (Kaha'ulelio 2006: 42–61).

Ruddle et al. (1992) suggest that the tabinau in Yap, the vanua in Fiji, and the puava in the Solomon Islands are variations on the Hawaiian system. Berkes (1999: 70) emphasizes the close connection between society and land and sea under each of these systems:

> In each, the term refers to an intimate association of a group of people with land, reef, and lagoon, and all that grows in or on them. This "integrated corporate estate" concept is effectively the "personal ecosystem" of the group in question: "puava is a defined, named area of land and, in most cases, sea. A puava in the widest sense includes all areas and resources associated with a butubutu (descent group) through ancestral rights, from the top of the mountains to the open sea outside the barrier reef (Hviding 1990:23)." The Fijian vanua is conceptualized in similar terms (Ruddle 1994; Ravuvu 1987). Vanua describes the totality of a Fijian community. Depending on the context, it may be used to refer either to a social group . . . or the territory it occupies, thereby expressing the inseparability of land and people in the Fijian ethos . . .

Such models are particularly useful in conceptualizing use and management of marine and terrestrial resources and their biophysical and human dimensions in the Pacific islands. It is essential to note that Native Hawaiians and other indigenous peoples in the Western Pacific region have long recognized and understood elements and dynamics of marine ecosystems, and there are numerous examples of customary marine tenure and various forms of marine resource management across the region (Johannes 1978).

In the case of Hawaii, for instance, an extensive postmissionary literature regarding use of the marine environment by Native Hawaiians indicates long-standing understanding of (a) complex biophysical relationships between land, reef, nearshore sea, deep sea, climate, and lunar phenomena; (b) effects of human activities on marine biophysical systems (no doubt including historically detrimental effects); (c) the benefits of specialization in knowledge and pursuit of marine and terrestrial resources; (d) the utility of or necessity for politically delineated bounds that related to geophysical, biological, and human considerations and needs; and (e) social organizational and customary means of managing marine and terrestrial resources in the island setting (Malo 1847, Fornander 1878, Beckley 1883, Lind 1938, Scobie 1949, Newman 1970, Handy et al. 1972, Titcomb and Pukui 1972, Kamakau 1976, Kirch and Hunt 1985, Goto 1986, Costa-Pierce 1987, Summers 1990, Abbott 1992, Maly and Maly 2003, Kaha'ulelio 2006).

In Hawaii, the historic ahupua'a system is a useful model for envisioning connections between components of the physical environment, and for effectively managing natural resources and their use by humans (e.g., Matsuoka et al. 1998). The system is reflective of the Hawaiian concepts of holo'oko'a (wholism) and pili'ana (connections therein).

But clearly, given the complex social, economic, and political contexts of contemporary life in the MHIs, such a system would be challenging to emplace in the manner it was used historically. The historic systems were developed in the context of well-established (though ever-evolving) forms of social and political organization, related customs and traditions, and modes of governance and enforcement, now significantly altered. This is not to say that some traditional form of ecosystem-oriented resource management could not be modified to fit modern conditions or vice versa. In fact, the old systems or related concepts may be particularly well suited for application in certain areas, and there is the possibility that aspects of traditional Hawaiian society may eventually be reestablished in certain areas. In any case, there is need for assessment of the pragmatic potential of traditional systems to succeed, given variable sociopolitical conditions and constraints across the region today.

Indeed, social, cultural, economic, and political aspects of life have changed radically since the ahupua'a system was used in Hawaii, and these factors continue to evolve in each of the island groups in the Western Pacific. Establishment of any new form or mode of management of marine resources in the Pacific islands clearly calls for assessment of those conditions that would influence its establishment in often rapidly changing modern settings. This begs questions about the historic challenges and successes of ever-evolving traditional forms of resource management; the nature of contemporary human pursuit, use, and management of marine resources in the archipelagos; and the capacity for success under the existing concepts and parameters of the ecosystem model now at the forefront of fisheries management across the United States.

Contemporary challenges notwithstanding, ecosystem concepts are in no way new to the Pacific islands. Johannes (1978:352) concurs, noting that "almost every basic fisheries conservation measure designed in the West was in use in the tropical Pacific centuries ago" (see Table 7.1). This is not to suggest that conservative use of marine resources was universal over time and space in the Pacific. Indeed, there were

Table 7.1. Forms of marine resource management used historically in the Pacific islands.[a]

Select management measure	Select places of usage
Closed fishing areas	Pukapuka, Marquesas, Truk, Tahiti, Satawal
Closed seasons	Hawaii, Tahiti, Palau, Tonga, Tokelaus
Allowing portion of catch to escape	Tonga, Micronesia, Hawaii, Enewetak
Holding excess catch in enclosures	Pukapuka, Tuamotus, Marshall Islands, Palau
Ban on taking small individuals	Pukapuka, Palau
Restricting some individuals for emergencies	Nauru, Palau, Gilbert Islands, Pukapuka
Restricting harvest of seabirds and/or eggs	Tobi, Pukapuka, Enewetak
Restricting number of fish traps	Woleai
Limited entry (by social structural arrangements)	e.g., Hawaii, Samoa
Aquaculture (fish ponds)	e.g., Hawaii

[a] After Berkes (1999: 70).

instances of problematic use and management of marine resources here (Johannes 1978: 355) as elsewhere (McGoodwin 1990: 57–59). In fact, radical modification of terrestrial ecosystems by early Polynesians had profound implications for the subsequent ordering of island societies (Kirch 1997; Kirch and Hunt 1997). As such, care is warranted in contemporary use of concepts regarding "traditional" forms of resource management in this region and others (see Pollnac and Johnson 2005).

Indeed, there were eons of trials and adaptation in the Pacific islands, as has been characteristic of societal interaction with the physical environment in all regions of the world. The assertion here is that, in the Western Pacific, small human societies gradually progressed through trial, error, and various social processes to enable broad expansion and growth of indigenous populations in each archipelago. Accumulation of ecological knowledge and mechanisms of social control were basic to this eventuality, as is made so clear, for instance, in the literature regarding life in protohistoric and early contact-era Hawaii. It should be noted that it was exogenous concepts, disease, and socioeconomic pressures that led to subsequent demographic decline of indigenous peoples in the larger archipelagos, such as Hawaii.

The story of indigenous people in the archipelagos has been one of change and adaptation, first to the unique environmental conditions of isolated islands and island groups, subsequently to environmental changes wrought by early settlers, and eventually to social and environmental pressures associated with the arrival and tenure of bearers of different cultures and economies. Localized knowledge of connections within and between biophysical and human elements of marine and terrestrial ecosystems and their use and management in the Western Pacific has been hard-won, characteristically dynamic, and perennially associated with mechanisms of social control. That knowledge and aspects of those mechanisms may be useful to contemporary resource managers who seek to adopt a "new" approach, elements of which have a long history of application in this region (Figures 7.2 and 7.3).

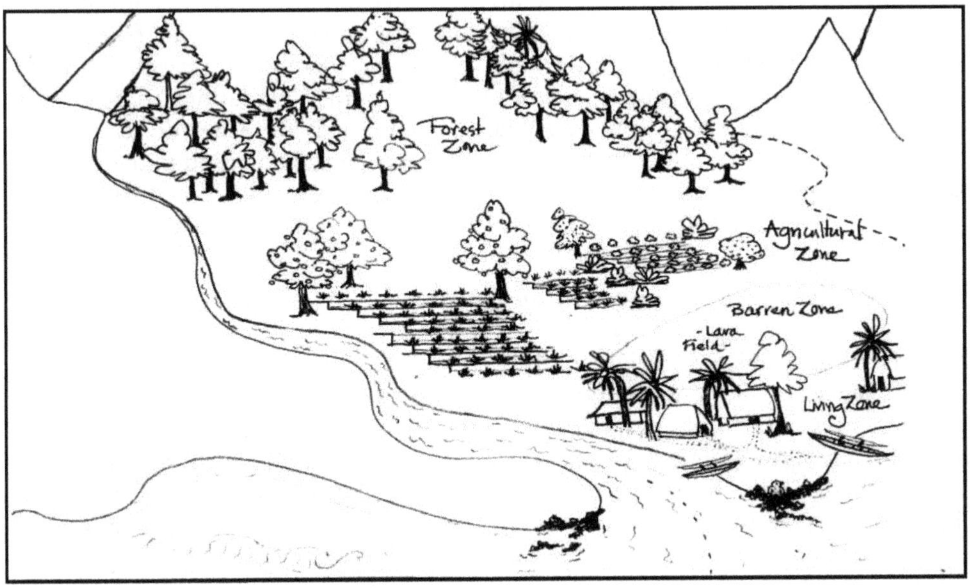

Figure 7.2. Illustration of a functional ahupua'a (rendering courtesy of Libby Stevens).

Figure 7.3. View of distant Ka'ena Point and the various ahupua'a of Waialua District; photo taken from ancient Pu'u o Mahuka Heaiau above Waimea Bay on O'ahu's North Shore, winter 2008. (For a color version of this figure, see Plate 12.)

Formal Conceptual and Policy Background

Social scientists have long recognized the importance of relationships between physical and human aspects of marine and terrestrial ecosystems (Berkes 1999). In fact, knowledge of such relationships and effective means of pursuing and using natural resources enabled the proliferation and broad geographic distribution of human societies over the course of time (Bentley and Ziegler 1999: 16).

In Europe, concepts about relationships between components of the natural world were gradually developed and modified through formalized scientific methods. These were communicated in literature through the use of a distinct terminology. Given the strong influence of "western" science on contemporary institutionalized management of natural resources, formalized concepts and principles regarding ecology and ecosystems have subsequently come to be widely recognized and used in the arena of fisheries management.

Ecological Principles Originating in Europe

Students of natural history in eighteenth-century Europe commonly drew upon theory and data regarding interconnections within and between components of the physical environment. Concepts about interrelationships between parts of a larger whole became popular and widely influential. Alexander von Humboldt used such principles in groundbreaking biogeographical observations described in *Kosmos* (1845), and Charles Darwin drew upon ecological principles and observation of exotic species and their manner of adaptation to the physical environment to formulate *The Origin of Species by Means of Natural Selection* (1859). The word *oekologie* (ecology) was

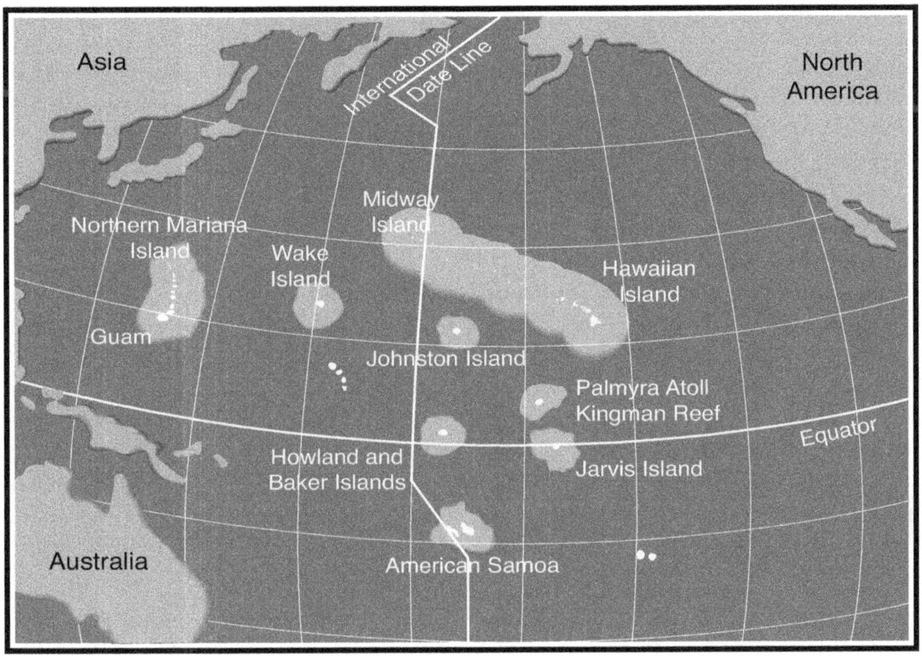

Figure 7.4. The archipelagos and remote islands under Council purview. (For a color version of this figure, see Plate 13.)

originated by German biologist and philosopher Ernst Haeckel in 1866. The term derives from the Greek words *oikos* (house) and *logie* (study). Baltic biologist and ecological pioneer, Jakob von Uexküll, reportedly used the word *umwelt* (environment) for the first time in 1909 (Capra 1996: 33).

The formal discipline of ecology, nascent in the 1930s, furthered the study of natural systems and relationships between components thereof. Significantly and ironically, botanist Sir Arthur Tansley first used the term "ecosystem" during a personal philosophical conversion to reductionism. Tansley (1935: 289) defined ecosystems as "wholes [that] are in analysis nothing but the synthesized actions of the components in associations."

Ecology stimulated interest in the study of whole systems, and it became more analytical, quantitative, and experimental with the passage of time. The discipline has become widely accepted and ecosystem concepts are now used in many environmental sciences. Definitions of ecosystems vary, but basic attention to linkages between components is typical, as in the definition provided by Mayhew (2004: 168):

> [Ecosystems are] communit[ies] of plants and animals within a particular physical environment which is linked by a flow of materials through the non-living (abiotic) as well as the living (biotic) sections of the system. Thus, ecosystems can range in size from the whole earth to a drop of water, although in current practice, the term ecosystem is generally used for units below the size of biomes...

Human Ecological Principles

Systems concepts were being applied in the social sciences as early as the 1920s. For instance, thinkers in the Chicago School used ecological principles to describe and explain social phenomena in urban settings where component parts of social systems such as individuals, families, modes of production and transportation, and government institutions were readily visible and tended to induce questions about how these functioned in totality. There was emphasis on the spatial distribution and explanation of social problems as these related to systemic processes of migration, economic problems, and coping mechanisms.

Other thinkers, such as Hawley (1950), used biological concepts such as natural selection, adaptation, and succession to characterize and explain the development of human communities in a range of settings. Yet others have used economic, sociocultural, geographic, and other human factors to explain social behavior in relation to various environmental contexts (e.g., Palinkas et al. 1985). As conceived in the contemporary context, the discipline provides a well-established conceptual framework for understanding human interactions with their physical environmental, social, and institutional surroundings.

Formal Development of EAFM

Application of ecosystem principles to fisheries management was initiated in the mid-1990s subsequent to increasingly common perceptions that management of single species through the principles of maximum sustainable yield and maximum economic yield were not producing optimal results. For example, the National Research Council (NRC 1999) asserted perspectives on the old and new paradigms in 1995 in *Understanding Marine Biodiversity*, and again in 1999 in *Sustaining Marine Fisheries*. The NRC perspective is clearly stated in the latter report:

> It is the perception of many observers that single-species fishery management has failed, and that a new approach, which recognizes ecosystem values, is required to achieve sustainable fisheries. A move toward fishing and management that recognizes the importance of species interactions, conserves biodiversity, and permits utilization only when the ecosystem and its productive potential is not damaged, is a worthy objective.

A precautionary approach to the new ecosystem paradigm was included in revisions to the Sustainable Fisheries Act (SFA 1996) and, as noted at the outset of this chapter, the Secretary of Commerce was authorized to form an Ecosystems Principles Advisory Panel to develop recommendations regarding application of ecosystem principles in the arena of fisheries management. These principles are as follows: (a) the ability to predict ecosystem behavior is limited; (b) ecosystems have real thresholds and limits that when exceeded can affect major system restructuring; (c) once thresholds and limits have been exceeded, changes can be irreversible; (d) diversity is important to ecosystem functioning; (e) multiple scales interact with and among ecosystems; (f) components of ecosystems are linked; (g) ecosystem boundaries are open; and (h) ecosystems change with time. The EPAP (1999) also identified basic elements of ecosystem-based management and developed concepts about ecosystem health and the need for indicators thereof:

> Ecosystem health refers to a balanced, integrated, adaptive community of organisms having a species composition, diversity, and functional organization that has evolved naturally. Provided that a healthy state can be determined or inferred, management should strive to generate and maintain such a state in a given ecosystem. Inherent in this management strategy would be specific goals for the ecosystem, including a description of "unhealthy" states to be avoided. (EPAP 1999)

The EPAP recognized that several legislative changes would be required to implement FEPs. It subsequently recommended interim measures to develop demonstration FEPs and, as noted above, called for voluntary adoption of ecosystem principles, goals, and policies by the nation's fishery management councils and NOAA Fisheries.

NOAA Fisheries' Marine Fisheries Advisory Committee established an Ecosystem Approach Task Force in 2001. Members identified five issues considered essential for implementing ecosystem-based fisheries management in the United States: (1) enhancement of intra- and interagency cooperation and communication; (2) delineation of geographic parameters of marine ecosystems; (3) preparation of quantified natural resource goals and objectives; (4) identification and application of specific indicators of ecosystem health; and (5) examination of socioeconomic data for the purpose of evaluating management tradeoffs. The Task Force recommended implementation of several pilot projects to illustrate the benefits and challenges to EAFM (Busch et al. 2003). These are currently being undertaken in New England, the Mid-Atlantic region, the South Atlantic region, and the Gulf of Mexico region.

Here we move beyond typical conceptualizations of the term "ecosystem," which are still primarily biophysically based, and emphasize the critical importance of human beings as pivotal elements of marine and terrestrial ecosystems. As such, the term "ecosystem" is specifically defined to encompass three basic elements: (1) a distinct biophysical realm; (2) people who are interested in or whose behavior affects or is affected by that realm; and (3) people who have authority or responsibility for developing and enforcing formal legal rules of human behavior with respect to that realm. The human and institutional ecology of constituent user and interest groups and governance entities must be considered an indispensable aspect of ecosystem research and ecosystem-based management.

Fishery Council Actions

Two influential reports indicating national dependence on marine resources and ecosystems also made clear the vulnerability of those systems to human activities. These are *America's Living Oceans* (Pew Oceans Commission 2003), and *An Ocean Blueprint for the 21st Century: Final Report of the US Commission on Ocean Policy to the President and Congress* (US Commission on Ocean Policy 2004). Both reports call for ecosystem-based approaches to fisheries management. A consensus statement signed by ocean science and policy experts followed release of the reports and also called for conservation and management of marine systems through an integrated ecosystem approach. The Bush Administration's *Ocean Action Plan* was released late in 2004 in response to the US Commission on Ocean Policy report and further advocates an ecosystem approach to marine resource management.

NOAA Fisheries has subsequently initiated exploration of regional governance structures to evaluate overall capacity for engaging in ecosystem-based management and to identify relevant management objectives, threats, options, and alternatives. The agency is developing quantitative decision support tools for evaluating management options, and it has conducted workshops to identify information and technical needs for scientists and managers seeking to examine and apply ecosystem principles in real-time settings.

Certain historic and current management strategies utilized by the nation's fishery management councils already incorporate ecosystem principles and considerations. But many of the councils are now initiating comprehensive ecosystem planning processes. The New England Fishery Management Council conducted 10 stakeholder meetings in 2005 to elicit public commentary on the nascent ecosystem approach. Preliminary workshop summaries are available at www.nefmc.org/ecosystems/index.html. The Mid-Atlantic Fishery Management Council has conducted stakeholder workshops for the same purpose. Results are pending. The South Atlantic Fishery Management Council developed an action plan outlining its ecosystem-relevant goals and objectives and is developing a Comprehensive Ecosystem Amendment (CEA) to convert its FMPs to FEPs. Information is available at www.safmc.net/ecosystem/Home/EcosystemHome. The South Atlantic Council has also initiated public scoping meetings on its action plan and CEA. The Gulf of Mexico Fishery Management Council has formed an Ecosystems Science and Statistical Committee to assist with development and implementation of FEPs in that region. It held stakeholder workshops in 2005; a report is available at www.gulfcouncil.org. The North Pacific Council has reconstituted its Ecosystem Committee and is moving forward with development of an Alaska FEP. Its Science and Statistical Committee held a multispecies ecosystem-modeling workshop in 2005. Relevant information is available at www.fakr.noaa.gov/npfmc.

Challenges to Effective EAFM

While the goal of improving management of marine resources through ecosystem-based approaches is laudable, requisite objectives and strategies are, as yet, largely uncertain. The process is continuing to unfold. Uncertainties notwithstanding, there is now much momentum, and actors in agencies and institutions in the United States are moving forward in response. Challenges associated with that response are reported further along in the book, as numerous workshop participants described them. Not the least of these is individual and collective adaptation to the new paradigm. Moreover, various management challenges persist while the new strategies are being formulated. Some amount of time and fiscal resources formerly applied to "traditional" management strategies are now being consumed by planning for and adopting the ecosystem approach. One function of the workshops being held by the Western Pacific Council is to identify such challenges early on and move toward meeting them in as efficient a manner as possible.

A perennial challenge is the desire, need, and mandate to balance the health or level of productivity of the physical environment with that of human user groups. Understanding relationships between biophysical components of marine and terrestrial ecosystems is a highly complex undertaking of vital importance to management needs and interests. But it must be kept in mind that the end goal of that understanding is to enhance the sustainability and utility of marine resources and environs *for the sake of human beings*. People both use and influence marine and terrestrial resources and environs in many ways, and understanding and appropriately managing those uses and influences are at the core of the management equation.

But ecosystem approaches in the present context will probably require increased attention to complex connections between humans and between humans and their environs. Complexity and challenges abound in that human behavior is multidimensional. It is cultural, social, economic, political, contemporary, historic, local, national, and global. Again, the intent of the ecosystem workshops being held by the Council is to move toward efficient identification and solution of the most critical of those complexities and challenges in the context of the Western Pacific.

Working Definition of Social Science and US Marine Resource Management Policy

Given that the ecosystem workshop described in this chapter is a *social science* workshop, it may be useful at the outset to provide a working definition of social science as relevant in this context. We offer the following: fisheries social science is the study and analysis of individual and/or collective human behavior associated with or affecting the pursuit, use, distribution, and management of marine resources and related environs. As noted above, when conceived in full, that behavior is complex and multidimensional. But research of human behavior in the context of marine ecosystems is not new. Complex dimensions of marine fisheries have been studied by social scientists for many decades. For instance, Bronislaw Malinowski published *Argonauts of the Western Pacific* in 1922, and Raymond Firth wrote his dissertation on Maori economics in 1927 and *We the Tikopia* in 1936. Policies requiring social research and analysis to meet the tailored information needs of resource management institutions and entities in the Pacific are relatively recent.

Passage of the National Environmental Policy Act (NEPA) in 1969 required that decision makers working in federal agencies adequately address the human dimension—the effects people have on the environment, and the way in which people are affected when some aspect of the environment is thereby altered. This was precedent for subsequent federal and state policies that recognized the place of human beings in the marine and terrestrial environment. The Environmental Impact Statement (EIS) was subsequently instituted as the standardized means for objective decision-making. NEPA required consideration of the human dimension in environmental analysis through use of a "systematic, interdisciplinary approach which will ensure the integrated use of the natural and social sciences . . . in planning and decision-making" [Section 102(2)(a)]. The Council on Environmental Quality clarified these terms in *Regulations for Implementing the Procedural Provisions of the National Environmental Policy Act*, expanding interpretation of the "human environment" to include the relationships of people with their natural and physical environment (40 CFR 1508.14).

NOAA Fisheries and the fishery management councils have addressed the human dimension in varying degrees since inception of the Magnuson Fishery Management and Conservation Act in 1976 (amended in 1996 as the Magnuson–Stevens Fishery

Figure 7.5. The ocean end of the ahupua'a of Halawa on Moloka'i area first inhabited about 650 A.D. (For a color version of this figure, see Plate 14.)

Management and Conservation Act or Magnuson-Stevens Act). In 1991, NOAA Fisheries provided interim guidance to the councils and its own regional offices regarding the need for social analysis to be included in EA and EIS documentation, and for Social Impact Assessment (SIA) to be a component of fishery plan amendments as stipulated by NEPA. SIA is particularly useful for assessing prospective or actual management actions in terms of their direct, indirect, and cumulative human effects, including aesthetic, historic, cultural, economic, social, and health effects (cf. Interorganizational Committee on Guidelines and Principles for Social Impact Assessment 1994).

Under stipulations in the Magnuson–Stevens Act, social impact assessment is to be linked to actions associated with FMPs and in some cases to a particular benchmark. For instance, when considering "a system for limiting access to the fishery in order to achieve optimum yield," the Secretary of Commerce and fishery management councils are to consider the social and economic effects of that decision [Magnuson–Stevens Act section 303(b)(6)]. Direct and indirect effects of management measures on fishery participants must also be included in fishery impact statements per amendments in the Sustainable Fisheries Act [Magnuson–Stevens Act section 303(a)(9)], and a significant part of the core definition of the Magnuson–Stevens Act, Optimum Yield, requires consideration of social and economic factors. Most recently, with the addition of National Standard 8, FMPs must incorporate assessment of management effects on fishing communities so as to assure sustained participation and avoidance of adverse economic impacts [Magnuson–Stevens Act section 301(a)(8)]. As the administrative and management parameters of FEPs are still under development in the Western Pacific region via planning and EIS processes, potential changes in the manner or extent of application of social science under the new paradigm remains uncertain. Again, the workshops are intended to assist in informing those processes (Figure 7.5).

References

Abbott, I. A. (1992) *La'au Hawaii—Traditional Hawaiian Uses of Plants*. Bishop Museum Press, Honolulu.

Amesbury, J. R. and Hunter-Anderson, R. L. (2003) Review of archaeological and historical data concerning reef fishing in the U.S. flag islands of Micronesia: Guam and the Northern Mariana islands. Prepared for Western Pacific Regional Fishery Management Council, Honolulu. Micronesian Archaeological Research Services, Guam.

Bentley, J. H. and Ziegler, H. F. (1999) *Traditions and Encounters: A Global Perspective on the Past*. McGraw-Hill, Boston.

Beckley, E. M. (1883) Hawaiian fisheries and methods of fishing with an account of the fishing implements used by the Natives of the Hawaiian Islands. Reprinted in U.S. Fish Commission. Bulletin No. 6 (1886), 245–256. Honolulu.

Berkes, F. (1999) *Sacred Ecology—Traditional Ecological Knowledge and Resource Management*. Taylor & Francis Group, London.

Busch, W.-D. N., Brown, B. L., Mayer, G. F. (eds). 2003. *Strategic Guidance for Implementing an Ecosystem-based Approach to Fisheries Management*. US Department of Commerce, National Oceanic and Atmospheric Administration, NMFS, Silver Spring, MD, 62p.

Cai, J., Leung, P., Pan, M., et al. (2001) *Linkage of Fisheries Sectors to Hawaii's Economy and Economic Impacts of Longline Regulations*. Pelagic Fishery Research Program. SOEST 05–01. JIMAR Contribution 05–355.

Capra, F. (1996) *The Web of Life*. Anchor Books, New York.

Costa-Pierce, B. A. (1987) Aquaculture in ancient Hawaii. *BioScience*, 37, 320–300.

Fornander, A. (1878) *An Account of the Polynesian Race, Its Origin and Migrations and the Ancient History of the Hawaiian People to the Time of Kamehameha I*, Vol. 3. Ballantyne, Hanson and Company, London.

EPAP (1999) Ecosystem based fishery management: A Report to Congress by the Ecosystem Principles Advisory Panel. US Department of Commerce, National Oceanic and Atmospheric Administration, National Marine Fisheries Service, Silver Spring.

Firth, R. (1967) *The Work of the Gods in Tikopia*, 2nd edn. Humanities Press, New York.

Firth, R. (1939) *Primitive Polynesian Economy*. George, Allen and Unwin, London.

Firth, R. (1936) *We the Tikopia*. George, Allen and Unwin, London.

Garcia, S. M., Zerbi, A., Aliaume, C., et al. (2003) *The Ecosystem Approach to Fisheries—Issues, Terminology, Principles, Institutional Foundation, Implementation and Outlook*. FAO Fisheries Technical Paper, (No. 443), Rome.

Gladwin, T. (1970) *East Is a Big Bird.* Harvard University Press, Cambridge.

Glazier, E. W. (2006) *Hawaiian Fishermen*. Wadsworth-Thomson Learning, Belmont, CA.

Goto, A. (1986) *Prehistoric Ecology and Economy of Fishing in Hawaii: An Ethnoarchaeological Approach.* Ph.D. dissertation in Anthropology. University of Hawaii, Honolulu.

Handy, E. S. C., Pukui, M. K., Handy, E. G. (1972) *Native Planters in Old Hawaii, Their Life, Lore and Environment*. Bernice P. Bishop Museum Bulletin 223. Bishop Museum Press, Honolulu.

Hawley, A. (1950) *Human Ecology: A Theory of Community Structure*. Ronald Press, New York.

Hviding, E. (1990) Keeping the sea: Aspects of marine tenure in Marovo Lagoon, Solomon Islands. In: *Traditional Marine Resource Management in the Pacific Basin: An Anthology*, (eds. K. Ruddle and R. E. Johannes). UNESCO/ROSTSEA, Jakarta.

Johannes, R. E. (1978) Traditional marine conservation methods in Oceania and their demise. *Annual Review of Ecological Systems*, 9, 349–364.

Kaha'ulelio, D. (2006) *Ka 'Oihana Lawai'a—Hawaiian Fishing Traditions.* Translated by M. K. Pukui, (ed. M. Puakea Nogelmeier). Bishop Museum Press, Honolulu.

Kamakau, S. M. (1992) *Ruling Chiefs*, revised edn. Kamehameha Schools, Honolulu.

Kamakau, S. M. (1976) *Na Hana a ka Po'e Kahiko* (The Works of the People of Old). Translated from the Newspaper Ke Au 'Oko'a by Mary Kawena Pukui, (ed. D. B. Barrere). Bernice Bishop Museum Special Publication 61. Bishop Museum Press, Honolulu.

Kirch, P. V. (1997) *Feathered Gods and Fishhooks*. Paperback version. University of Hawaii Press, Honolulu.

Kirch, P. V. (2000) *On the Road of the Winds—An Archaeological History of the Pacific Islands before European Contact*. University of California Press, Berkeley.

Kirch, P. V. and Hunt, T. L. (1997) *Historical Ecology in the Pacific Islands—Prehistoric Environmental and Landscape Change*. Yale University Press, New Haven.

Kirch, P. V. and Hunt, T. L. (1997) Microcosmic histories—Island perspectives on global change. *American Anthropologist*, 99 (1), 30–42.

Kirch, P. V. and Hunt, T. L. (1985) *Feathered Gods and Fishhooks: An Introduction to Hawaiian Archaeology and Prehistory*. University of Hawaii Press, Honolulu.

Lewis, D. (1972) *We the Navigators: The Ancient Art of Landfinding in the Pacific.* University of Hawaii Press, Honolulu.

Lind, A. W. (1938) *An Island Community: Ecological Succession in Hawaii*. University of Chicago Press, Chicago.

Malinowski, B. (1922) *Argonauts of the Western Pacific*. E.P. Dutton and Company, New York.

Malo, D. (1847) *Ka Mo'olelo Hawaii*. (Translation by Chun, 1987). First People's Productions, Honolulu.

Maly, K. and Maly, O. (2003) Ka Hana Lawai'a A Me Na Ko'a O Na Kai 'Ewalu. A history of Fishing Practices and Marine Fisheries of the Hawaiian Islands. Kumu Pono Associates. Prepared for the Nature Conservancy and Kamehameha Schools, Hilo.

Matsuoka, J., McGregor, D., Minerbi, L. (1998) Molokai: A study of hawaiian subsistence and community sustainability. In: *Sustainable Community Development: Studies in Economic, Environmental, and Cultural Revitalization*, (ed. M. Hoff). CRC Press, Boca Raton.

Mayhew, S. (2004) *Oxford Dictionary of Geography*. Oxford University Press, Oxford.

McGoodwin, J. R. (1990) *Crisis in the World's Fisheries*. Stanford University Press, Stanford.

National Marine Fisheries Service (2005) Towards an Ecosystem Approach for the Western Pacific Region: From Species-Based Fishery Management Plans to Place-Based Fishery Ecosystem Plans. Draft Programmatic Environmental Impact Statement. NMFS Pacific Islands Region, Honolulu.

National Marine Fisheries Service (2008) Fisheries of the United States 2008. Current Fishery Statistics No. 2008. US Department of Commerce, National Oceanographic and Atmospheric Administration, National Marine Fisheries Service, Silver Spring.

National Research Council (1999) *Sustaining Marine Fisheries*. National Academy Press, Washington, DC.

Newman, S. (1970) *Hawaiian Fishing and Farming on the Island of Hawaii: A.D. 1778*. State of Hawaii, Department of Land and Natural Resources, Honolulu.

O'Meara, T. J. (1990) *Samoan Planters—Tradition and Development in Polynesia*. Case Studies in Cultural Anthropology. Holt, Rhinehart, and Winston, Fort Worth.

Palinkas, L. A., Harris, B. M., Petterson, J. S. (1985) *A Systems Approach to Social Impact Assessment: Two Alaska Case Studies.* Westview Press, Boulder, CO.

Pan, M. and Pooley, S. G. (2005) Tuna price in relation to economic factors and sea surface temperature. Proceedings of International Institute of Fisheries Economics and Trade (IIFET) 2004 Conference, Japan.

Pew Oceans Commission (2003) *America's Living Oceans: Charting a Course for Sea Change* Pew Oceans Commission, Arlington, VA.

Poepoe, K., Bartram, P., Friedlander, A. (2003) The use of traditional knowledge in the contemporary management of a Hawaiian community's marine resources. In: *Putting Fisher's Knowledge to Work*, (eds. N. Haggan, C. Brignall, L. Wood). *Fisheries Centre Research Reports*, Vol. 11 (1), 328.

Pollnac, R. and Johnson, J. C. (2005) Folk management and conservation of marine resources: Towards a theoretical and methodological assessment. In: *Indigenous Use and Management of Marine Resources*, (eds. N. Kishigami and J. M. Savelle). *Senri Ethnological Studies*, Vol. 67, pp. 33–50. National Museum of Ethnology, Osaka.

Ravuvu, A. D. (1987) *The Fijian Ethos*. Institute of Pacific Studies, University of the South Pacific, Suva, Fiji.

Ruddle, K. E. (1994) A guide to the literature on traditional community-based fishery management in the Asia-Pacific tropics. FAO Fisheries Circular No. 869. FAO, Rome.

Ruddle, K. E., Hviding, E., Johannes, R. E. (1992) Marine resources management in the context of customary tenure. *Marine Resource Economics*, 7, 249–273.

Sahlins, M. (1992) *Anahulu: The Anthropology of History in the Kingdom of Hawaii*, Vol. 1, Historical Ethnography. University of Chicago Press, Chicago.

Scobie, R. (1949) The Technology and Economics of Fishing in Relationship to Hawaiian Culture. M.A. Thesis. London School of Economics.

Severance, C. (2006) Personal Communication. Department of Anthropology, University of Hawaii at Hilo.

Severance, C. and Franco, R. (1989) Justification and Design of Limited Entry Alternatives for the Offshore Fisheries of American Samoa, and an Examination of Preferential Fishing Rights for Native People of American Samoa within a Limited Entry Context. Final Report 89–1 prepared for the Western Pacific Regional Fishery Management Council. Honolulu.

Sibert, J. and Hampton, J. (2003) Mobility of tropical tunas and the implications for fisheries management. *Marine Policy*, 27, 87–95.

Summers, C. C. (1990) Hawaiian cordage. *Pacific Anthropological Research*, Vol. 39. Bishop Museum Press, Honolulu.

Tansley, A. G. (1935) The use and abuse of vegetational concepts and terms. *Ecology*, 16, 284–307.

Titcomb, M. and Pukui, M. K. (1972) *Native Use of Fish in Hawaii*. University of Hawaii Press, Honolulu.

Tuilosega, H. (2005) A Review of the Land Use Planning Process and Proposals to Assist Environmental and Developmental Planning in American Samoa. University of Hawaii at Mānoa. Department of Urban and Regional Planning, Honolulu.

U.S. Commission on Ocean Policy (2004) An Ocean Blueprint for the 21st Century. Final Report, U.S. Commission on Ocean Policy, Washington, DC.

Vitousek, P. M. (1995) The Hawaiian Islands as a model system for ecosystem studies. *Pacific Science*, 49, 2–16.

Western Pacific Fisheries Management Council (2003) *Managing Marine Fisheries of Hawaii and the U.S. Pacific Islands—Past, Present, and Future. Honolulu.*

Western Pacific Fisheries Management Council (2005) *Fishery Ecosystem Plan for the Hawaii Archipelago Honolulu*.

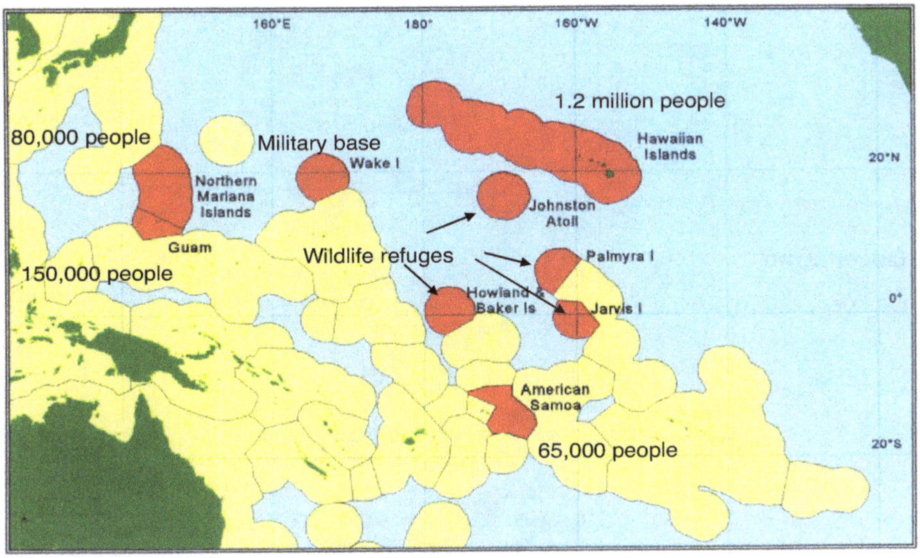

Plate 1. US and adjacent EEZs of the Western Pacific.

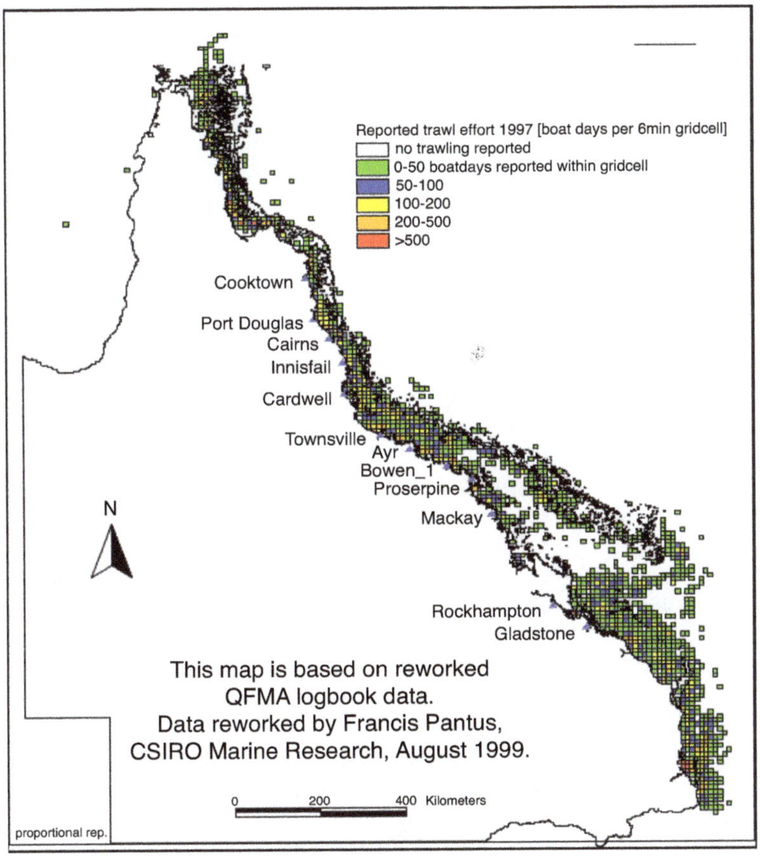

Plate 2. Estimated spatial distribution of effort for the prawn trawl fishery on Australia's Great Barrier Reef.

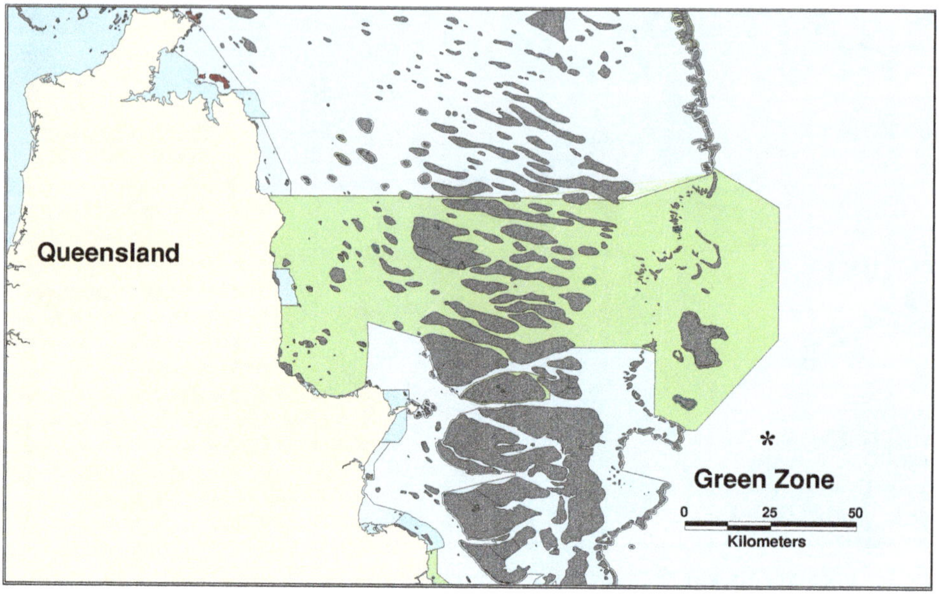

Plate 3. Great Barrier Reef ecosystem model, depicting regional prawn trawl closure.

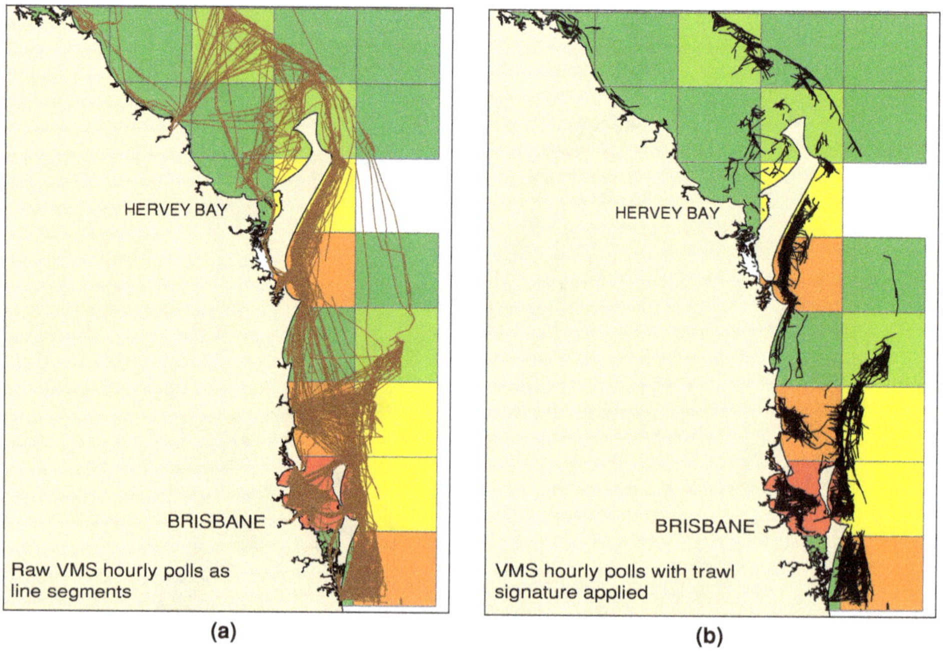

Plate 4. Map depicting VMS signatures of prawn trawlers overlaid over compulsory daily logbook 30-minute effort grids. (a) Raw VMS hourly polls as line segments. (b) Trawl signature filter applied (i.e., speed dependent net deployment).

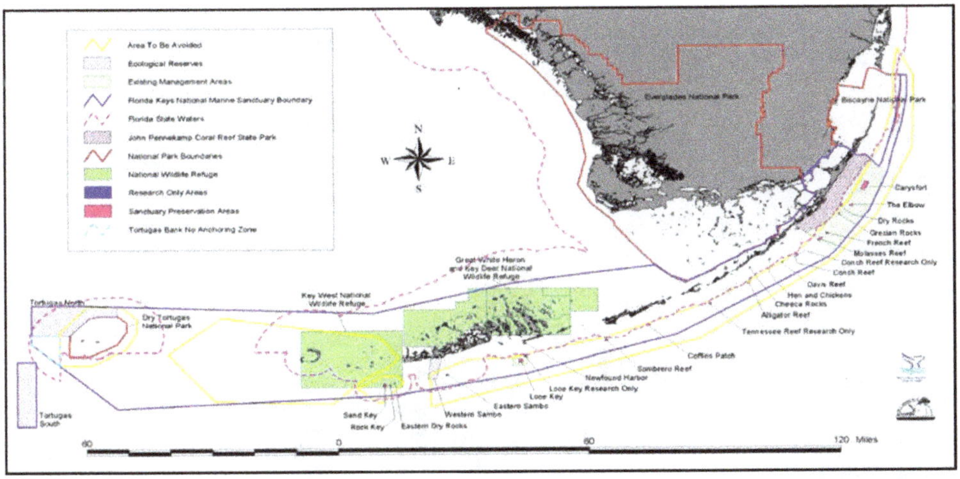

Plate 5. Multiple management jurisdictions in the Florida Keys and South Florida.

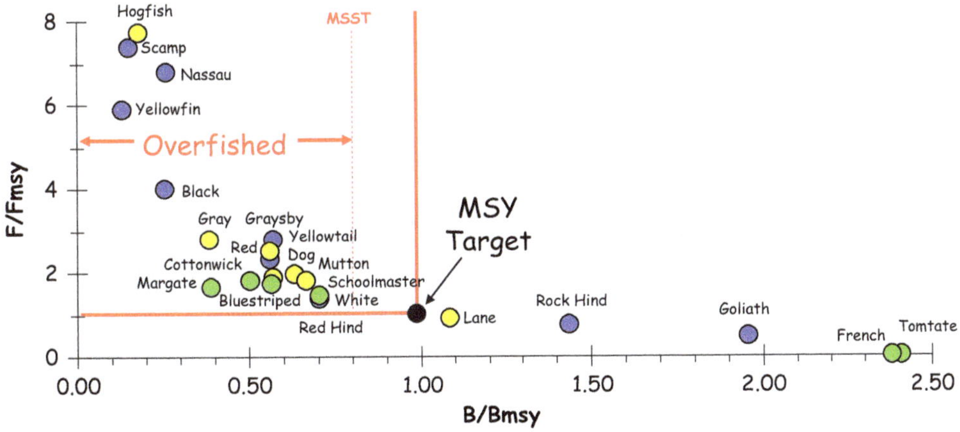

Plate 6. Management benchmarks for coral reef ecosystems in the Florida Keys and South Florida.

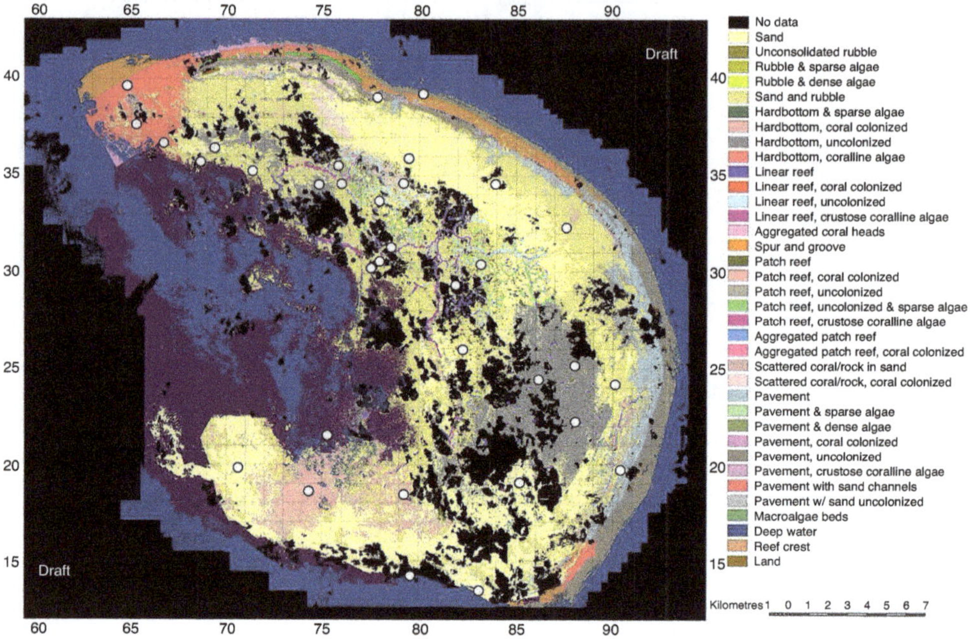

Plate 7. Habitat information for French Frigate Shoals, as derived from IKONOS imagery.

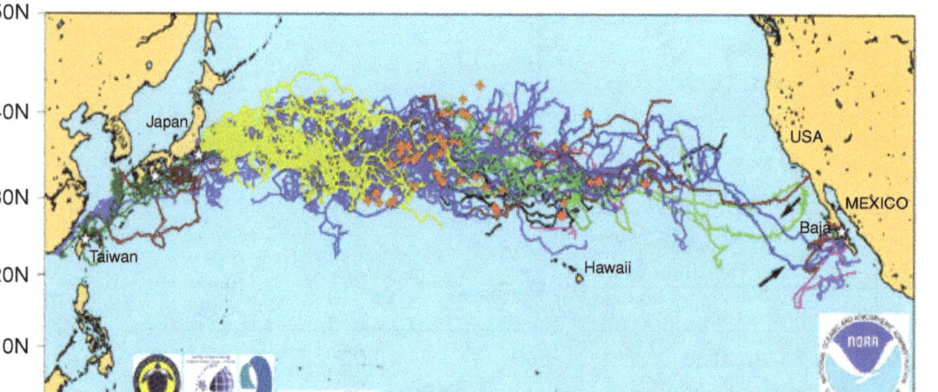

Plate 8. Distribution and movement of loggerhead turtles in the North Pacific Ocean. (Data consist of both loggerheads found in the 1990–1992 high-seas driftnet fishery, and track lines indicating movement of free roaming turtles fitted with satellite tracking transmitters between 1997 and 2005. Prepared by D.M. Parker in collaboration with J.J. Polovina, G.H. Balazs, I. Cheng, P.H. Dutton, N. Kamezacki, W.J. Nichols, and I. Uchida.)

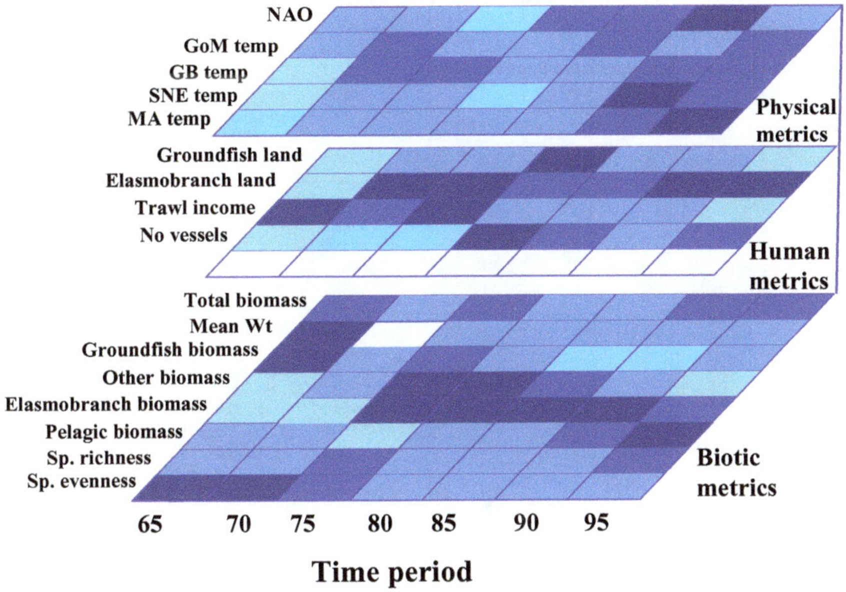

Plate 9. Ternary matrix of ecosystem indicators for US fisheries on Georges Bank.

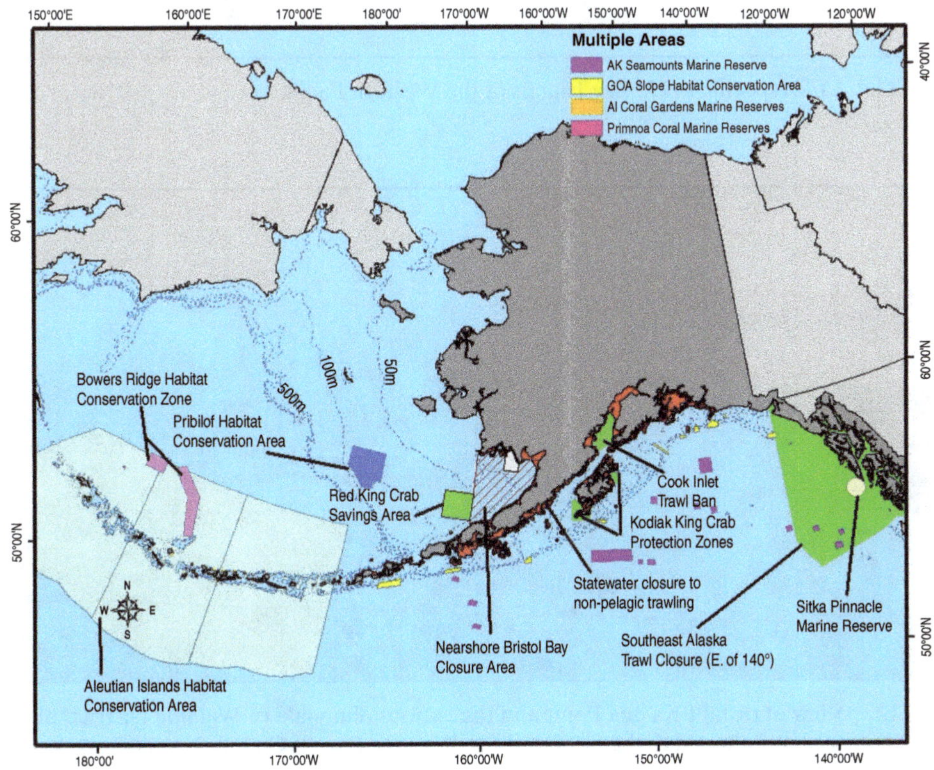

Plate 10. Conservation zones and area closures affecting Alaska commercial fisheries.

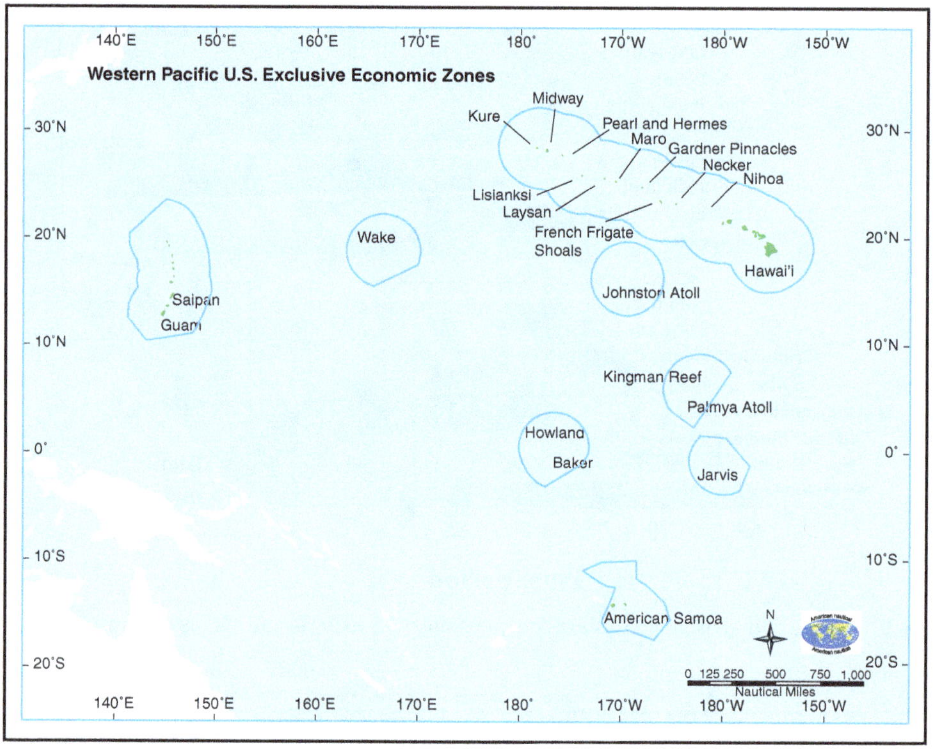

Plate 11. EEZs specific to US flag Islands of the Western Pacific.

Plate 12. View of distant Ka'ena Point and the various ahupua'a of Waialua District; photo taken from ancient Pu'u o Mahuka Heaiau above Waimea Bay on O'ahu's North Shore, winter 2008.

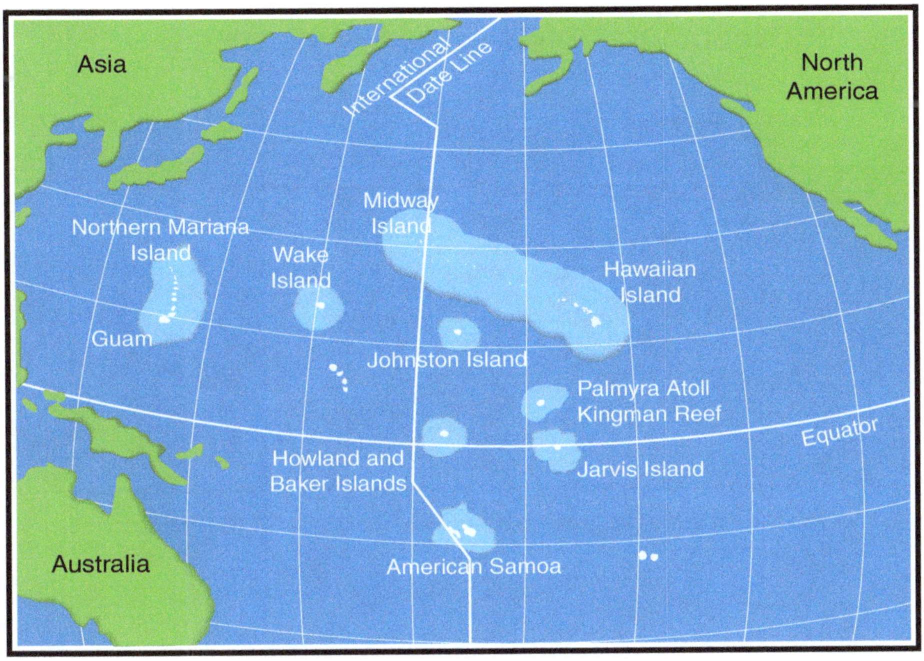

Plate 13. The archipelagos and remote islands under council purview.

Plate 14. The ocean end of the ahupua'a of Halawa on Moloka'i area first inhabited about 650 A.D.

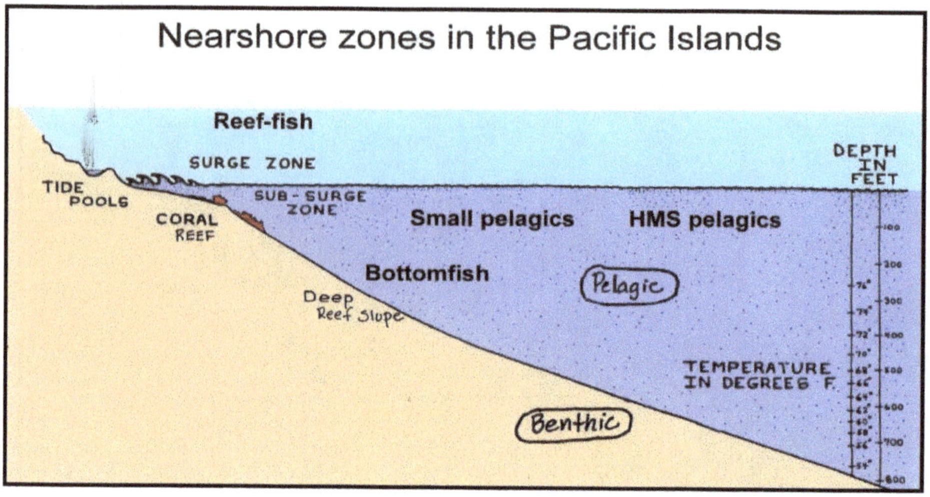

Plate 15. Distribution of species types in typical nearshore zone of the Pacific Islands.

Plate 16. Fisherman prepare for Guam Lunar Festival.

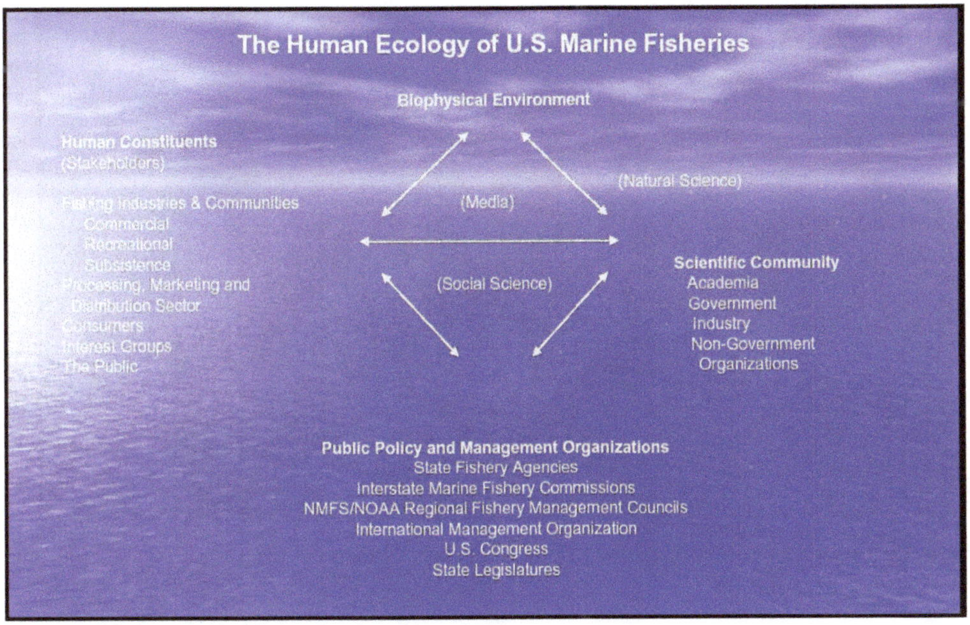

Plate 17. The human ecology of US marine fisheries.

Plate 18. Roundtable discussion during the first day of the workshop.

Plate 19. Littoral and riverine zones along the Pacific Northwest Coast.

Plate 20. Employment options related to commercial fishing are important in many coastal communities of the U.S.

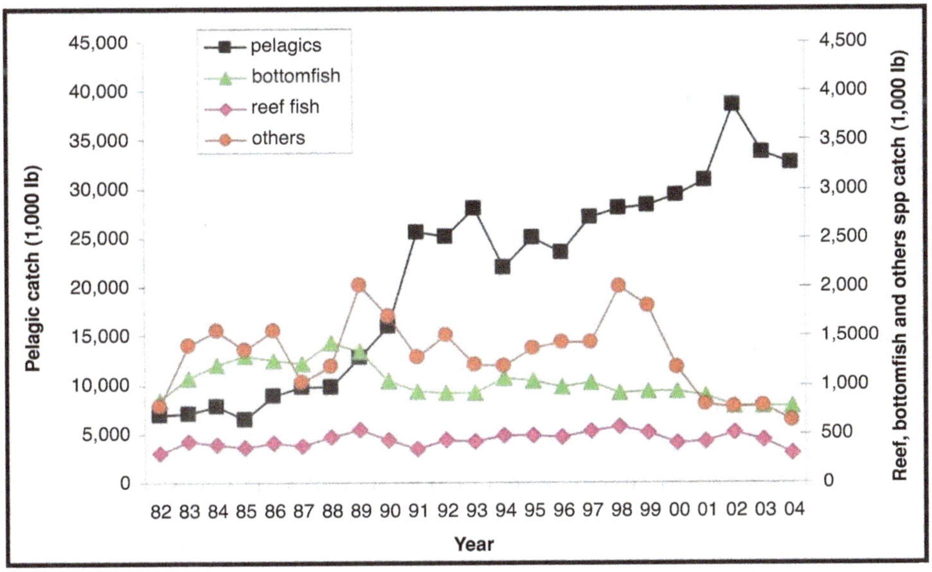

Plate 21. Annual domestic fishery production in the Western Pacific region.

Plate 22. Alia moored in small cove in American Samoa.

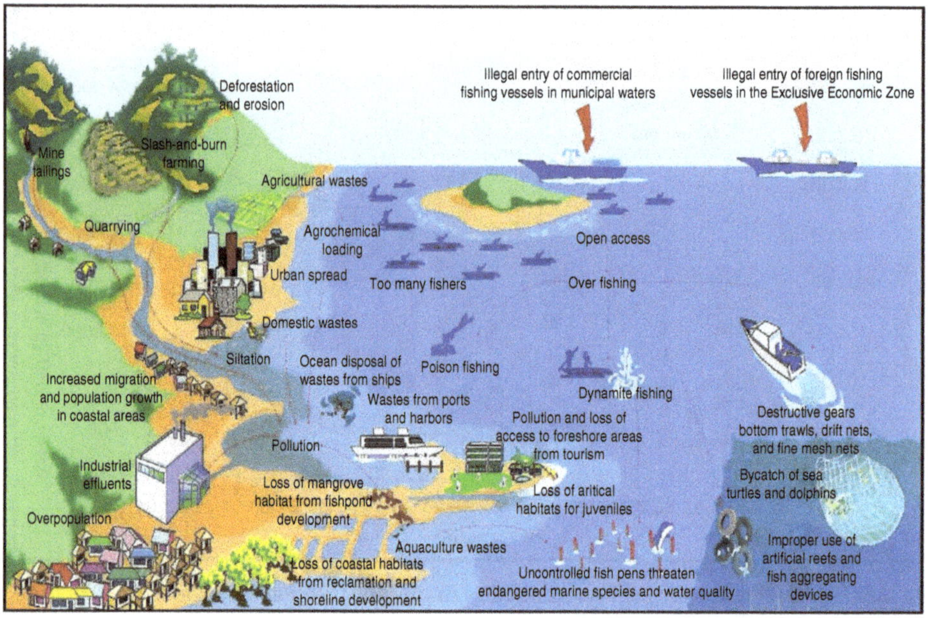

Plate 23. Direct and indirect ecosystem relationships (Coastal Resource Management Project).

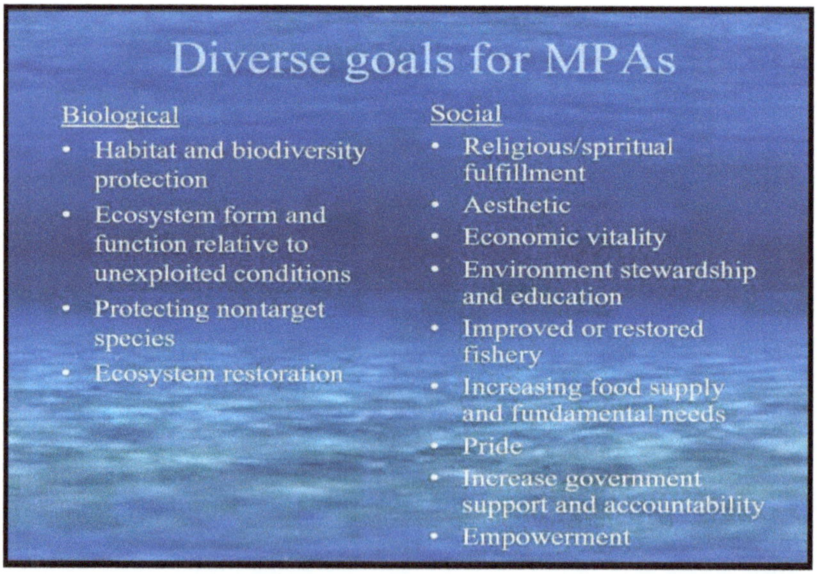

Plate 24. Diverse biological and social goals of the MPA management approach.

Plate 25. Community involvement imparts local knowledge.

Plate 26. View of traditional Fijian settlement.

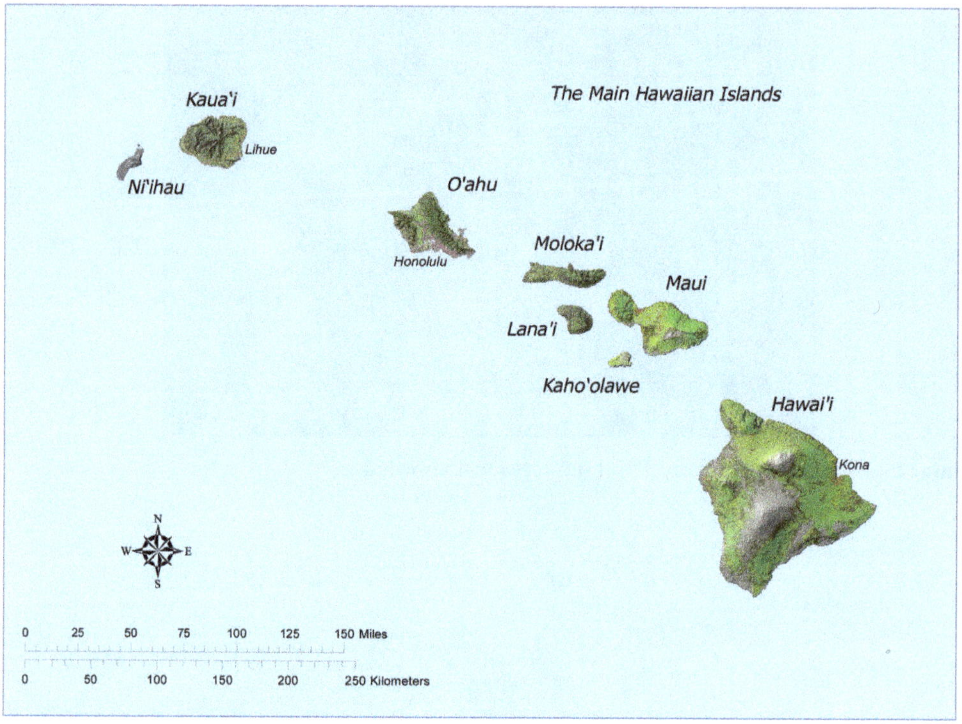

Plate 27. Kūʻula rising from the ocean.

144°40'E
144°50'E
13°40'N
13°30'N
13°20'N
13°10'N
145°E
Philippine Sea
Pacific Ocean
Pati Point MPA
Andersen Air Force Base
Yigo
Dededo
Tamuning
Tumon Bay MPA
Asan
Hågatña
Piti
Barrigada
Sasa Bay MPA
Chalan
Mangilao
Camp Bright
Apra Heights
Yona
Agat
Santa Rita
Talofofo
Umatac
Inarajan
Merizo
Achang Reef Flat MPA
Aga Point
Guam
Map by Impact Assessment, Inc.
UTM Projection, Zone 55
N
0 5 10 15 Kilometers
0 5 10 Miles

Plate 28.

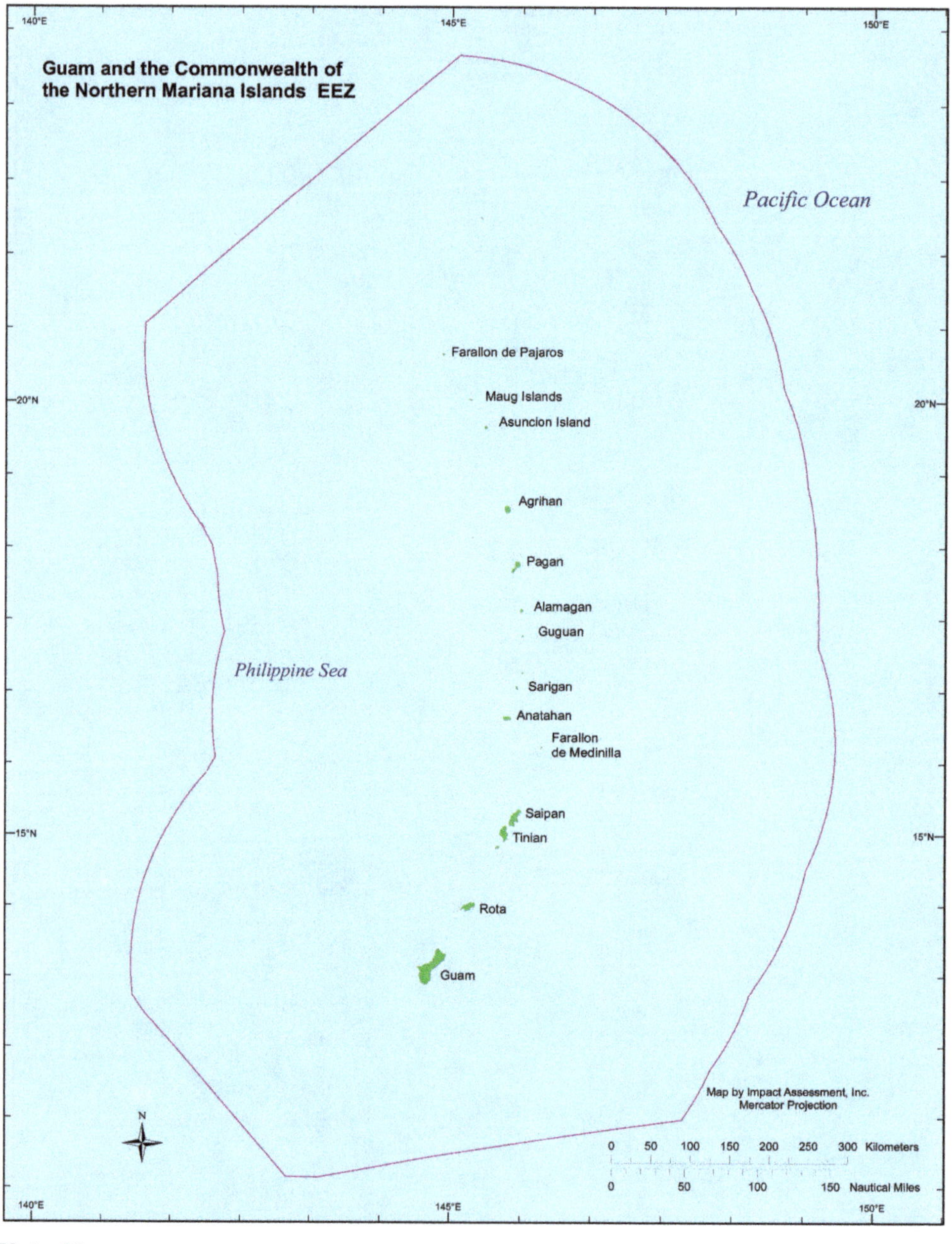

Guam and the Commonwealth of
the Northern Mariana Islands EEZ
Pacific Ocean
Philippine Sea
Farallon de Pajaros
Maug Islands
Asuncion Island
Agrihan
Pagan
Alamagan
Guguan
Sarigan
Anatahan
Farallon
de Medinilla
Saipan
Tinian
Rota
Guam
Map by Impact Assessment, Inc.
Mercator Projection
0 50 100 150 200 250 300 Kilometers
0 50 100 150 Nautical Miles
140°E
145°E
150°E
20°N
15°N
N

Plate 29.

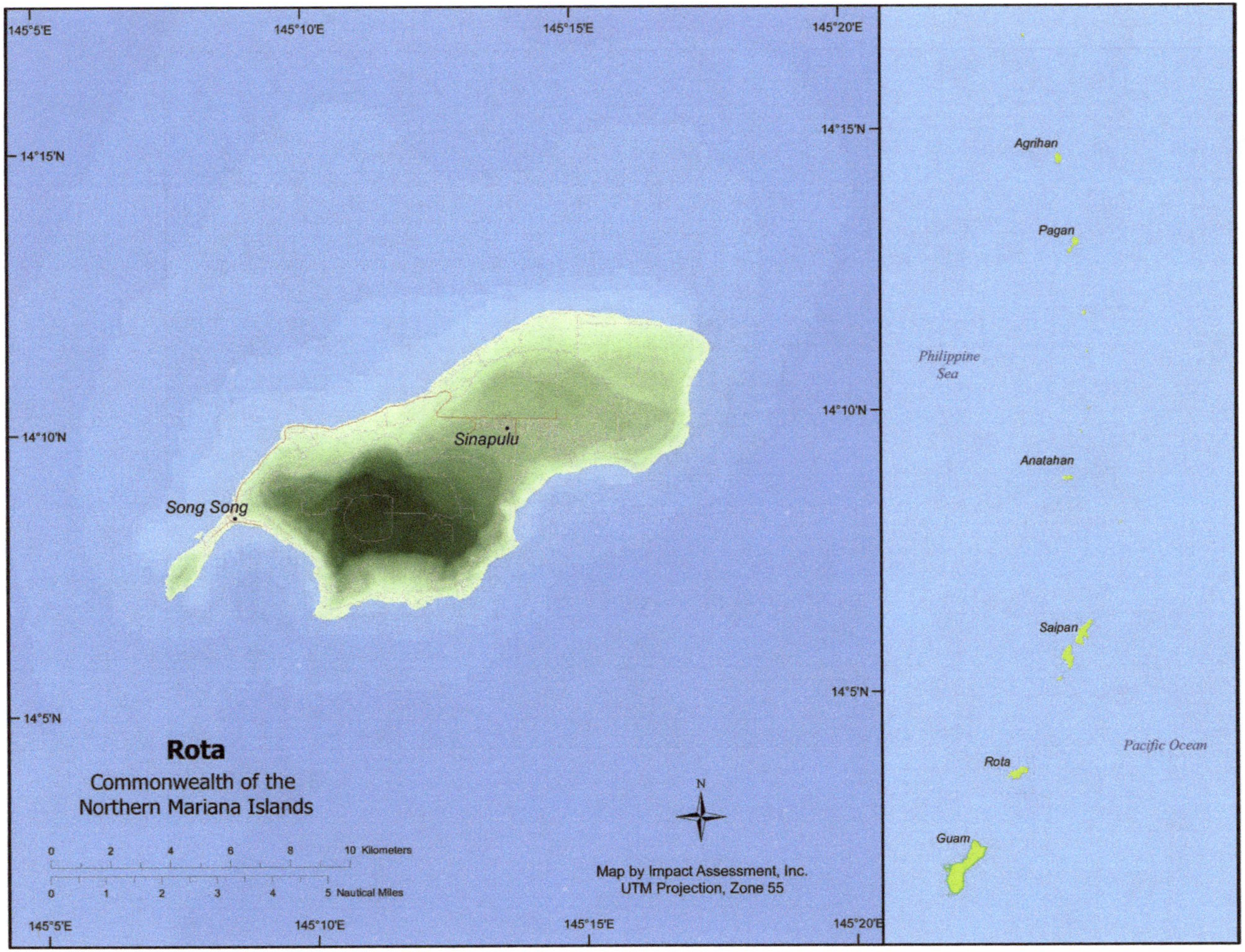
145°5'E
145°10'E
145°15'E
145°20'E
14°15'N
14°10'N
14°5'N
Song Song
Sinapulu
Rota
Commonwealth of the
Northern Mariana Islands
0 2 4 6 8 10 Kilometers
0 1 2 3 4 5 Nautical Miles
N
Map by Impact Assessment, Inc.
UTM Projection, Zone 55
Agrihan
Pagan
Philippine Sea
Anatahan
Saipan
Pacific Ocean
Rota
Guam

Plate 30.

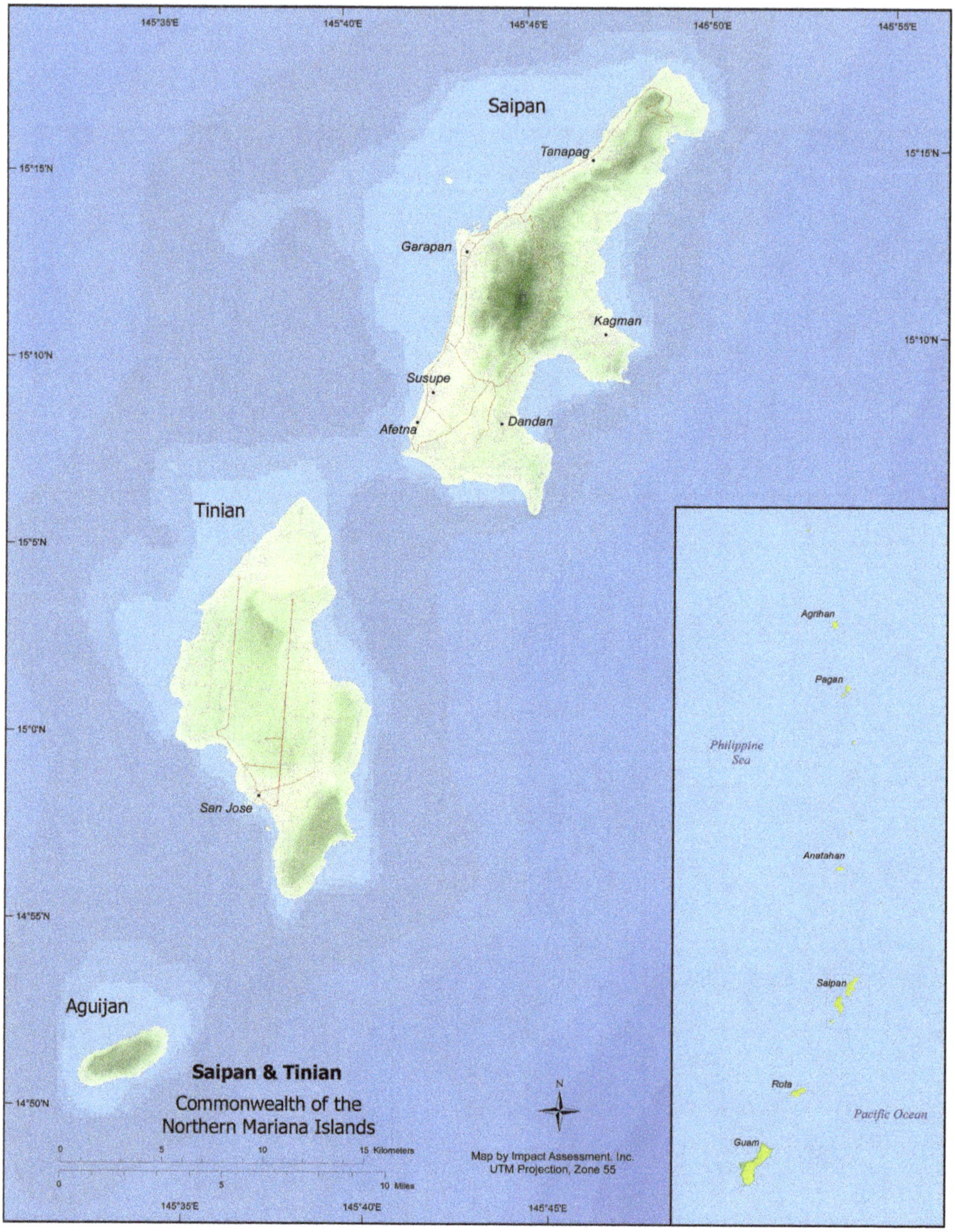
Saipan
Tanapag
Garapan
Kagman
Susupe
Afetna
Dandan
Tinian
San Jose
Aguijan
Saipan & Tinian
Commonwealth of the
Northern Mariana Islands
Map by Impact Assessment, Inc.
UTM Projection, Zone 55
Agrihan
Pagan
Philippine Sea
Anatahan
Saipan
Rota
Guam
Pacific Ocean

Plate 31.

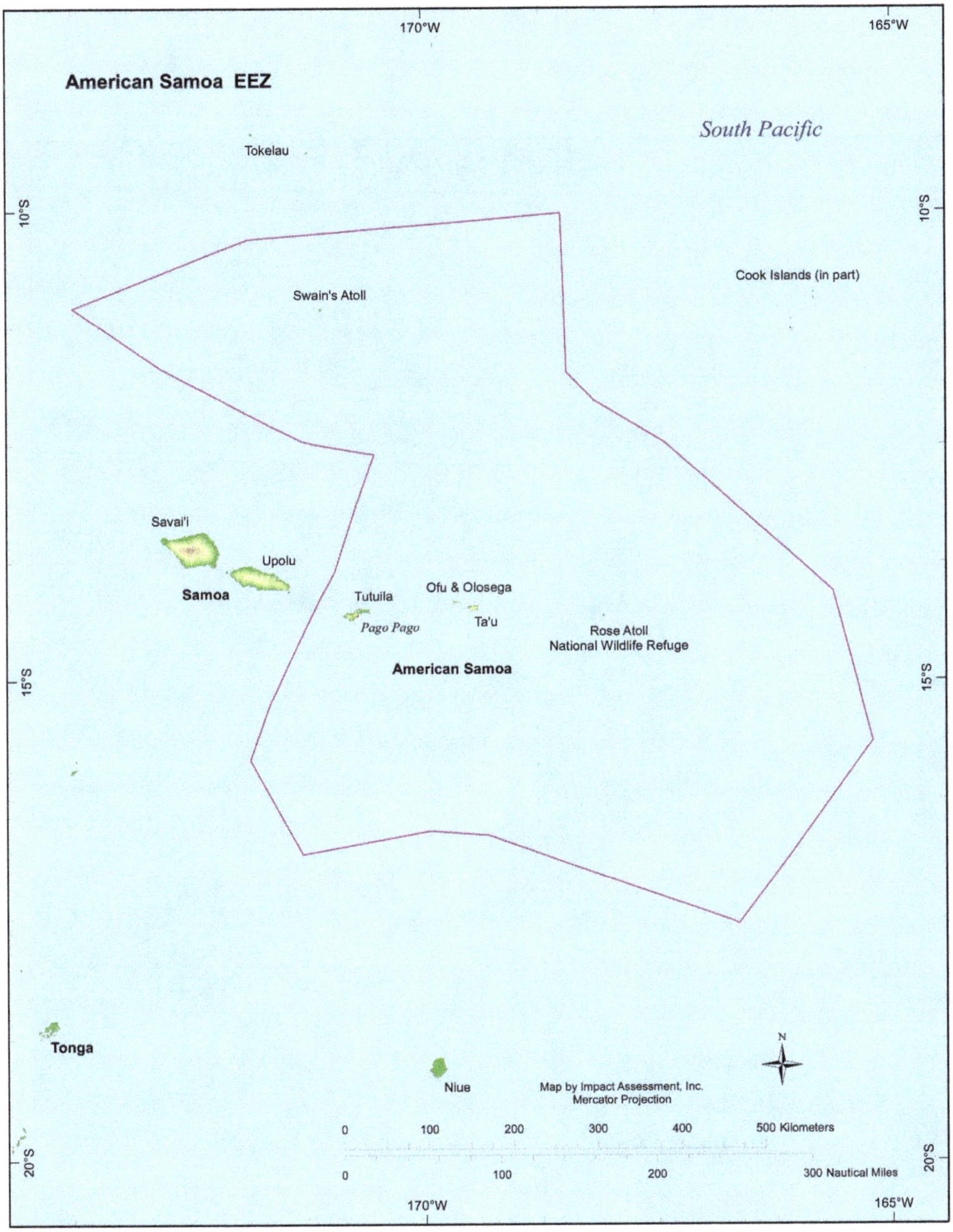

170°W
165°W
American Samoa EEZ
South Pacific
Tokelau
10°S
Swain's Atoll
Cook Islands (in part)
Savai'i
Upolu
Samoa
Tutuila
Pago Pago
Ofu & Olosega
Ta'u
Rose Atoll
National Wildlife Refuge
American Samoa
15°S
Tonga
Niue
N
Map by Impact Assessment, Inc.
Mercator Projection
0 100 200 300 400 500 Kilometers
0 100 200 300 Nautical Miles
20°S

Plate 32.

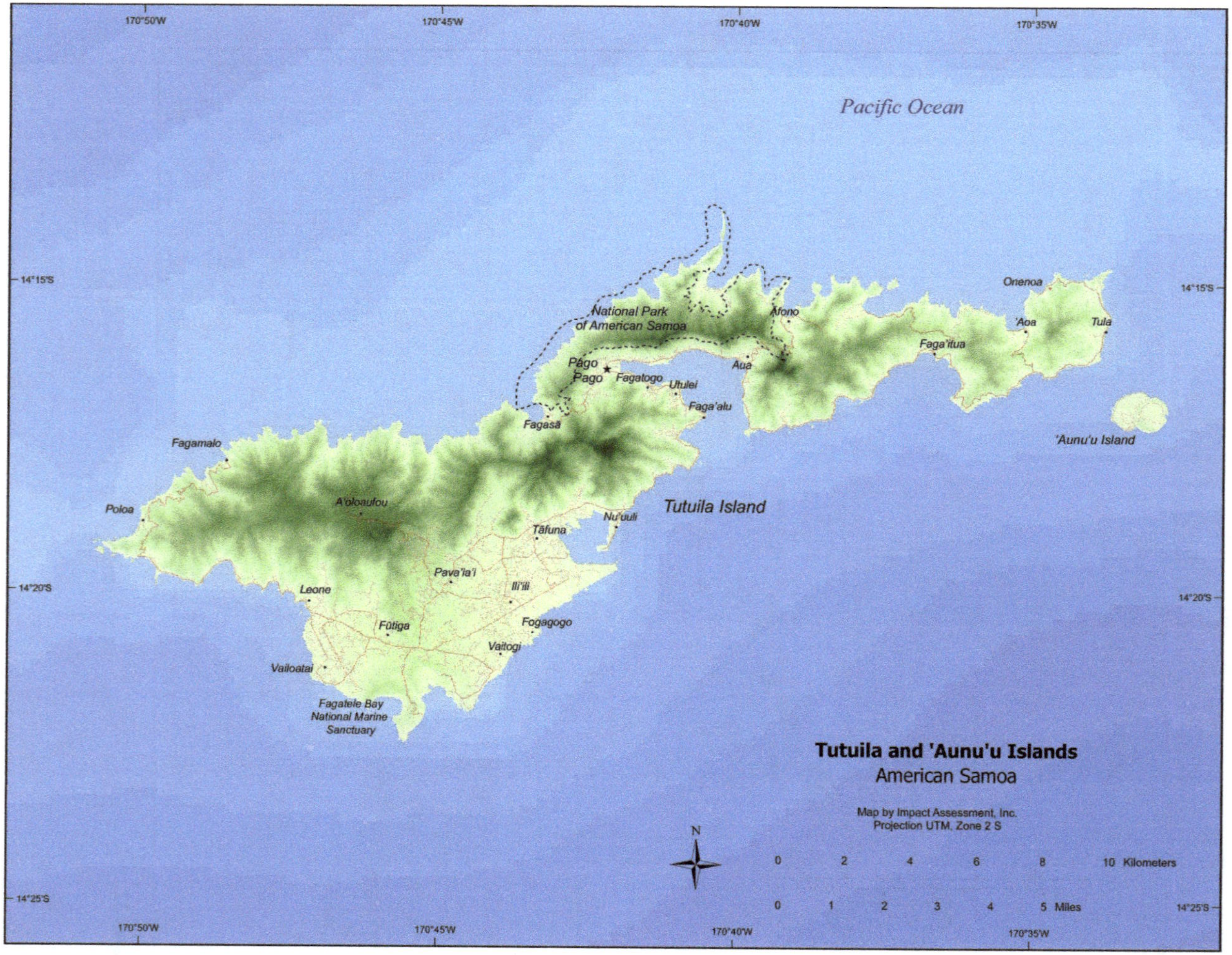
170°50'W
170°45'W
170°40'W
170°35'W
14°15'S
14°20'S
14°25'S
Pacific Ocean
National Park
of American Samoa
Afono
Onenoa
'Aoa
Tula
Faga'itua
Aua
Pago
Pago
Fagatogo
Utulei
Faga'alu
Fagasa
'Aunu'u Island
Fagamalo
Poloa
A'oloaufou
Nu'uuli
Tutuila Island
Tāfuna
Pava'ia'i
Ili'ili
Leone
Fogagogo
Fūtiga
Vaitogi
Vailoatai
Fagatele Bay
National Marine
Sanctuary
Tutuila and 'Aunu'u Islands
American Samoa
Map by Impact Assessment, Inc.
Projection UTM, Zone 2 S
N
0 2 4 6 8 10 Kilometers
0 1 2 3 4 5 Miles

Plate 33.

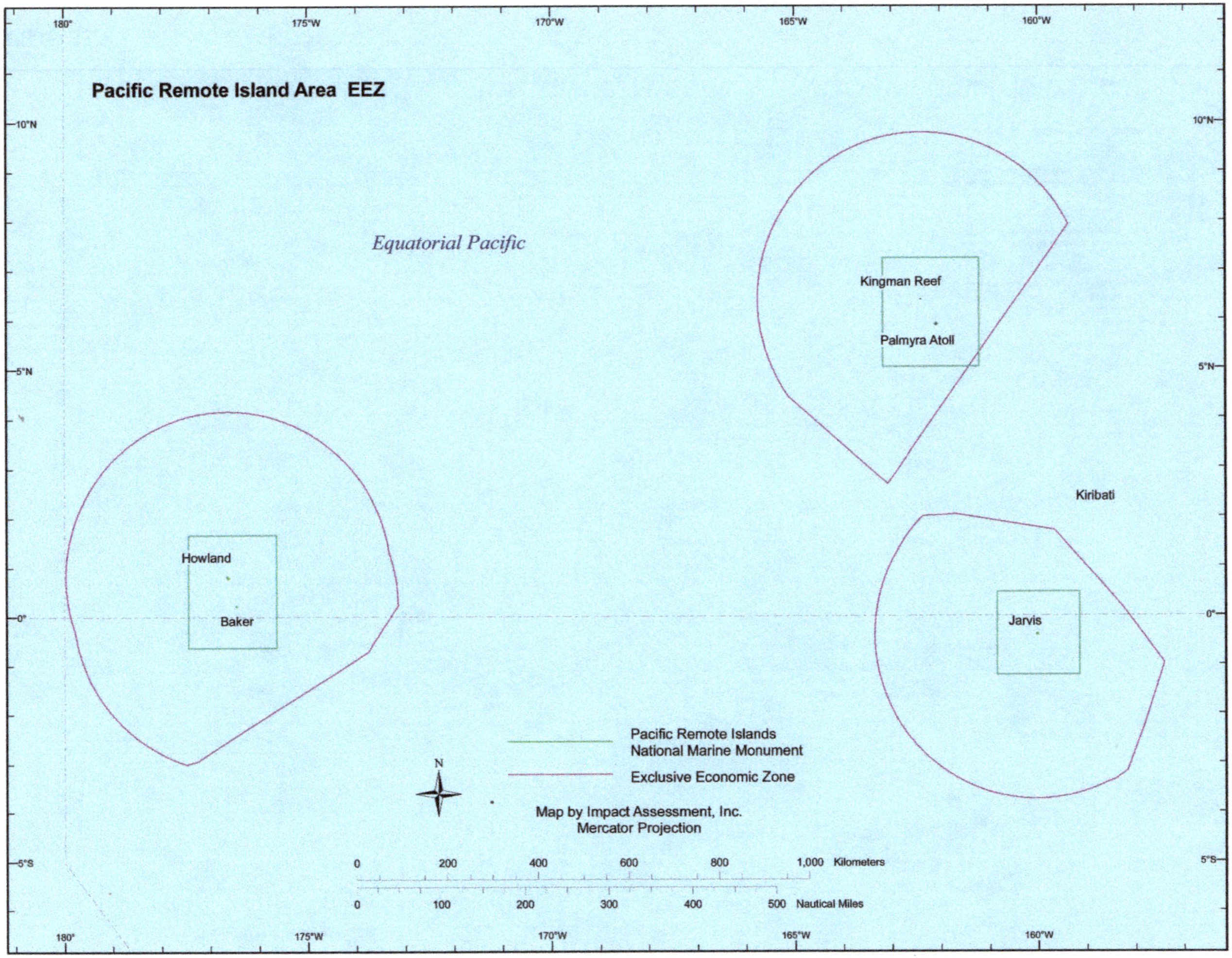
Pacific Remote Island Area EEZ
Equatorial Pacific
Kingman Reef
Palmyra Atoll
Kiribati
Howland
Baker
Jarvis
Pacific Remote Islands
National Marine Monument
Exclusive Economic Zone
N
Map by Impact Assessment, Inc.
Mercator Projection
0 200 400 600 800 1,000 Kilometers
0 100 200 300 400 500 Nautical Miles
180°
175°W
170°W
165°W
160°W
10°N
5°N
0°
5°S

Plate 34.

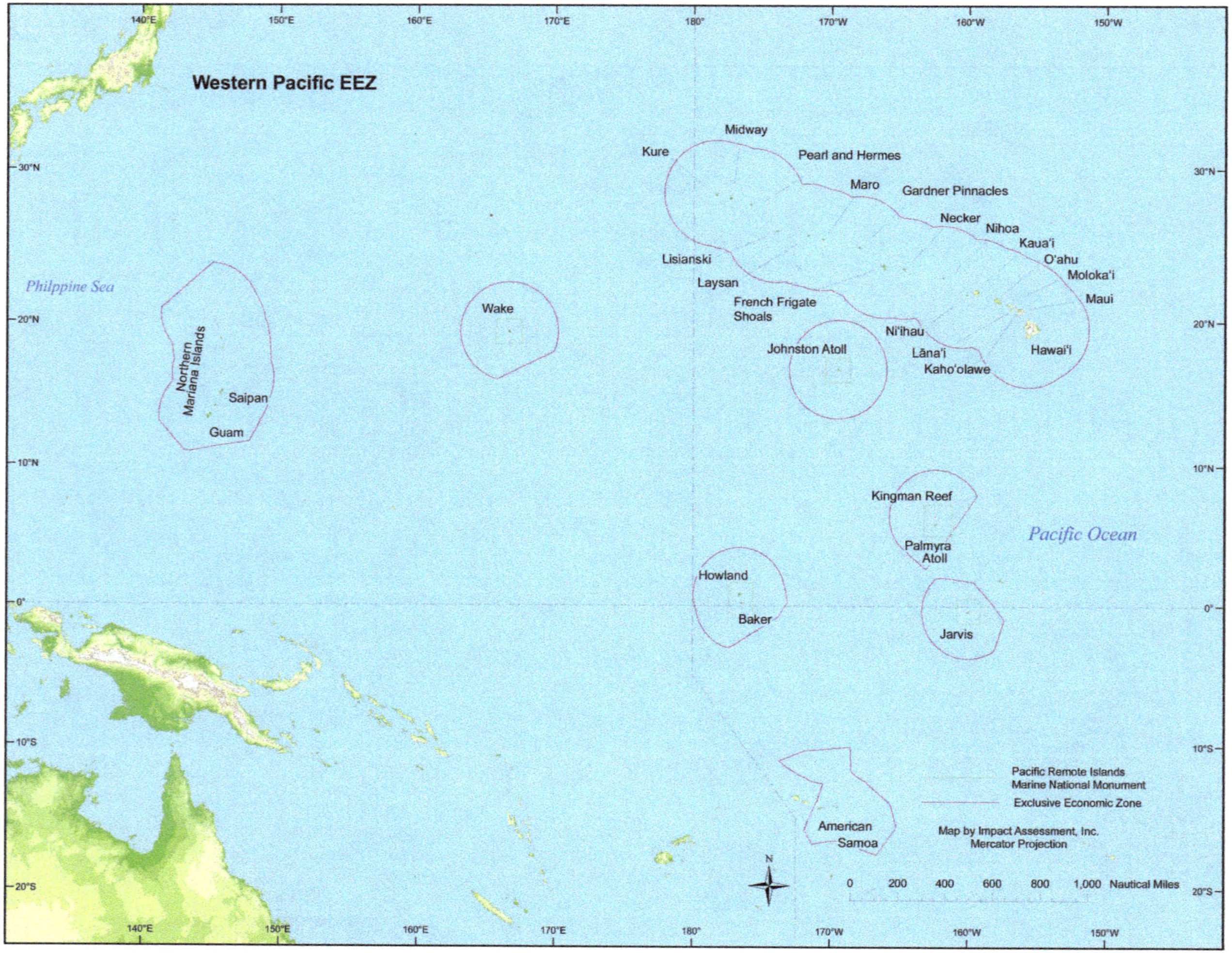
Western Pacific EEZ
Philppine Sea
Pacific Ocean
Northern Mariana Islands
Saipan
Guam
Wake
Kure
Midway
Pearl and Hermes
Maro
Gardner Pinnacles
Necker
Nihoa
Kaua'i
O'ahu
Moloka'i
Maui
Hawai'i
Ni'ihau
Lāna'i
Kaho'olawe
Lisianski
Laysan
French Frigate Shoals
Johnston Atoll
Kingman Reef
Palmyra Atoll
Howland
Baker
Jarvis
American Samoa
Pacific Remote Islands Marine National Monument
Exclusive Economic Zone
Map by Impact Assessment, Inc.
Mercator Projection
0 200 400 600 800 1,000 Nautical Miles
N
140°E 150°E 160°E 170°E 180° 170°W 160°W 150°W
30°N 20°N 10°N 0° 10°S 20°S

Plate 35.

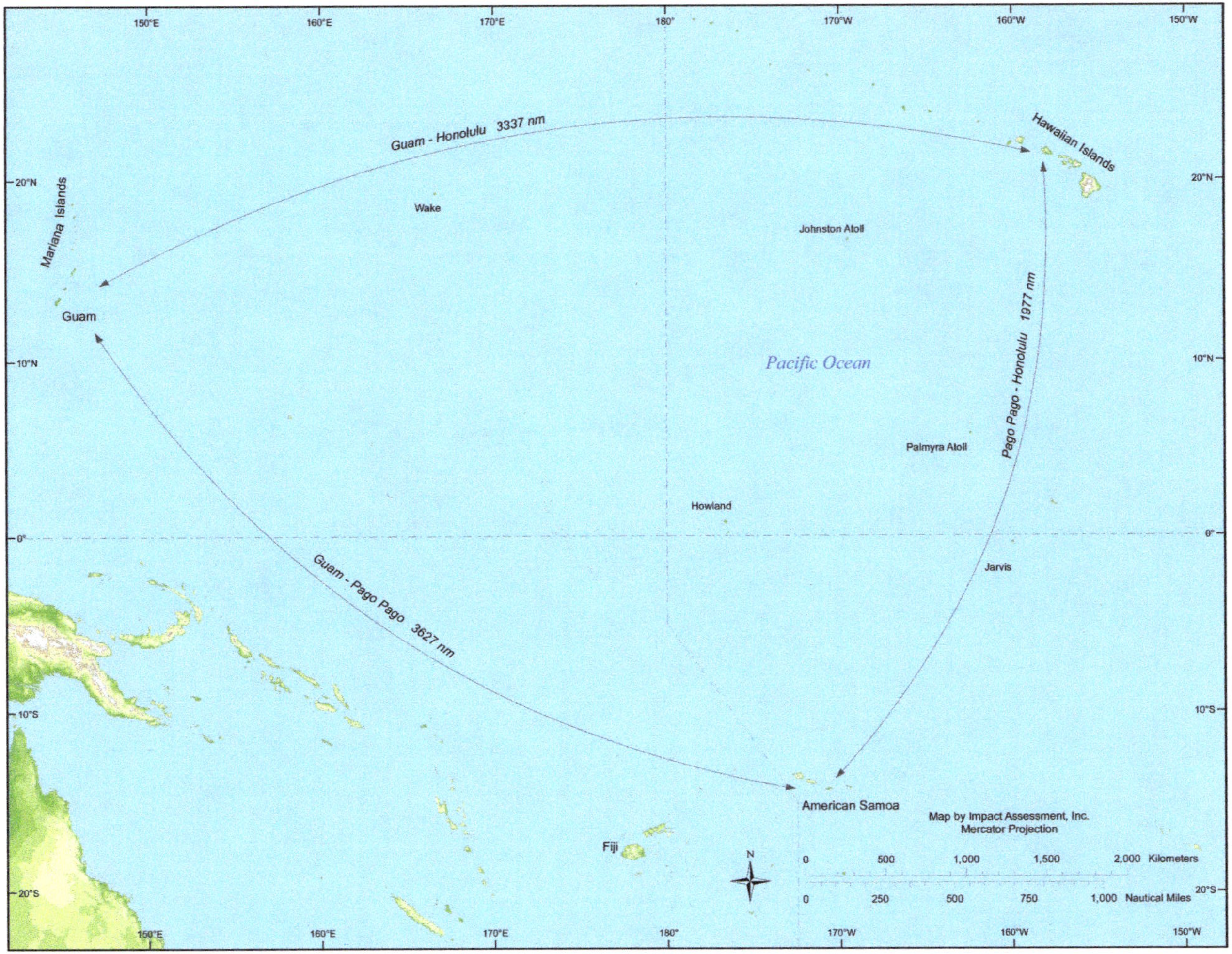
Guam - Honolulu 3337 nm
Hawaiian Islands
Mariana Islands
Guam
Wake
Johnston Atoll
Pacific Ocean
Pago Pago - Honolulu 1977 nm
Palmyra Atoll
Howland
Jarvis
Guam - Pago Pago 3627 nm
American Samoa
Fiji
Map by Impact Assessment, Inc.
Mercator Projection
0 500 1,000 1,500 2,000 Kilometers
0 250 500 750 1,000 Nautical Miles
N
150°E 160°E 170°E 180° 170°W 160°W 150°W
20°N 10°N 0° 10°S 20°S

Plate 36.

Chapter 8
Introduction to Ecosystem Social Science and Planning

The focus of this section is a summary presentation of the organized discussions held during the course of the workshop. We precede this with reiteration of the rationale for the workshop and discussion of its purpose and manner of conduct.

Workshop Goal and Objectives

The social science workshop was initiated based on recognition of the pivotal importance of humans in marine ecosystems in the Western Pacific and elsewhere. The Council recognizes the utility of an ecosystem approach in island settings with extensive historical precedent in what essentially were indigenous forms of ecosystem-based management. The approach is also considered highly practical in that it may reduce administrative burdens associated with management of single species by multiple fleets across highly divergent and distant archipelagos.

The overarching goal of the social science workshop was to facilitate informed discussion of social science requirements for effectively supporting ecosystem-based approaches to marine resource management in the Western Pacific region and its island subregions (archipelagos). A series of interrelated objectives was developed to meet this overarching goal, as follows:

1. Convene nationally recognized social scientists and regional experts to review social science applications relevant to ecosystem-based marine resource management.
2. Review resource management requirements and pertinent issues in the Western Pacific and its subregions.
3. Identify the best suite of ecosystem indicators related to the human and institutional ecology of marine ecosystems in the Western Pacific and its subregions.
4. In the short term, and within the parameters of existing mandates, identify the most effective ecosystem-based approaches to marine resource management that incorporate the human dimension and that can be implemented based on current data.
5. Explore what new social and policy science data or models would be needed to advance ecosystem-based approaches to marine resource management in the Western Pacific region and its subregions.
6. Explore changes in policy or social and policy science administration that would be needed to more effectively implement ecosystem-based approaches to marine resource management in the Western Pacific region and its subregions.

Objectives (2) through (6) above involve complex issues and challenges that have been addressed to varying degrees by fisheries social science and marine policy experts

Ecosystem-Based Fisheries Management in the Western Pacific, First Edition. Edward Glazier.

in other regions of the country. As such, it was decided that much would be gained by inviting such experts to discuss their experiences and explore the issues at the Council offices in Honolulu.

Social, cultural, economic, demographic, and political conditions and factors vary widely in the Western Pacific region (see Appendix B). Moreover, they differ in many ways from those of the Continent. Therefore, it was decided that persons knowledgeable of both social science applications and social conditions in each island group should also be invited to participate in the workshop.

Continent-based participants were invited to inform generalized discussion about: (a) fisheries social science methods and models as applicable in the context of ecosystem-based management, (b) fisheries social science data challenges and solutions in that context, and (c) social and economic indicators of potential utility for managers engaging the ecosystem approach. The workshop was so organized in part to parallel the conduct of the previously conducted biophysical workshop, and in part because these are indispensable elements of a comprehensive social science approach to implementing the new strategy. Island-based participants were also invited to the workshop to provide their perspectives on methods and models, data challenges and solutions, and prospective indicators, but with the additional dimension of local knowledge and experience.

Given time limitations, a subset of participants was asked to conduct presentations. Time was allotted by the entire group for discussion of the presentations, and all participants were encouraged to provide input as desired. The workshop was led by Michael Orbach of the Duke University Marine Lab. Dr Orbach is a widely recognized authority on marine policy and application of social science to issues surrounding management of marine resources.

Introductory Presentations

Following are summaries of speaker discussions, provided in sequence as presented during the course of the workshop. While presentations were often followed by group discussion, this varied in nature and extent. Discussion is limited to cases in which it afforded particularly useful insight into the primary points being made by the speakers. We do, however, synthesize additional group discussion in later analytical sections of this chapter. The summaries were developed through review of observer notes, transcripts, and written presentations (in most cases, PowerPoint presentations). The summaries are consistently presented in a third-person narrative form so as to minimize use of quotations and redundant shifting between person and tense. Interpretive-artistic license was taken in certain cases with the intent of clarifying points being made by the presenters.

Overview of Ecosystem Approaches to Management in the Region

Kitty Simonds, Executive Director, Western Pacific Regional Fishery Management Council

Following pule (prayer) by the Council's Indigenous Coordinator, Charles Kaaiai, Executive Director Simonds welcomed the participants to Hawaii and to the Council offices and, as presented in a summary form below, subsequently offered an overview of the Council and its mission, and the purpose, nature, and utility of the shift to ecosystem management planning in the region.

Ecosystem-based planning efforts are now being undertaken by the nation's fisheries management councils. The Western Pacific Council has been moving toward adoption of ecosystem-based management for some time, and the conduct of the biophysical and social science workshops is intended to facilitate informed discussion of the approach and related issues of importance to resource management in the region.

Ecosystem-based management should be seen as highly appropriate and practical in the island settings that comprise the region. The ocean surrounds island residents, and has long been and continues to be a critically important aspect of social, cultural, and religious life. The ocean provides sustenance, and thus the well-being of islanders is directly related to the status of its resources.

As noted previously, the Council is developing fisheries ecosystem plans (FEPs) for each archipelago. The new FEPs will combine the formerly distinct fisheries management plans for coral reef ecosystems, precious corals, bottomfish, and crustaceans, while pelagic species will be managed under a separate FEP.

The archipelago-based strategy is intended to accommodate and address the unique biophysical, social, and cultural attributes characteristic of each island area. It will also enable more equitable attention to the needs and issues specific to each region.

The workshop series is a proactive effort to enable expert discussion of the ecosystem approach to fisheries management. The intent is to enhance planning efforts in advance of full adoption of the FEPs. The Council views federal mandates to adopt ecosystem-based management as following rather than driving its own ecosystem planning efforts. Indeed, the Council interacted with NMFS to hold an ecosystems workshop in 1986, and planning for a coral reef ecosystem approach to management was first undertaken by the Council in the 1990s.

The social science workshop was convened to identify and review concepts and models of how the social sciences may contribute to understanding about the role of human beings in fisheries ecosystems, and in ecosystem-based fishery management. It was also intended to generate discussion about the manner and sources of information needed to facilitate that understanding, and research methods needed to acquire it. Finally, it was intended to facilitate discussion of indicators of social conditions and human behavior as these are relevant to understanding the dynamics of marine ecosystems and their effective management.

The social science workshop involved the convening of experts from academia, government agencies, and nongovernment organizations to discuss social science in the island context of the Western Pacific. Again, this is the second of a three-part series. A third workshop will incorporate results of this and the biophysical workshop to address marine policy issues under the new ecosystem-based approach. The final meeting convenes biophysical scientists, social scientists, resource managers, and stakeholders.

Two Hawaiian proverbs provide guidance for the social science workshop and for ecosystem planning in general. The first translates as "no breadfruit can be reached when the picking stick is too short." That is, success requires preparation and the acquisition and use of the appropriate tools. The second translates as "gird the loincloth and sharpen the spear." Historically, this was a call to prepare for war. In this case, it was a call to prepare for the project at hand.

NOAA Fisheries, WP Council, and Managing Fisheries in the Region

Paul Dalzell, Senior Scientist, Western Pacific Regional Fisheries Management Council

Prior to 1976, representatives of coastal states were responsible for managing marine fisheries in their respective nearshore and offshore areas. Federal representatives resolved disputes among the states and addressed matters associated with international fleets, then pursuing fishery resources in territorial waters. Subsequent to the Magnuson–Stevens Act and delineation of state and federal jurisdictional boundaries, the eight fishery councils were established to assist in governing fisheries in the EEZ. The principal goals were to conserve and manage domestic marine fishery resources, phase out the activities of foreign fleets in territorial waters, and enable development of domestic fisheries in the EEZ. Those original goals have been met to greater and lesser degrees, though the sustainable use of marine resources obviously continues to challenge managers throughout the nation. Landings by foreign fleets in US territorial waters declined from 71 percent of overall offshore landings in 1977 to near 0 percent in 1992.

But the subsequent expansion of domestic fishing activities may have been too successful in some ways, as it tended to leave certain stocks in peril. Thus, when amended in 1996, the Magnuson–Stevens Act focused primarily on conservation-related issues. The amendments focus on overfishing, rebuilding stocks, minimizing bycatch, protecting habitat and, in the social realm, establishing means for assessing and protecting fishing communities. The amendments also set the stage for an ecosystem approach to resource management.

The "National Standards" (Section 301) are the principles by which management efforts are assessed under the Magnuson–Stevens Act. The latter three were added as part of the 1996 amendments. These involve the following: (1) prevent overfishing while achieving optimum yield; (2) use the best scientific information available in decision-making processes; (3) manage stocks as units; (4) do not discriminate between residents of different states; (5) consider efficiency issues; (6) take into account variations and contingencies; (7) minimize costs and avoid duplication; (8) consider the effects of decisions on fishing communities; (9) minimize bycatch and mortality; and (10) promote safety of human life at sea.

The Magnuson–Stevens Act called for the establishment of eight regional fishery management councils to be funded through Congressional appropriations. Today, the council system provides stakeholders with a substantial opportunity for involvement in managing fisheries and resources in their respective areas. Council members are appointed by the Secretary of Commerce for 3-year terms. The Western Pacific Council has 16 members, three of whom are nonvoting federal agency representatives from the State Department, US Fish and Wildlife Service, and US Coast Guard. Current voting membership on the Council includes three commercial fishery representatives, four recreational fishery representatives, and one cultural representative. Since 1976, the composition of voting council members has remained relatively constant, with government representing 35–55 percent of votes, recreational industry comprising 25–35 percent, and the commercial industry representing 20–35 percent of votes over time. Local and federal fishery agency representatives hold the remaining five voting seats.

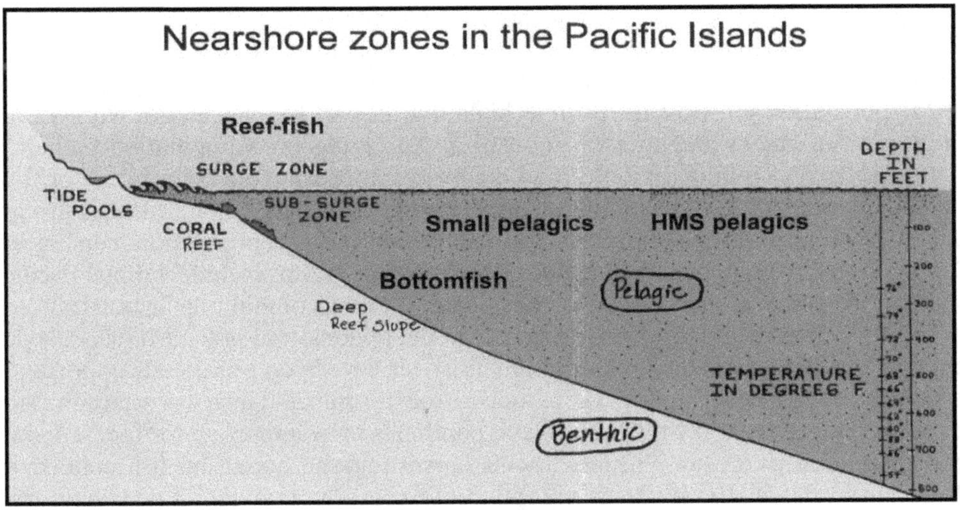

Figure 8.1. Distribution of species types in typical nearshore zone of the Pacific Islands. (For a color version of this figure, see Plate 15.)

The EEZ in this region extends far beyond what is popularly conceived to be the western limit of federal jurisdiction in Hawaii. The MHIs actually comprise merely the eastern edge of the Western Pacific Council area of jurisdiction. That area actually continues westward for many thousands of miles.

The equatorial and subtropical islands in this region are characterized by narrow fringing reefs, precipitous bathymetric slopes, a deep nearshore zone, and vast open ocean areas. Many of the fisheries in the region tend to occur in the reef zone and along the steep slopes of the islands, with pelagic fisheries occurring farther offshore (Figure 8.1).

As regards contemporary sociodemographic conditions in the region, the MHIs are relatively highly populated, with the year 2000 Census figure surpassing one million persons. Guam has the second largest population with some 155,000 persons enumerated in 2000, followed by CNMI with some 69,000 persons, and finally American Samoa with some 57,000 persons in 2000. Various military bases contribute to the population.

In American Samoa, fishing is dominated by pelagic longlining (with volume of landings currently approaching that of the Hawaii-based longline fleet), pelagic trolling, bottom fishing, and reef fishing. In sociocultural terms, this is a relatively homogenous island area, with nearly 90 percent of residents reporting Samoan ancestry in 2000. Independent Western Samoa is immediately adjacent.

Note that Guam is much closer to Asia than North America. As such, a burgeoning Japanese-based tourism industry has developed here. Moreover, most seafood landed in Guam is marketed in Asia. Guam fisheries include troll fishing, bottomfish fishing, a short-line fishery for sharks, and an emergent pelagic longline fishery. There is considerable ethnic and cultural diversity on Guam. Persons of Chamorro ancestry comprise 37 percent of the population, Filipinos 26 percent, Palauans 14 percent, Caucasians 7 percent, and persons of Chuuk ancestry 4 percent.

The Northern Mariana Islands form a half-moon shaped island-arc chain to the north and west of Guam, also in close proximity to Asia. Fisheries here include pelagic trolling, bottom fishing, and various forms of fishing along the reef ecosystems. The CNMI population growth rate is quite high, due in part to an influx of workers in the growing garment and tourism industries. The indigenous population includes Chamorro and Carolinian peoples and comprises approximately 24 percent of the total population of 70,000. The Filipino ethnic group is the largest single ethnic group in the CNMI, comprising approximately 26 percent of the total population. Persons of Chinese ancestry comprise 21 percent. Other resident groups include various Pacific Islanders, Japanese, Caucasians, Koreans, and individuals of multiple ancestries.

The most diverse and productive fisheries in the region occur in the MHIs . Pelagic longline, handline, troll, and pole-and-line fisheries have been historically important, as have the bottomfish fisheries. There are also lobster trap and mixed crustacean trap fisheries. Fishing along the reef ecosystems continues to be important for recreational and subsistence purposes. Precious corals harvesting and aquarium fish collecting also occur in the MHIs. With the exception of Oahu, the population density in the MHI is relatively low. As per the 2000 census, some 9 percent of persons reporting one race reported being Native Hawaiians or other Pacific Islanders, some 42 percent reported a single Asian background, 24 percent reported being Caucasian, and 2 percent reported being African American. Significantly, over 21 percent of Hawaii residents reported having two or more ancestral backgrounds (the national average is about 2.4 percent), and nearly 7 percent reported three or more backgrounds. Alu Like, Inc., a nonprofit organization for Native Hawaiians, reports that as much as 20 percent of the population in Hawaii may be Native Hawaiian or part Native Hawaiian (Severance 2006). The per capita gross domestic product in Hawaii is much higher than in American Samoa, Guam, and CNMI.

Reported domestic landings of pelagic species in the Western Pacific region increased from about 7,000,000 pounds in 1982 to about 32,500,000 pounds in 2004. Landings of other species remained static or declined over the same time period. Hawaii-based fleets contribute the vast majority of landings volume and value. In 2004, Hawaii-based fleets landed 35.7 million pounds for an ex-vessel value of $67.9 million. This was 73 percent of total landings and 85 percent of total value in the region. American Samoa-based fleets landed about 25 percent of total landings during 2004. American Samoan tuna canneries process between 150,000 and 200,000 tons of skipjack, yellowfin, and albacore tuna each year, generating some $250–$300 million annually. Guam is a major point of air-transshipment for Japanese, Chinese, and Taiwanese-based longline operators. Between 5,000 and 12,000 tons of sashimi-grade yellowfin and bigeye are shipped from Guam to Tokyo yearly. Exports were worth about $43 million in 2004.

As regards management of these fisheries and resources, the Council devotes much energy to informed decision-making processes in association with its Fishery Management Plans (FMPs). NOAA Fisheries staff provides the Council with extensive scientific information and consultation, and is continually advised by various standing committees, a Scientific and Statistical Committee, five Plan Teams, four Advisory Panels, and various ad hoc committees and review boards.

Nationwide, 40 FMPs have been instituted by the councils to manage a total of 535 species. The Council has established five FMPs, all of which are multispecies plans, described below. One such plan, the Bottomfish Fishery FMP was established for the purpose of managing snappers, groupers, emperors, jacks, and seamount ground fish, such as armorheads, alfonsins, and ratfish. The principal management measures involve a ban on bottom-trawls and longlines used to pursue bottomfish species. The FMP also includes a permanent moratorium on fishing for groundfish at Hancock seamount; and a limited entry program, vessel size cap, and observer program for the NWHI bottomfish fishery. The Council is currently implementing area closures to large (>50 ft) vessels pursuing bottomfish in certain areas in the region.

Spiny lobsters, slipper lobsters, and Kona crab are covered by the Crustacean Fishery FMP. Its principal management measures involve a harvest guideline model based on a 10 percent risk of overfishing the stocks; this was exceeded and the entire fishery has been closed since the late 1990s. A major closure was previously instituted for Laysan Island in the NWHI in order to provide a spawning refuge and protect monk seals. Prior to closure, the plan included trap specifications to permit maximum escapement of juvenile lobsters and incidental species.

The Precious Corals Fisheries FMP established management measures for pink, gold, black, and bamboo corals. It involves establishment of quotas, and establishment of a NWHI mega-refugium to protect monk seals. There is no active black coral fishery at the present time.

As noted previously in this chapter, the Coral Reef Ecosystem Fishery FMP was the first ecosystem-based fishery management plan established in the United States. The plan addresses over 250 species of coral reef fish, invertebrates, and a diverse range of potentially harvestable species. Principal management measures involve permitting and reporting requirements for coral reef fisheries occurring in the EEZ, a ban on the use of destructive gear types (including SCUBA-assisted spearfishing), and a ban on the collection of certain corals and live rock. The FMP also includes a network of no-use and low-use Marine Protected Areas (MPAs).

The Pelagic Fishery FMP involves the regulation of fisheries involving tunas, mackerels, billfish, pelagic sharks, wahoo, mahimahi, gempylids, and pomfrets. It will also include squid fisheries. Principal management measures include limited entry programs for longline fisheries in Hawaii and American Samoa; closed areas for longline fisheries in the NWHI to protect turtles, seabirds, and monk seals; and areas closed to longline fishing around the MHIs and American Samoa to enhance small vessel pelagic fisheries and obviate potential gear conflicts. A protected species program supports several turtle conservation programs at nesting beaches and foraging grounds along the Pacific Rim. [In reiteration], management of pelagic species in the Pacific relates to decisions made by the Inter-American Tropical Tuna Commission (IATTC), the Western and Central Pacific Fisheries Commission (WTFCC), and other entities and conventions. A longline quota for bigeyed tuna is now being addressed by the Council.

The shift from FMPs to FEPs will augment the efforts of the Council to effectively manage fisheries in the region. The Council has been working toward this end since 2003. Revising the Pelagics plan may be the simplest process, as the current Pelagics FMP is a *de facto* ecosystem plan. Transforming the other current FMPs into a series of FEPs is more challenging.

The Council is working with NOAA Fisheries to implement a two-step approach to establishing the FEPs. Boundaries and institutional structures are being determined initially through draft plans, and a programmatic DEIS has been prepared. Information needs and indicators will be determined, and fishery regulations will ultimately be finalized.

Some Perspectives on Marine Ecosystems, Governance, and Science

Steven Murawski, Chief Scientist, NOAA Fisheries (Presented by Sam Pooley, Director, PIFSC)

Although marine ecosystem science is still relatively young, significant progress has been made in theory, observation, and experimentation. The application of ecosystem science to marine resource management in the EEZ and adjacent marine and coastal areas has been the focus of the NOAA Ecosystem Goal Team since its inception. The task is a particularly challenging one given the complexities of temporal scope and spatial scale associated with the physical processes of marine ecosystems and the larger climatic and oceanographic systems that influence those processes. But given the goods and services marine ecosystems can provide to human beings, and the importance of understanding the interconnections of physical and social systems, application of ecosystem science is critically important. But significantly, a clear path of reference for building and implementing such an approach is not yet available.

An ecosystem involves a geographically specified system of organisms (including humans) that interact in dynamic fashion in the physical environment. EAFM must also be geographically specified and dynamic. It must be adaptive to both knowledge and uncertainty, and it must be capable of accommodating multiple external influences and diverse social objectives. Clearly, such an approach must address the human dimension—the influence of culture, economics, and other factors specific to the relationship between humans and the marine environment. It must consider the adequacy and competence of existing institutions to effectively manage and the prospect that existing institutions may need to be reformulated or reconstructed in their entirety to enable the new approach. The need for effective governance is a critical consideration in establishing an ecosystem approach to management, considered at all levels of analysis, from local to regional to international.

The Large Marine Ecosystem (LME) approach adopted by NOAA Fisheries involves efforts to manage coastal systems that exhibit a wide range of physical characteristics and challenges to effective governance. This is a perennially complex process, especially given the need to assess and understand changes occurring across space and over the course of time.

The Integrated Ocean Observing System (IOOS) has been employed as a scalable system for observing marine ecosystems. The system involves coordinated sampling of relevant data at fixed locations, and the management and communication of data products and analysis for use in the context of management. The PICES Reports prepared by the Alaska Science Center for use by the North Pacific Council are IOOS products.

An effective ecosystem approach to fisheries management involves three principal objectives: (1) development of a broad stakeholder-based governance system; (2) conservation of essential parts of the ecosystem in question; and (3) conservation of essential ecosystem processes. Effective ecosystem approaches to management should be seen as accommodating the input of multiple stakeholders in developing options that accomplish specific management goals. These should be linked with decision support systems, and assisted by an observing system (such as IOOS) that provides feedback on the effectiveness of those decisions and the nature of effects resulting from other sources of change.

The development of an ecosystem approach to management will require tradeoffs between conflicting interests. The approach should be transparent, equitable, and highly adaptable. Moreover, it may be necessary to broaden the scope of attention beyond fisheries so as to address coastal or terrestrial processes that affect or are related to marine ecosystems. Development of standard definitions, objectives, and requirements may enhance the EAM framework. Investment in good science (e.g., the IOOS), and outreach efforts that increase the coordination, efficiency, and outcome of ecosystem goods and services should be aggressively pursued.

Pacific Islands Ecosystem and Social Science Research

Samuel Pooley, Director, NOAA Pacific Island Fisheries Science Center

Islands in the Western Pacific are exceptional in many ways. They are remote and often rugged islands, those in the Hawaiian Archipelago being an extreme case. Some are pristine and others are relatively degraded, with such variability sometimes occurring in close proximity. They are situated on narrow topographical bands of volcanic origin amidst the vast Pacific Ocean. They are profoundly impacted by cyclical oceanographic and atmospheric influences such as the ENSO cycle. Relative to other marine regions in the Pacific, marine resources are not superabundant. Yet there is extensive cultural diversity in the region, and while local social and economic processes obviously are important, in reality many of the most profound economic decisions affecting people here are made elsewhere in the United States and in Asia.

One way in which the ancient Hawaiians adapted to environmental conditions in this region was through the ahupua'a system. Responsibility for managing resources of land and sea was based in part on the geography of the islands, wherein a given ahupua'a typically encompassed land between mountain peaks, down through the upland area, into the broadening lower valley, across the inshore reef system and into the offshore zone. The approach tended to integrate human and biophysical systems and marine and terrestrial systems. This ancient form of environmental interaction and economic production is not explicitly used by NOAA Fisheries in Hawaii, but it does serve as a valid means for conceptualizing island ecosystems and related management issues.

Contemporary management and conservation issues and challenges abound and intersect with an array of institutions and regulatory efforts in the Western Pacific region. The WPFMC must address a range of issues while maintaining a focus on sustainable fisheries across a broad region: sea turtles and marine mammal interactions, community development, international management conventions, and regulations such as the Endangered Species Act, the Marine Mammal Protection Act, the Coral Reef Protection Act, and the National Marine Sanctuary Act. Analysis of the distribution of costs and benefits resulting from management strategies is also a subject of concern as per the Regulatory Flexibility Act and Executive Order 12866 (Figure 8.2).

Social science research activities at the Pacific Islands Fisheries Science Center (PIFSC) are diverse and include extensive application of social science in support

Figure 8.2. Fisherman prepare for Guam Lunar Festival. (For a color version of this figure, see Plate 16.)

of effective resource management throughout the region. Fisheries Monitoring and Socioeconomics Division staff undertakes socioeconomic research with an implicit ecosystem approach. Led by David Hamm, this includes an Economics Program, a Human Dimensions Program, and two Fishery Monitoring Programs. The Economics Program supports NOAA Fisheries' conservation and management goals, and based on historical development of the program, remains largely utilitarian in nature. The program, led by Minling Pan, involves collection and compilation of economic data, assessment of changes in economic indicators, and conduct of applied research and analysis of the economic impacts of alternative management measures for the range of fleets active in the Western Pacific. Stewart Allen leads the Human Dimensions Program, which addresses social and cultural aspects of fisheries, fishing communities, and ecosystems. A Fishery Monitoring Program undertakes various ecosystem-relevant missions and is associated with the observer program.

It is anticipated that future social science research undertaken by the PIFSC will address the needs of NOAA's Fisheries Economics and Social Science Coordinating Committees, the Council's Social Science Research Committee, and priorities set by the University of Hawaii Pelagic Fisheries Research Program. Addressing linkages between the terrestrial, near-shore, and open ocean marine environments is a priority. Plans also include heightened social and economic research in other parts of the region, such as American Samoa, Guam, and the Northern Marianas. Relevant topical areas for which increasing attention will be applied include governance processes, noncommercial fisheries, marine aquaculture, and invasive species issues.

As an interagency initiative, the Hawaii Archipelagic Living Marine Ecosystem Research Initiative is being established to involve a broad range of partner agencies and participants in ecosystem research, monitoring, and management efforts in the region. This multiyear, multidisciplinary research program is designed to enhance understanding of the range of factors affecting marine ecosystems in the Hawaiian Islands, including those associated with terrestrial and nearshore zones. Major partners include the PIFSC, the University of Hawaii, the University of Guam, the National Ocean Service, and the Fish and Wildlife Service.

An important question raised by this process, and still an open one, is whether the social and economic sciences would need to be applied differently within the nascent ecosystem-based management context. In moving toward an answer to this question, it should be noted that: human beings may be seen as a pivotal aspect of marine ecosystems; ecosystem approaches to resource management may require consideration of broader range of participants, stakeholders, and jurisdictions; and, thus, striving to attain balance between the needs and interests of a larger field of constituents may be concomitantly challenging. It can also be suggested that efforts to apply social science in this context may benefit by examination of historic social scientific approaches to management of terrestrial ecosystems.

New issues are inevitably raised in this context. These include a range of important topics that may need to be addressed as the Council moves toward the new approach: prospective regulatory change; the potential need for and utility of ecosystem-related education and outreach; need for management-derived community development and optimization of benefits; and potentially competing social science paradigms.

Non-Economic Social Science and Ecosystem-Based Management

Michael Orbach, Director, Duke University Marine Laboratory

People have long migrated to the Pacific islands from around the Pacific basin and beyond. This has resulted in a highly diverse and complex mosaic of human societies and cultures. Such diversity forces social scientists to adopt an array of methods and models appropriate for understanding that context. Such researchers also have an opportunity to apprehend the perspectives of those who are part of that context. In acquiring this emic perspective in the specific context of human–environmental interaction in the marine realm, a better understanding of marine ecosystems may be achieved and communicated to managers, scientists, and others.

It is important to recognize the diversity of perspectives, needs, and expectations of the many individuals and groups involved in the use and management of marine resources in this region and elsewhere. More often than not, the nominal or behavioral "bounds" between categories of user groups are permeable. They are, in reality, fluid and mixed. For instance, persons fitting in a commercial category may best be described in a recreational or subsistence category, or in both. It is important to consider such individuals and groups, because they are part of marine ecological systems. Ecological systems involve interrelated biophysical and human dimensions. The human dimension is critically important—human beings and social groups, relationships between individuals and groups, and relationships between those persons and groups and the biophysical environment. The study of these relationships is known as human ecology. As a subset of human ecology, institutional ecology involves the study of relationships between institutions involved in governing marine resources and resource users (Figure 8.3).

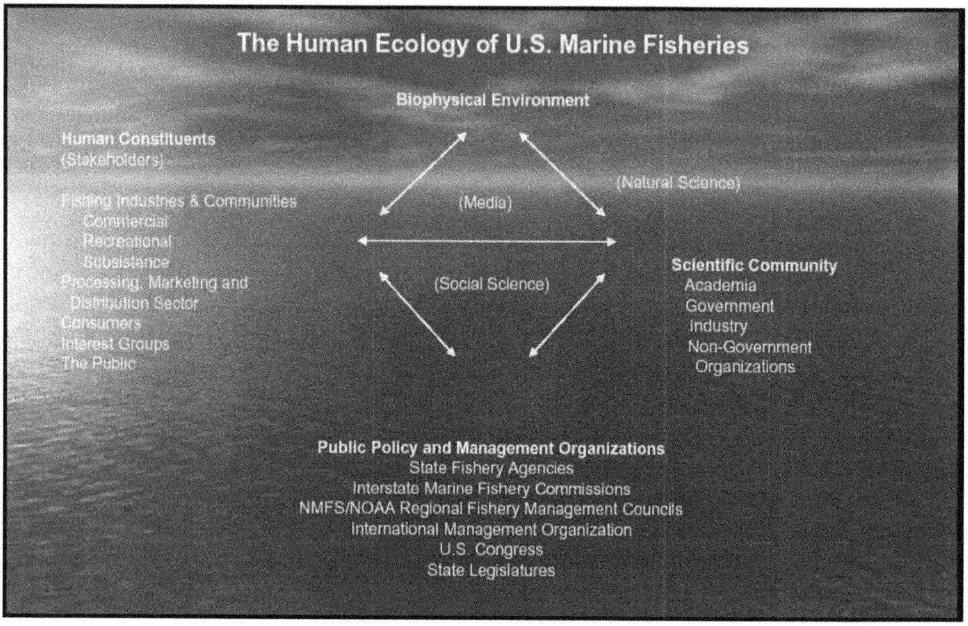

Figure 8.3. The human ecology of US marine fisheries. (For a color version of this figure, see Plate 17.)

Thus, when looking at the ecology of a fisheries system, one must consider its biophysical aspects—typically the primary interest of the scientific community, academia, government, industry, and NGOs. One clearly must also consider and address the human component as part of the ecology of fishing systems. This is identifiable as a system of stakeholders or constituents: the fishing industries and communities, those who process and distribute seafood, the consumers and interest groups, the recreational user groups, the general public. From the institutional subset, public policy and management organizations should also be considered. Natural scientists tend to focus on the nonhuman components of the ecosystem, whereas social scientists focus on the interplay between human user groups and governance entities, and between these components and the biophysical environment.

When considering these components and relationships in the context of resource management, inevitably, there is need for a certain detachment from objective science. Science is about objectivity, reliability, and validity. It can serve to describe and explain what was, what is, and the range of possibilities and probabilities for what might be. But it cannot determine what management *choices* should be made. That is a subjective process involving human judgment, guided by human values, assisted by science and scientific findings. Governance is ultimately a human value-based decision-making process. This fact is critically important in that human values and choices are malleable, with important implications for the management of ecosystems, including marine ecosystems.

It follows that ecosystem management may be defined as the management of human behavior against specific objectives, defined and executed through systems of governance behaviors, which influence and are influenced by the physical world. Contrary to managing biophysical factors to achieve certain social changes, the process ideally involves managing human–social behavior to achieve certain biophysical effects. Governance systems by definition involve a variety of acceptable options and approaches, formal and informal, from modern institutions such as the NMFS and Council, to the ancient *ahupua'a* or *matai* systems in the Pacific islands. One important role of social science in this context involves the production of a better understanding of systems of governance, how they have historically functioned to work in any given place, and how they might be applied in the future.

Important issues emerge from a perspective that prioritizes management of human behavior above management of the resources they pursue and the environment in which they interact. First, if ecosystem management has not prioritized human behavior, how would such change if human beings and their behaviors were prioritized? Might it be simply a matter of exchanging existing concepts, labels, and perceptions, or are truly significant differences in approach called for? Clearly, better integration of human behavioral considerations is sorely needed to improve the current system(s) of management, especially under an approach that emphasizes the importance of relationships between aspects or components of environmental systems. Can social science contribute to this task and what changes are called for to better accommodate human–social considerations in marine ecosystems?

Again, given the broad range of relationships people have with the marine environment, it is clear that social science does have the capacity to contribute to more effective understanding and management of marine ecosystems. In fact, given that human beings have the capacity to modify and regulate their own behavior, and positively affect and influence the biophysical environment, the social sciences may be applied not only to study of ecosystems, but also more proactively to the *design* of healthy marine ecosystems. There is much promise in the application of social science in this arena, and many challenges ahead (Figure 8.4).

Figure 8.4. Roundtable discussion during the first day of the workshop. (For a color version of this figure, see Plate 18.)

Economics of Organization and Ecosystem Management

Susan Hanna, Professor of Marine Economics, Oregon State University

The US Commission on Ocean Policy explicitly stated that institutional challenges underlay many of the problems associated with fisheries management and efforts to maintain the sustainability of ocean resources. Commission members believe that humans are an integral and influential part of ecosystems and that problems stem from failures to build robust institutions to solve those problems. Better integration of agencies and institutions is central to the challenge. The Commission recommends enacting regional ocean governance strategies for which it gives primarily structural recommendations. Should management efforts extend beyond marine fisheries to address marine-terrestrial relationships, this approach may prove particularly beneficial.

The FAO Code of Conduct provides guidance on issues of direct relevance to ecosystem management. It calls for attention to issues of biodiversity, endangered species, species interdependence, fishery impacts, nonfishery impacts, waste, uncertainty, and risk. It also mandates integration with coastal zone management objectives. Accommodating each of these factors has clear institutional implications (Figure 8.5).

Ecosystem management entails an expanded perspective of biophysical and human interactions within and between marine ecosystems. The approach involves a change in degree over the way fisheries have been managed thus far. It will require a more extensive examination of a range of interactions: climate-oceans, oceans-ecosystems, within-ecosystems, people-ecosystems, people-communities, and user group-user group. All of this also relates, in varying degrees, to broader social and economic spheres of consideration.

Figure 8.5. Littoral and riverine zones along the Pacific Northwest Coast. (For a color version of this figure, see Plate 19.)

As we expand the degree to which management addresses more and more complex interactions, the institutional decision environment becomes similarly complicated. More and more complex interactions between scarcity, tradeoffs, impacts, reversibility, integration, compatibility, decentralization, devolution, and equity will need to be considered.

The issue of tradeoffs becomes increasingly complicated under conditions of numerous interests, goals, and objectives. Maintaining flexibility and the ability to reverse decisions enables learning and subsequently reduces uncertainty. Issues of compatibility among components of the institutional environment are critical, as is the option to decentralize decisions or devolve them down to their most effective level. Questions of equity, which are always difficult to address, become even more so when expanding the scope of interests and actors under the parameters of ecosystem management. Greater institutional complexity may be inevitable.

But assuming institutions will simply expand naturally to address the wider scope of considerations is spurious. Institutional expansion raises design issues relevant to important economic principles. A number of incentive problems can limit the effectiveness of complex organizations and increase transaction costs. One problem involves a failure to understand fully the implications of expanding parameters as these relate to incentives and behaviors, and as the decision environment is made more complicated and complex. Incentive problems have very direct implications for transaction costs; in this case, the costs of implementing the new system of management.

The Williamson Puzzle accurately frames the incentive problem. It asks, "Why can't a large firm do all that a collection of small firms can do, and more?" In truth, we cannot realize gains from integration without experiencing losses. In institutional environments, such losses result from impaired incentives. As transactions are transferred from smaller organizational entities and melded together into a more complex institutional environment, all of the ways the smaller entities developed to deal with transactions are misfit—problems inevitably result. These are all fairly typical institutional problems. As fisheries management expands to broaden its institutional scope, it needs to pay attention to typical kinds of incentive problems, anticipate and design around them, and thus prevent costlier outcomes.

A typical problem is power ambiguity, or uncertainty about the distribution of power. A second is the failure to make credible commitments, a situation that can occur in the absence of clear delineations of responsibility. Manifesting in uncertainty, instability, or the absence of property rights, this problem can make it hard to deliver on promises. Similarly, low-intensity incentives can reduce accountability in the system. Opportunism is another typical institutional expansion problem. Increased opportunity for unobservable actions arises, because transparency decreases as complexity increases. The issue of bounded rationality should also be considered. Under expanding parameters, full scientific information or constant information feedback is increasingly less likely, and uncertainty and inconsistency can lead to unintended consequences. Truncated learning relates to bounded rationality. It suggests that opportunities for learning-by-doing and adaptation can be limited by heavily regulated decision environments.

All of these issues lead to design questions. How can incentive problems and transaction costs be minimized? Unifying goals and objectives across widely disparate interests will clarify direction. Uncertainties need to be recognized, with management occurring in a manner sufficiently conservative to accommodate that recognition.

Limits to scale will also need to be determined in terms of institutional structure, decision environments, and the unique contexts that are characteristic of marine fisheries.

Given what is known about incentive problems, it is rational to be proactive in crafting solutions as the institutional environment is expanded. We must pay careful attention to the design of institutional structures, with explicit consideration of incentive problems and transaction costs. As these design requirements are known, measurable indicators may be identified through which the performance of institutional environments can be monitored and evaluated. Studies of governance and institutions may contribute to understanding of institutional or organizational issues likely to surface under conditions of increasing complexity, and hence to design programs that minimize social problems.

Group Discussion

Dr Hanna's presentation led to discussion about whether studies of institutions were relevant to FEPs. Dr Hanna and others emphasized that studies of institutions and governance were critical to effectively implementing ecosystem management, but not necessarily material for a FEP. Dr Leanne Fernandes concurred, describing how her agency experienced institutional disincentives in rezoning the Great Barrier Reef Marine Park, both in working across institutions and with communities. The agency had institutional limits regarding the degree to which it could involve other agencies and stakeholders in decision making. But arriving at a common goal, clearly articulating authority, and not making promises that could not be delivered were important for success. While this information did not appear in the plan, it was critical in the decision-making process.

Workshop participants struggled to understand what product the Council needed or wanted with regard to social science aspects of ecosystem planning. Dr Hanna's presentation highlighted important institutional process considerations that should accompany implementation of complex regulatory structures, such as ecosystem management. Some participants considered it important to provide the council with social science principles and instructions about how these would best contribute to implementation of FEPs. Executive Director Simonds asserted that information about institutional processes and appropriate application of social science would both be appreciated by the council.

Reference

Severance, C. (2006) Personal communication. Department of Anthropology, University of Hawaii at Hilo, July.

Chapter 9
Ecosystem Social Science Models

Incorporating Humans in EAFM

Jeffrey Johnson, Senior Scientist, Institute for Coastal and Marine Resources; Professor, East Carolina University

Scientists are often too narrowly specialized to address the complexities of modern environmental problems. Solutions thus often require the contribution and interaction of persons with different perspectives and expertise. The National Academy of Sciences has recognized the importance of the multidisciplinary approach. For instance, the National Science Foundation Advisory Committee for Environmental Research and Education recognized the need "to meet these complex challenges as well as urgent human needs, [by moving toward] environmental synthesis to frame integrated interdisciplinary research questions and activities to merge data, approaches and ideas across spatial, temporal and societal scales." Coastal problems are particularly amenable to such an approach; several large-scale research programs in the United State currently address these, with particular attention to the interface between human and natural systems.

Despite the fact that humans are components of natural systems, they often see themselves as external to nature. This is evident in management designs and approaches that have no or little embedded human component. As defined by USCOP, effective approaches to ecosystem-based management are to be understood the following way:

> . . . ocean and coastal resources should be managed to reflect the relationship among all ecosystem components, including human and non-human species, and the environments in which they live. Applying this principle will require definition of relevant geographic management areas based on ecosystem, rather than political, boundaries.

A review of the current literature would indicate, however, that humans are to be accounted for primarily in terms of the extractive actions and disturbances they exert on natural systems. Research tends to focus on biological communities, trophic structures, habitat issues, and so forth. A preferable approach is one that accepts humans as critically important components of natural systems and examines interactions within and between social groups and the environment—a total ecosystem. Consequently, it becomes important to examine the relationships of human systems. The complexity and richness inherent in the human component need to be recognized. This

Ecosystem-Based Fisheries Management in the Western Pacific, First Edition. Edward Glazier.

includes behavioral networks, the trophic effects of human activities, the structure of communities of user groups and institutions, incentives and well-being, nutrient output behaviors, and others.

Human behavior can be seen from multiple perspectives: behavioral structures or networks; behavioral incentives or dependencies occurring in systems with diverse degrees of robustness and adaptability; terrestrial-based activities that impact marine ecosystems, such as nutrient loading. There are clear and measurable relationships between human and ecological networks. Regarding these, the conventional view is focused on how humans impact the ecosystem, either directly or indirectly, through cascading effects on trophic systems. Less attention has been given to the way human systems are affected by ecosystems; how these fare in terms of robustness, flexibility, and adaptability; and the nature of feedback between humans and surrounding natural systems.

This difference in treatment is further evinced by the number of indicators that have been developed for the natural system, befitting conservation objectives such as ecosystem diversity, species diversity, genetic variability within species, directly impacted species, ecologically dependent species, and trophic balance. But analog indicators can also be developed for the human counterpart, as for example: fishery diversity, fishing constituency diversity, ecologically dependent communities, social and economic balance, directly impacted groups, indirectly impacted groups, human system diversity, and so forth. It stands to reason that it would be useful to acquire knowledge of these, and then proceed to link indicators from both sides in order to obtain a more comprehensive understanding of the overall ecosystem. Doing so will eventually require the establishment of causation between indicators. One of the ways of doing so is through experimental research.

In the context of marine fisheries milieu, human behavioral networks can be depicted in terms of the structural relationships between different types of species and gear combinations in relation to those persons who use them. These can be expressed as either two-node or one-node networks (or relations), depending on the number of variables that are being used (2 or 1, respectively). Such networks may exhibit certain structural characteristics that can be linked to elements of the natural system in which the behavior in question occurs.

Characteristics of social networks include density (connectivity); graph centralization (the extent to which a given node of a certain behavior dominates and influences the entire system); and fragmentation (the number of network components that are produced when removing a key player from the system). The latter proves useful in revealing key behaviors in fishing behavioral networks, as a change of behavior among those persons who play a central role in structuring the system will incur a significant change in the structure of that system. Identified mathematically are these keystone behaviors; related to the robustness of the network are changes in its preponderance (by reduction or elimination). Depending on the density and centralization around a node or a group of nodes, the removal of one or more behaviors from the network will have different effect(s), revealing varying degrees of resilience, adaptability, and so forth.

This approach is exemplified in the project titled "Incorporating Humans in Ecosystem-Based Models of Fishery Management," with funding from UNC Sea Grant. A multidisciplinary team of scientists developed a framework that includes humans as important components of coastal food webs and that enables modeling of the behavior of fishermen in that system. In terms that are more detailed, the objectives are as follows:

1. Controlling for ecological and environmental factors through standard sampling techniques and analyses of gut contents and stable isotopes, characterize trophic networks (food webs) in Core Sound with: (a) little human input (Primary Nursing Areas), (b) moderate human input (Secondary Special Nursing Areas), and (c) high human input (Core Sound).
2. Characterize behavioral networks of fishery participants (harvester) in the study area.
3. Characterize the interactions between human behavioral networks (fishing networks) and corresponding trophic networks (food webs).
4. Work to understand and measure the direct and indirect effects of fishing effort on benthic, plankton, and fish components of the food web, and vice versa.

Six unique food webs are to be modeled for the Core Sound area, each including the fishery pressure component. By testing various hypotheses, comparing and contrasting the effects of fishing behavior, it will be possible to evaluate whether and how fishing pressure and types of fishing pressure ultimately impact the food webs.

This is a unique project in that it involves simultaneous collection of social and natural science data that is directly relevant to understanding of complete ecosystems. It applies network methods to model both food webs and the shifting behaviors of the harvesters. Finally, it is capable of assessing species richness and predator/prey linkages resulting from changing fishing pressure.

The study of Traditional Ecological Knowledge (TEK) may be particularly useful for ecosystem-based management of marine fisheries. Much of the work on TEK has been of an anecdotal or qualitative nature, though TEK approaches have been increasingly systematic and quantitative. When approached systematically, this form in inquiry can enable understanding and comparison across different knowledge systems or time gradients, between types of knowledge such as TEK and Scientific Ecological Knowledge (SEK), and within particular groups. Regarding the latter, it is possible to model variations of knowledge and how these relate to factors such as social roles, experience, social status, and gender.

Recent work with Iñupiaq people in Kotzebue Sound has been effective for enhancing understanding of human connections with the physical environment. Kotzebue Sound is home to about 3,500 persons, 80 percent of whom are Iñupiaq. Traditional ecological knowledge is critically important in this setting where hunting, fishing, and gathering are common means of survival. Two hypotheses were tested in this case, one relating to the existence of a shared cultural ecological model of the Ugruk (bearded seal) among Iñupiaq hunters, and the second relating to the level of agreement between TEK and SEK regarding the structure of the trophic web. It was found that agreement among hunters increased when the topic of interest involved higher levels of the food such as the Ugruk, and that TEK and SEK were most closely correlated when the topic involved higher trophic levels.

This approach allows for direct comparisons of different systems of knowledge, detecting where similarities and differences lie. This may assist scientists and resource managers in assessing TEK for potential use in frameworks of ecosystem-based management. Also examined by this study were problems associated with communication of TEK to scientists; the methodology may assist in structuring that information for use in management, as in the formulation of ecological boundaries, seasons, and gear use. Furthermore, TEK may be useful in formulating valid research topics. Future use of more advanced techniques, such as Markov chain models of trophic network dynamics, would open the way for more powerful analytical approaches. The application of true statistical models would allow the measurement of changes in some network parameters while controlling others, and in conducting tests of significance on structurally based food web statistics. Changes in food web structure could be taken to more detailed levels of analysis, and modeling could be extended to include true changes in ecosystem structure and no longer be limited to macrolevel summary statistics, such as species diversity.

In sum, there is a need to conceptualize ecosystems in terms of interactions both within and between human and natural subsystems. A promising avenue of understanding is through analysis of human and ecological networks. Recognition that impacts occur not only from the human to the biophysical but also the other way around is important, as is the need to incorporate and relate indicators from both components. Finally, modeling TEK may help in developing a comparative interface between various forms of human knowledge, so as to assist in an ecosystem approach to management processes.

Modeling NonEconomic Aspects of Human Behavior for Managing Marine Ecosystems

Richard Pollnac, Professor of Anthropology and Marine Affairs, University of Rhode Island

Taking human behavior into full consideration in the management of marine ecosystems is of paramount importance. But it is hampered by the complex and delicate task of identifying and analyzing the many variables that influence and structure behavioral interaction between humans and marine ecosystems. In addressing noneconomic aspects of human interaction with the marine environment, a very broad set of variables could potentially be addressed. Of particular relevance in any such analysis of relationships between participants in the harvest sector of marine fisheries is the dependent variable "happiness" or "well-being," envisioned as an expression of occupational satisfaction. By extension, one could examine management options in terms of their capacity to enable continued job satisfaction or well-being.

The complexity of relationships between factors associated with the well-being of human beings calls for analytical tools that enable conceptualization and analysis of those relationships. Several types of models are relevant and useful: (a) the heuristic model, (b) the causal model, and (c) the pattern-oriented model. Heuristic models are useful during early stages of research. They assist in developing theory and structuring decision-making processes associated with data collection. They can be constructed in various ways and involve an iterative approach. Causal models are useful for testing hypotheses, for building theory, and for making decisions. A typical sequence involves use of a heuristic model to collect quantitative data about predefined variables, and use of that data to test hypotheses about causal relationships. Pattern-oriented models are helpful for development of theory about complex agent-based human systems.

Building such models requires initial review of the previous relevant research and associated data, as available. Fieldwork and collection of new data ensue. Models are built and tested and their elements and configuration confirmed or dismissed.

If findings lead to reformulation or abduction of a model, a new structure emerges. This cycle of compilation of existing information, field observation, induction, deduction, abduction, and trial and error tends to generate myriad variables purportedly connected to the dependent variable under consideration.

Study of a given dependent variable involves examination of a range of known and unknown factors. A good sampling strategy is fundamental to proper evaluation of the contribution of known factors and the validity of the research hypothesis. Case studies may also yield important information and suggest unseen relationships. Attention to feedback between quantitative and qualitative research is often helpful. It is important to remember that if these considerations are to be applied in an ecosystem-based management context, the known and unknown factors potentially relating to the dependent variable will derive from both the human and natural environment.

An example of the complexities inherent in modeling the interface between human groups and the marine environment is noted of a coral reef project conducted in the Philippines. It was thought that high human population density would correlate with coral reef mortality. But a model constructed with both coral reef mortality data and several socioeconomic variables indicated that increasing population density was actually positively correlated with high-quality reefs, while decreasing density was inversely correlated. A hypothesis that sound reef areas were attracting migrants was confirmed by fieldwork; new communities were being settled where the reefs were still in good shape. The findings clearly suggest a need for nuanced interpretation of the situation and perhaps longitudinal assessment of population-related fishing pressures in the area. This has obvious implications for managers seeking to predict changes in coral ecosystems in the region.

Human factors, such as fishing pressure, environmental activism, and demographic change, will tend to affect management strategies, as will biophysical factors such as oceanographic or climatic regime shifts. Changes in management strategies may, in turn, shape the attributes of fishing occupations by, for instance, shifting the allowable intensity of activity. This may affect job satisfaction and, subsequently, job performance, mental and physical health, family and social relations, and so on. However, job satisfaction appears to involve more than gainful employment. Several variables appear to contribute to satisfaction, but the experience of hunting and catching fish in a challenging environment appears central, and a corresponding personality type is implicated. An independent, self-reliant, risk-taking personality is common, and issues of identity come into play. In sum, job satisfaction relates in part to identity and affinity with a way of life.

External forces will influence and be influenced by management strategies. Management will influence occupation attributes, which, in turn, will be interrelated to job satisfaction. Job satisfaction will influence individual attributes and the occurrence of social problems. Social problems are interlinked not only to individual attributes but also to the contextual social structure. These cascading factors affect the well-being of the fishers.

An example is the Individual Fishery Quota (IFQ) system in Alaska. External forces led managers to implement IFQs; this influenced crew size and structure: having no need to maximize effort during a limited season, the owner could rely on a few family members. On the other hand, the high costs of the IFQ led to the relatively few, relatively wealthy in the fishery. Job satisfaction increased for those able to stay in the fishery, but diminished for those unable to participate. Social stratification in certain communities was amplified, with unemployed crewmembers on the one side and IFQ holders on the other. Participants simultaneously improved access to and influence in the management arena. As such, there was unequal distribution of effects, with implications for the well-being of the participants.

Another approach to modeling involves inductive use of statistics and working backward from the dataset to develop heuristic models that can be tested with causal analysis of another dataset. A correlation matrix with predictor variables and a measure of performance is developed, and the analyst, using a predetermined criterion of strength of relationship, works backward from the dependent variable through the predictor variables to determine patterns of interrelationships between strongly related variables.

An example of this procedure was applied to evaluation of 45 no-take MPAs in the Philippines. A measure of MPA success was created by incorporating three biological parameters, assessed by direct observation with snorkel surveys. A coral mortality index was created, and a number of fish families and top predators were assessed to generate an aggregate indicator of MPA success (the dependent variable). A previous survey of literature was used to identify independent variables suggested to be related to MPA success. The dependent variable was examined in terms of its relationships with 83 independent variables: 12 environmental and demographic variables, 29 socioeconomic and cultural variables, and 42 project activity and project output variables. The resulting matrix had 3,486 entries. It was found that independent variables such as adaptive management, monitoring by community, and successful alternative form of income were linked to the dependent variable. By replicating and extending this process in several steps, it was possible to construct a model, including a total of 15 independent variables and infer causal relations that could later be tested.

In conclusion, different types of models can be of great assistance at different stages of the research process. These can serve to highlight relationships that might go unnoticed and further understanding about relations between multiple human and biophysical factors, thereby enabling a better grasp of systems and related management decisions of great complexity.

Bioeconomics and EAFM

Lee Anderson, Professor, University of Delaware

An initial challenge to a regime shift in fisheries management involves the clarification of definitions of terms related to ecosystems and ecosystem-based management. The goals and objectives of an ecosystem approach to management must also be made clear at the outset. A valid approach would involve managers and stakeholders generating a list of objectives, goods and services desired for whom, and what types of trade-offs would be allowed. Management plans or planning exercises could then be designed according to this prioritized list, and implemented carefully to ensure that scientists, managers, and the public understand and have reached some level of consensus on objectives and intentions.

Ecosystem approaches to fisheries management may be seen as differing from single species management primarily in terms of degree rather than radically different approach. The ecosystem approach signifies a paradigm shift. But by considering it as primarily a change in degree, we can eliminate some uncertainty about its true nature. Certain commonly held beliefs about ecosystems-based approaches may at this time be inappropriate or not fully developed and should be subject to assessment and potential revision. These include the following: (1) we need to conserve the form and structure of the ecosystem; (2) we need to develop a series of ecosystem indicators; (3) we need to develop a series of MPAs (no-take areas) over wide sections of the ocean; and (4) we need to maximize species diversity. Clearly, these are important concepts of pertinence to management. Rather than instigating argument between biologists, social scientists, and economists, it may be possible to engage in free and open discussion about these concepts.

Ecosystem approaches are geographically specified and adaptive approaches that take into account levels of current knowledge, uncertainty, and multiple external influences. Ideally, they balance diverse social objectives and are implemented incrementally and collaboratively. The process of developing and implementing such approaches would start with a clear statement of the objectives of management while specifying acceptable trade-off rates between conflicting objectives; that is, what types of goods and services do we want to flow in and out of the system over time? Persons developing such approaches would then select a range of potential management options and seek to determine the likely effects of those options. Finally, those options that most nearly achieve the objectives of management would be selected and implemented.

In prioritizing formulation of management objectives, it is necessary to know what types of goods and services are desired from the ecosystem. This should include the widest conceivable range of goods and services, not simply those that are market-related. For example, we might be interested in fish for sustenance, to catch, to observe, and other related ecosystem services, such as the ability of estuaries to protect against storm surges. In the end, this would be a long list.

Determination of trade-offs will necessitate stakeholder involvement. It may be a messy process and likely will generate conflict. At this stage, diverse social objectives may be balanced through a collaborative process. There will be degrees of winning and losing. Some may be unwilling to make trade-offs, perhaps because they view the problem as more of a moral imperative than a basket of goods and services, no matter how broadly the basket is defined. But while stakeholders may have strong feelings about what they want, clear leadership and thinking are needed because not all stakeholders possess complete understanding of all ecological interactions. For instance, some will have desires for directed catch but may not have bycatch on their radar. In sum, we must determine what we know, do not know, cannot know, and what risks society is willing to take with what we do not or cannot know.

Compared to the conventional approach, ecosystem approaches will involve a broader range of potential outputs, interested stakeholders, and a greater likelihood that moral imperatives will be brought to the table. Bioeconomic Management Strategy Evaluations (MSEs) may be useful in that they can enable determination of which management strategies will most closely achieve stated objectives. These tools can simultaneously reveal the effect of a management strategy on (a) likely future changes in important elements of the ecosystem over time, (b) likely future pattern of ecosystem-related flow of goods and services over time, and (c) distribution of gains and losses over time [examples are available at the following link: http://www.st.nmfs.noaa.gov/st7/ecosystem/workshop/2005/index.html].

A bioeconomic MSE model for the Western Pacific would incorporate standard stock assessments and relationships between patterns of fishing effort, fishing mortality, effects on outputs, and welfare of participants. It could help provide answers to questions about the effect of various patterns of fishing mortality on stock size, age class, distribution, and spawning biomass. It could also address issues about how or what regulations could affect patterns and distribution of profits, exit and entry, and gear changes among various participants. In order to create an effective model, we must understand the incentive structure of participants and be ready to accommodate new factors, such as cultural assessment components. The Fulton model contains many ecological and human dimensions. But models will be useful only if they take into account the objectives of management.

While we may seek to conserve the very complex and ever-changing form and structure of the ecosystem, in a very real sense it is not the ecosystem we are interested in *per se*, but rather the associated flow of goods and services over time. Emphasis on form and structure may be misplaced. As such, biological indicators should also be able to tell us about the ability of the system to provide the desired goods and services over time (e.g., status of structural habitat biota, environmental fluxes, and seabird population trends). Design of MPAs and no-take areas over wide sections of the ocean should account for patchy distribution of goods and services and relate clearly to management objectives. In seeking to maximize species diversity, once again, the real goal is maximizing the ability of the ecosystem to provide goods and services associated with those species.

In sum, movement toward the ecosystem approach needs to proceed incrementally, patiently, and carefully. Clarification of definitions and management objectives is critical, as is recognition that ecosystems may be conceived and managed in terms of their long-term capacity for bearing goods and services for society.

[Commenting on the presentation in heuristic spirit, Dr. Jim Burchfield reversed the logic, pointing out that goods and services may also be seen as following from ecosystem *functionality* as an objective. Thus, by prioritizing functional marine ecosystems as a management objective, assurance of extractive marketable products and nonmarket goods and services may follow].

The Ecology of Governance: Policy Perspectives on EAFM

Tim Hennessey, Professor of Economics, University of Rhode Island

From a broad policy perspective, the ecosystem model can be considered a paradigm shift driven by mounting problems and perceptions that the existing management structure has been largely ineffective. The shift was instigated, in part, by depleted fish resources in national and international waters, lawsuits brought against NOAA Fisheries, and an institutionalized process of developing solutions without truly linking them to problems.

Yet, is it not entirely clear what an ecosystem approach to management truly entails. As a solution generated long ago in the absence of a specific problem, ecosystem-based management has recently been called upon to deal with the problems at hand. But its ambiguity has already surfaced in the current setting. Even in the current context, while many workshop participants have expressed ideas about ecosystem-based fisheries management, no clear definition or objectives have been clearly formulated.

Throughout the policy process, ecosystem managers must cope with the uncertainty and changing nature of the organizational and institutional environment. These structures and processes may be referred to as "the ecology of governance." Players in such a system are connected in some way from a central node of governance structure and process. In our example, one can envision NMFS headquarters as the central point of structures and processes. Within the system, some connections work well, others do not, and yet others do not exist. Moreover, the system changes as personnel move to other positions, as policy shifts, and as new administrations enter and exit. Uncertainty characterizes the setting. While adaptive management may be useful, it is difficult to achieve in the policy arena.

Adaptive management is "learning by doing." Implementing programs involves an opportunity to test and improve the scientific basis for action. Adaptive implementation means there is active participation by relevant actors. Those involved in the delivery of policy learn by doing rather than mechanically following standard operating procedures. This requires adjustment of policy based on the situation at hand. Generally speaking, the implementation phase enables clarification of policy. Only at this time can the words comprising the policy in question be understood in pragmatic terms. But significant challenges may remain even after a modified policy is indoctrinated. For instance, when the Clean Water Act was enacted several decades ago, those delivering the policy dealt with point source pollution first. After some initial adjustments and successes, the full implications of a comprehensive policy began to be recognized and numerous agencies are now addressing the massive challenge of implementing nonpoint, source pollution programs around the nation.

The Organizational Conservatism Hypothesis describes additional constraints to a new ecosystem management paradigm. For example, imagine a bureaucrat at Headquarters. A brand new management system is indoctrinated. You say to yourself, "wait a minute, we devoted 20 years to this other system, there are massive sunk costs, I don't think I'm going to like this, I think I'll just drag my feet." This may be termed organizational conservatism.

In the context of the nascent ecosystem approach, a significant challenge involves institutional integration among as many as 16 agencies. Yet, historically, NOAA and the EPA could not effectively integrate when explicitly required by the Coastal Zone Management Act Amendments of 1990. This is problematic in that ecosystems must be subdivided for management purposes into goals, subdivided into programs, and again into projects. Assuming five goals are established, for each of these there will be 20 programs and 100 projects. The challenge is to relate the measures for these goals into an assessment of the overall progress toward achieving the broad goal of ecosystem health.

For example, according to an expert review panel on the Chesapeake Bay Program, the overall progress toward ecosystem health in the Chesapeake Bay should have been assessed using an integrated approach with broad scale measures. Because ecosystem elements are no longer to be viewed separately, integration and unifying concepts become critically important issues. This has relevance for the establishment of boundaries, selection of policies and issues, designation of the lead agency, and the establishment of partnerships between levels of government.

[Drs Hennessey and Anderson were asked to clarify whether the extent of challenges associated with ecosystem-based management could be terminal to the effort. Neither believed implementation of the approach to be impossible. Dr Anderson reiterated that defining and developing goals for an ecosystem approach to management requires ongoing clarification and careful deliberation. Dr Hennessey asserted that when the short list of goals is developed for the Council FEPs, these need to be integrated into one final, overarching measure, such as productivity, sustainability, resilience, or ecosystem health since funding and accountability agencies tend to respond best to a single overarching measure.]

Chapter 10
Social Science Data Sources

People and Marine Protected Areas: Research Priorities

Bryan P. Oles, Social Scientist, National MPA Center-Science Institute

Persons and groups potentially affected by marine resource management decisions are often highly diversified. People have a wide range of experience with and perspectives on the marine environment and what it may afford them. The capacity to assess what is valuable and important and to whom is an important aspect of the management process. The role of the social scientist in the formulation, monitoring, and evaluation of MPAs and other forms of resource management is crucial to success.

The policy basis for the establishment of MPAs in the United States is Executive Order 13158, signed by President Clinton in 2000. The three principal goals of the order are as follow: (1) strengthen the management, protection, and conservation regime of existing MPAs; (2) develop a comprehensive system of MPAs based on scientific advice that represents the diversity of ecosystems, and natural and cultural resources; and (3) develop federally funded activities to protect existing MPAs from any harm. The link between Executive Order 13158 and social science derives from some of the conditions stated therein: the need for science-based criteria and protocols for monitoring and evaluating the effectiveness of MPAs; the identification of potential threats and user conflicts that pertain to MPAs; the identification of appropriate management solutions, including enforcement strategies that could mitigate, reduce, or eliminate those threats; and the assessment of economic effects of management solutions.

The National MPA Center's Science Institute has developed a social science research strategy for strengthening the application of social science in MPA planning, management, and evaluation, while catalyzing the interest in human dimensions as they apply to MPAs. The strategy addresses the following: (1) governance, referring to the structure and function of relevant institutions and processes; (2) patterns of resource use, relating to the intensity, significance, and spatial and temporal patterns of relevant activities; (3) attitudes, perceptions, beliefs, cultural models, and systems of meaning associated with human-environmental relationships; (4) market and non-market values and economic trends; (5) socioeconomic and cultural characteristics of both place-based and identity-based communities; and (6) cultural heritage and resources as socioeconomic dimensions of maritime heritage.

Several regional social science workshops have been held to address these themes and issues, and to discern regional variation. The resulting reports include recommendations and guidelines for enhancing regional capacity for social science research

Ecosystem-Based Fisheries Management in the Western Pacific, First Edition. Edward Glazier.

and documenting existing research, regulations, institutions, and information sources pertaining to MPAs.

This effort relates to the outreach and coordination role assumed by the Center. Similarly, public dialog sessions have been held with stakeholders, and online regional information centers have been established. The Center also supported and engaged several research projects to develop baseline data needed for MPA planning, management, and evaluation. This includes development of the Marine Managed Area (MMA) descriptive inventory of MPAs around the US and its territories.

Other programs involve an MPA decision support tool being developed in cooperation with the Monterey Bay National Marine Sanctuary and the California Marine Life Protection Act Team. *OceanMap* was designed to collect and depict, in spatial terms, ecological knowledge of fishery participants for use in MPA planning processes. Reports such as *Enforcing MPAs* and *Lessons Learned from MPA Designations* have focused on stakeholder processes and the success and failures of models for stakeholder participation. As regards the future of the Institute, the West Coast Pilot Project will coordinate identification of priority conservation areas at a regional level.

Identification, implementation, and assessment of MPAs bear lessons for the establishment of ecosystem-based management strategies around the nation. The issues and data needs that apply to the MPA context may be scaled to inform the nascent ecosystem management process. The challenge lies in defining which data needs are truly essential.

NOAA Fisheries National Social Science Research Programs

Susan Abbott-Jamieson, Senior Social Scientist, NOAA Fisheries

The recent history of social sciences such as geography, anthropology, political science, social psychology, and sociology emerged with the Magnuson-Stevens Act in 1996. Implementation of National Standard 8 (NS8) and appropriation of research funds in FY 1999 allowed NMFS to begin developing its Social Science Research Program. Its Sociocultural Analysis Component now employs 10 full-time staff nationwide and operates on a budget of approximately $300,000 across six regions. Staff members are developing the Sociocultural Practitioners Handbook, community profiles and databases, a national community port database, and a Social Impact Assessment Conceptual Model. Various social science research is ongoing in a variety of topical areas, including the dynamics of fishing crews, women in fisheries, and local and traditional ecological knowledge, among others (Figure 10.1).

Perhaps most relevant to ecosystem management considerations are the research program's Community Profiles Databases and Indicators. The Community Profiles Databases identify and profile communities and ports where fishing-related activities occur. Given the specific requirements of NS-8, fishing community analysis is place-based. NOAA Fisheries evaluates extent of community involvement in fishing-related activities, including those associated with commercial, subsistence, and recreational fishing.

The research program is creating and maintaining regional and national databases to support research and monitoring at the community level of analysis. These data

Figure 10.1. Employment options related to commercial fishing are important in many coastal communities of the U.S. (For a color version of this figure, see Plate 20.)

may ultimately be of use in the development of fishery ecosystem plans. The databases incorporate a wide range of information relevant to fishing activities and local socioeconomic and demographic conditions.

Key indicators that will help determine extent of involvement and engagement in fishing include pounds and value of landings attributable to the community, base economic activity generated by fishing or related services, and history of involvement in marine fisheries.

Once completed, the federal database will allow for comparative sociocultural analyses of fishing communities and activities. Since ecosystem approaches to management are likely to require the spanning of local and regional boundaries, the data will support analyses that extend beyond the community.

Participants in the ESSW should pay attention to a parallel effort being developed by agency and academic social scientists. The Social Impact Assessment (SIA) Conceptual Model project aims to make social impact assessments more compatible with those of biologists and economists. The effort corresponds with visions of integrating social science into fisheries ecosystem plans. The model incorporates data on community demographics; community jobs related to fishing and associated industries; crew, owner–operator, processor worker information; characteristics of fishing-related businesses; subsistence participation, landings, and consumption; species; governance and institutions; cultural heritage and resources; community resilience; public health and social problems; perceptions of the future; and perceived community identity.

National Ocean Service Social Science Program

Peter Wiley, Economist, NOAA; National Ocean Service

The National Ocean Service (NOS) is one of the several NOAA line offices, as are the National Marine Fisheries Service (NMFS), the National Weather Service (NWS), the National Environmental Satellite, Data, and Information Services (NESDIS), the Office of Oceanic and Atmospheric Research (OAR), and the Office of Program Planning and Integration (PPI). NOS personnel are responsible for handling matters related to coastal zone resources and ecosystems, through the Office of Coastal Resource Management. Other NOS offices include the Office of Coast Survey, the Center for Oceanographic Products and Services, the Coastal Services Center, and the Sanctuaries Program, among others.

The development of an NOS Social Sciences Plan (SSP) dates back to work by Leah Bunce and a Social Science Review Panel. Several recommendations were issued by the Panel to the NOAA Science Advisory Board as a way to build social science capacity inside the institution. One of the proposed measures was that each line office should develop a social science plan. A Social Science Team (SST) was thus set up within NOS to determine the status and direction of social science in the agency. NOS social science needs are diverse and encompass distinct areas of inquiry: characterization of sanctuary resource use, evaluation of MPA use and impacts, socioeconomic monitoring, and providing assistance in management planning and technical support.

Several goals were defined to aid in initiating a coordinated effort to build social science capacity in NOS that would explicitly support both NOAA and NOS missions. The underlying vision aimed at strengthening program planning, management decision-making and performance measures to better integrate the biophysical and social sciences in NOAA, NOS, and outside organizations. The guiding goals were to (1) enhance NOAA's ability to monitor, understand, evaluate, and communicate socioeconomic benefits; (2) acquire more accurate decision-support tools by integrating social science, biophysical science, and monitoring results; (3) increase models and methods for assessing the impact of human and natural disturbances; and (4) improve the understanding of the needs, knowledge, perceptions, and values of NOAA's partners and their constituents.

Because of the complexity of this process and a need to harmonize the interests of social scientists and program directors, the SSP was constructed in several steps. Issues and needs for social science that would meet and support NOS and NOAA social science-related missions were identified. NOAA directors and key personnel then prioritized these. From this point forward, the social science team identified areas with better integration potential. Since an increase in efficiency, communication, and coordination between social scientists and users of social science within NOS was one of the major objectives, a database was assembled to better manage existing information. Because of specific demands, a dual-system database format was compiled, separately describing the personnel and projects components. This allowed for deeper and more flexible analysis of information at several levels (geographic, functional, chronological, etc.).

Members of the social science team were linked to a NOS Office and a NOAA program, working with them to develop social sciences-relevant missions within an integrated vision. The degree of effort required varied in part based on the existence of previous social science programs. One challenge related to the current structure of NOAA. NOAA programs define strategic planning and budgeting, while the line offices provide the operational and functional framework. Development of a social sciences plan for NOS had to take both perspectives into account. Later, NOAA Office of Program Planning and Integration would assume the task of integrating the NOS Social Science Program with those from other NOAA agency social science plans.

Important questions arose early in the process regarding the definition of social science. The following definition was used as a reference point for the SSP: "the process of describing, explaining, and predicting human behavior and institutional structure and change in interaction with their environments, to include the fields of economics, anthropology, sociology, geography, political science, social psychology, and history."

Connection of social science with NOS Ecosystem Approach to Management (EAM) was the next step. In accordance with an NOS EAM Action Plan, this connection was established at the Ecosystems Goal Team (EGT) level. This would correspond to the lowest level of decentralization possible in NOAA, following the "bottom-up" approach defended by the Action Plan for the programs of interest: the Coastal and Marine Resources Program, the Corals Program, and Ecosystem Research and Habitat Restoration.

The Coastal and Marine Resources Program focuses on promoting healthy and productive ecosystems and incorporates socioeconomic and demographic factors in its management processes. There are 11 full-time social scientists in this program, distributed across the MPA Science Center, National Marine Sanctuaries Program, the Coastal Services Center, the Office of Coastal Resource Management, and the International Programs Office.

The Corals Program aims to preserve, sustain, and restore coral reef ecosystems. These bring numerous benefits to society through the tourism and fishing industry, bio-prospecting for pharmaceutical research, protection of the shoreline, and so forth. With only one fully funded social scientist, this program relies extensively on NOS staff. Total NOS investment in social science amounts to $594,000 ($614,000 with partnerships), distributed by the Offices of Response and Restoration, International Programs Office, the National Marine Sanctuaries Program, and the Coastal Services Center.

Ecosystem Research focuses on providing scientific information and decision-support tools by integrating research from the biophysical and social science perspectives to advance understanding of ecosystems. The integration process is performed through assessment of information needs of coastal managers, coordination of biophysical and social science research, facilitating use of said research by coastal managers, and building local capacity and environmental literacy.

Finally, the Habitat Restoration Program is designed to improve the quality and quantity of coastal habitat restoration. The main role of social science in this program is in the area of damage assessment. This is accomplished via National Research Damage Assessment (NRDA).

Two specific examples of projects being developed by these programs include the Regional Priorities for Research on MPA's project and the Northwestern Hawaii Islands Reserve Commercial Bottomfish study. The priorities project established to detect needs for social science research were at a regional level. Methodologies have included the development of several workshops, with focus groups, targeted discussions, and identification of regional priority projects. The bottomfish fishing study originated as a means for enhancing the Environmental Impact Statement and management plan for the Sanctuary Designation Process of the Northwestern Hawaii Islands Reserve. The methodology was based on the assessment of existing information, an iterative survey of the fishers, and spatial analysis of logbook data.

Island Context and Social Science Data Challenges and Solutions

Fisheries Management in the Western Pacific EEZ

Paul Dalzell, Senior Scientist, Western Pacific Regional Fishery Management Council

The US purse seine fleet is an important aspect of the economy and fisheries in the Western Pacific region. A treaty between the US and 16 Pacific island nations provides regulatory structure and authority over the fleet and other purse seine operations in the region. The seine fleet contributes substantially to the fisheries-related economy in the region, in large part because it supplies tuna to canneries in American Samoa.

Troll fishing is the most widely practiced form of fishing in American Samoa. Bottomfish fishing and diverse reef fishing practices also occur here. Longline operations in American Samoa are subject to area closures. Large pelagic fishing fleets are subject to 50-mile closures around the islands, while the nearshore waters are open to small-scale longline vessels. The latter employ outboard motors and gear is deployed and retrieved without the use of hydraulics. American Samoa tuna canneries process more fish than any other in the world. Between 150 and 200,000 metric tons of skipjack, yellowfin, and albacore valued at between $250 and $300 million are processed on an annual basis.

Trolling is the most important fishery for Guam. Bottomfish fishing occurs predominantly on the southern banks. A small fleet specializes in deployment of short lines (less than 1 mile in length). Guam is a major point of air transshipment of seafood destined for markets in Japan, China, and Taiwan. Guam-based distributors typically send between 5,000 and 12,000 metric tons of large sashimi grade tuna to Tokyo. Exports were valued at $43 million in 2004. Various regulations such as the Shark Finning Act have impacted the transshipment industry, as have shifts in home–port preferences by longline operators.

The principal fishing methods in the CNMI are troll fishing and bottomfish fishing. A long distance bottomfish fishery operates to the north. Small-scale coral reef fisheries are prolific. There is an extensive skipjack fishery in Saipan.

The longline fishery has a long and productive history in Hawaii. An extensive small boat pelagic troll fishery operates within about 20 miles of the Main Hawaiian Islands. Participants in a smaller handline fishery target offshore banks, nearshore ko‘a, seamounts, and weather buoys. Once popular as a source of fresh skipjack, the pole-and-line fishery is now almost defunct. The lobster trap fishery in the Northwestern Hawaiian Islands is under a moratorium. Trap fishing for crustaceans also occurs in the region, as does black coral harvesting. Trap, hook and lines, and spear fishing

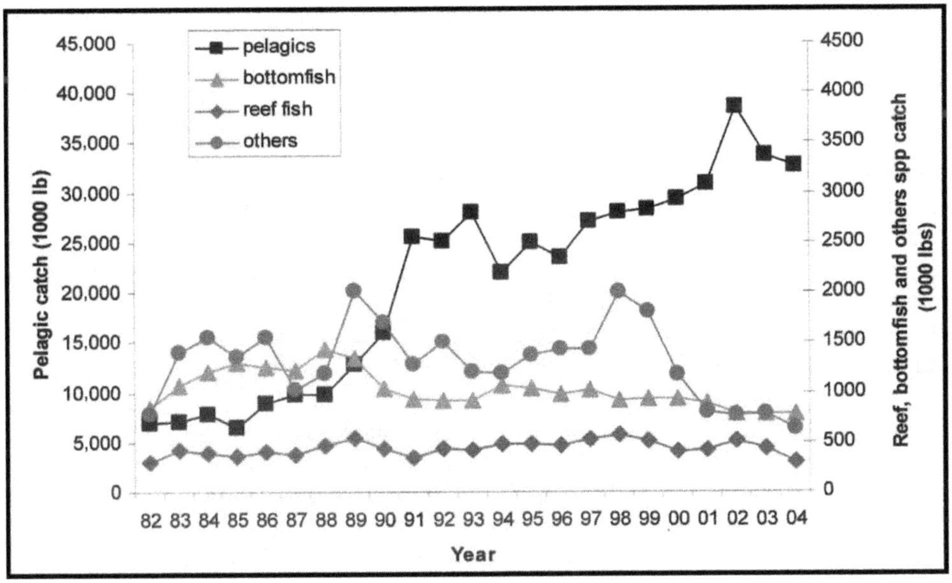

Figure 10.2. Annual domestic fishery production in the Western Pacific region. (For a color version of this figure, see Plate 21.)

are some of the gear used to target reef fishes in the islands. Tropical aquarium fish collection has been increasing in intensity.

Area closures have been established to separate longliners from protected species in the Northwestern Hawaiian Islands, and in the Main Hawaiian Islands to separate them from small boats. Most fishing that occurs near Hawaii occurs around the main islands (~10,000 metric tons per year), while relatively little occurs around the Northwestern Hawaiian Islands (~100 metric tons annually; primarily bottomfish species) (Figure 10.2).

Recreation-oriented fishing is also important in Hawaii, in per capita terms perhaps more than anywhere in the country. Most people in the islands have a strong relationship to the sea and love to fish for a variety of reasons. But differentiating commercial, recreational, and consumptive-oriented fishers can be rather difficult in Hawaii, and nearly impossible in places like Samoa. Another factor that makes the region unique is the strong affinity for marlin. Marlin fishing is important for the charter vessel fisheries and it is also commonly consumed, sold, and otherwise distributed. There are close to 200 active charter vessels based in Hawaii. Smaller charter fleets are based in Guam and CNMI.

The vast majority of landings in the region derive from pelagic stocks. While coral reef fishes (small pelagic species such as *'ōpelu* and *akule*) and various bottomfish have been subject to static or somewhat declining harvest, the harvest of pelagic species has increased by 1 percent or more each year. The Samoa longline fleet is increasingly productive. It should be noted that fisheries in the region are characterized by relatively low volume of landings and relatively high ex-vessel value. Virtually all seafood landed in the region is sold fresh.

Overview of Social Science Data Issues in the Western Pacific Region

Craig Severance, Department of Anthropology, University of Hawaii at Hilo; Member of WPFC-SSC

Despite political bifurcation, American Samoa and Independent Samoa are bound by a single culture. As regards marine fisheries, although only American Samoa has a seat at the Council, there are significant interactions between both Samoas. There is much interaction and exchange. For instance, a shared albacore stock may present an opportunity for international comanagement.

Understanding *Fa'a Samoa*—the Samoan way—is critical to understanding society, culture, and human interaction with the physical environment in Samoa. The chiefs or *matai* hold great authority, and cultural identity and resilience among Samoans relate to respect for the matai system, and adherence to customs and traditions. Oversight of fisheries occurs at the village level and permission to fish in a given reef area is granted by local chiefs. Fishing and fish are central to local society and culture, and fishing-related matters are taken seriously.

Commercial fishing is particularly important in economic terms, facilitated in part by the presence of the canneries. In American Samoa, commercial permits and licenses are granted by the Department of Marine and Wildlife Resources.

The pursuit, capture, distribution, and consumption of seafood in this context are critically important. Seafood circulates or flows across the community from point of extraction to point of sharing or consumption, and important social processes and cultural meanings are associated with each exchange. Fish and fishing are pivotal in the cultural identity and vitality of the Fa'a Samoa and its resilience, integrity, and continuity. Analysis of the flow of seafood is useful in this setting not only for enhancing understanding of its role in Samoan society, but also as a means for guiding management in a manner that is culturally appropriate and that ensures the well-being of local societies.

Chamorros are deeply involved in the pursuit of seafood in the Mariana Archipelago, which includes Guam and the CNMI. Although Chamorros are a minority population in the CNMI, the group tends to dominate the political landscape. Other ethnic groups are involved in fishing as well, including Filipinos, Micronesians, and Carolinians.

The social and cultural importance of seafood in the Marianas is exemplified in *fiestas*--festivities held in honor of the patron saints of the villages. Fresh fish is of paramount significance at these events. Captains and crew in the local fleets are seasonally focused on fishing for such events, raising questions about whether management scenarios can accommodate such cultural considerations.

Multiple cultures and social groups are characteristic of life in the Hawaiian Archipelago, and seafood is significant throughout. There is extensive demand for seafood products with distinct characteristics at different times of the year and for diverse ends—from the *'ohana* setting to Japanese and Chinese New Year celebrations to the mixed commercial-recreational fishery sectors. Again, analysis of the flow of seafood is an appropriate way of conceptualizing these culturally significant patterns of distribution and use (Figure 10.3).

Description and analysis of social and cultural systems associated with the pursuit, distribution, and use of seafood requires the capacity to conduct a specific brand of

Figure 10.3. Alia moored in small cove in American Samoa. (For a color version of this figure, see Plate 22.)

social science. There is limited potential in this regard in Samoa and CNMI. Generally speaking, there are relatively few social scientists trained in research and analysis of fishing cultures and ecosystems. Limitations to such investigation are further constrained by logistical factors, including long travel distances and costs, and language and cultural challenges.

Given that Pacific island societies have undergone significant changes following the arrival of Europeans, resistance to research undertaken by outsiders is common. Incorporating local community members as full and paid participants and as interviewers on the research team has been effective in mitigating such resistance and in enhancing meaningful input in social science research conducted in American Samoa and CNMI. For example, a Samoan research team member recently developed a particularly useful social network approach for tracing the culturally important flow of fish backward from the event of presentation and consumption to the source of the fish. This method is now being applied elsewhere in the region. Adequate representation of the perspectives of Pacific Islanders in the arena of marine resource management should be seen as more than just a diplomatic gesture. Rather, the necessary steps should be taken to enhance fisheries social science capacity throughout the region. One such step toward capacity building efforts in the region should involve implementation of properly focused graduate and undergraduate programs and internship programs.

Importance of Traditional and Local Ecological Knowledge in the Hawaii Context

Paul Bartram, Cultural Practitioner and Scientist from Moloka'i

While it is less than 50 miles from the island of O'ahu, Moloka'i is a very different island. It is rural and undeveloped. It is an ideal location through which to practice and illustrate traditional principles of Native Hawaiian interaction with the marine environment. Gauging and living in rhythm with local environmental cycles was and is critically important to the kanaka maoli (indigenous Hawaiians) (Figure 10.4).

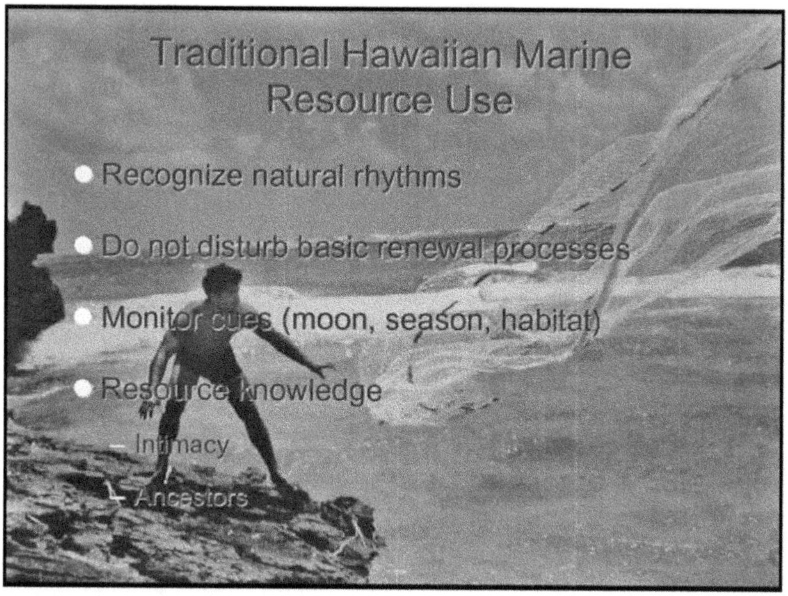

Figure 10.4. Principles of Hawaiian traditional ecological knowledge.

Hawaiians conducted and conduct research in the daily practices of fishing and through other extractive and observation-based forms of interaction with the physical environment. They also used and use models about that environment. Cognitive models of environmental cycles have been developed through observation of very long-term patterns over the course of many generations. Hawaiians traditionally monitored the moon, seasonal changes in resource behavior and abundance, and the nature of habitats. Such observation-based models formed the basis for regulating fishing pressure in certain times and places. Such models and related regulatory rationale were communicated between generations and modified based on intimate and ongoing contact with the ecosystems and resources.

To the Hawaiians, the primary objective of fisheries management is to enable the sustainability of marine resources so that they may be used for purposes of consumption, sharing, celebration, and so forth. Consumptive or subsistence-oriented fishing is critically important in the Hawai‘i context. In some rural locations, fishing may provide as much as 30 percent of the local diet.

Ancient Hawaiians developed a code of conduct to regulate fishing. As nearshore fish aggregate in favorable locations called ko‘a, the Hawaiians monitored these areas and decisions were made to open or close fishing based on environmental cues. A management action might be taken, for example, if one ko‘a seemed to be bearing an excessive burden such as carrying an unusually large proportion of spawning fish as compared to other nearby ko‘a (Figure 10.5).

Hawaiians also regulated fisheries by seasons and by the phase of the moon. There are two general seasons. The wet season, *Ho‘oilo*, typically occurs between November and April. The dry season, *Ka‘u*, generally sets in between May and October. Nighttime fishing activities tended to be undertaken in the wet season and daytime fishing

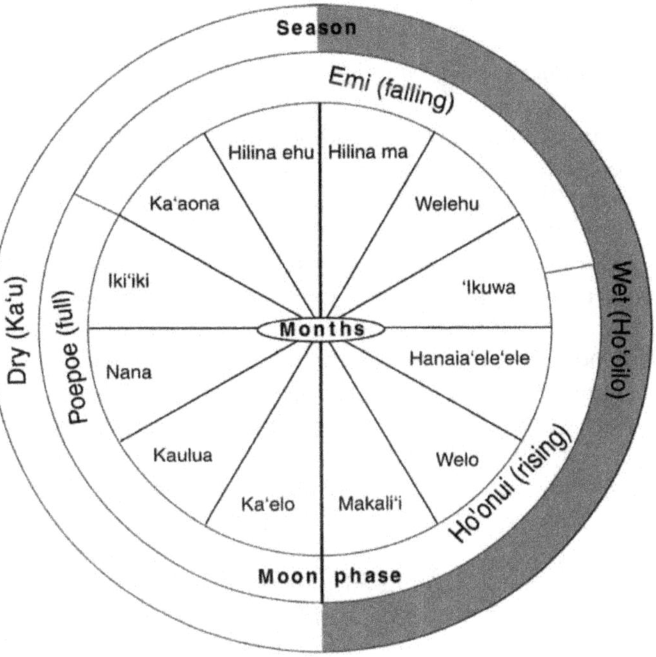

Figure 10.5. Hawaiian lunar and seasonal calendar.

activities in the dry season. The lowest of the low tides in the winter months occur in the dark, and during the day in the summer.

Seasonality in tidal flux has profound effects on tropical shallow water reefs. Accordingly, under the Hawaiian system, certain fish could be taken at certain phases of the moon and left alone during others. The effects of lunar phase on fishing were and are believed to be profound and complex. In general terms, nights prior to the emergence of the moon are the best nights for nearshore fishing, while the Whole-Day Nights or full moons are thought to present poor conditions for nearshore fishing. Moons like egg drops, or those before the full moon, are considered good for pursuit of offshore pelagic species.

[Paul Bartram was asked to respond to the suggestion that Western science is not needed in this context, but rather a good understanding of local culture and a way to facilitate it in the resource management context. He responded that this may hold true in some places, but less so in those areas where indigenous residents have been displaced. The Hawaiian system requires that participants have been continually associated with ecosystems and that a social and cultural system is extant to enable its conduct and enforcement. In historic times, the timing and nature of marine resource management varied across the Hawaiian Islands depending on local conditions, knowledge, resource migration patterns, and so on. The key was and is localized knowledge and an established system of social interaction with and informed awareness of ecosystems and associated resources.]

Fishing, Culture, and Data Collection in American Samoa

Fini Aitaoto, Western Pacific Council On-Site Coordinator for American Samoa

American Samoa tuna canneries employ about one-third of the roughly 15,000 documented employees in the country. Another several hundred persons supply the fish.

Subsistence-oriented fishing is also important for American Samoans in various ways—dietary, cultural, social. The 660,300 pounds landed in 1994 by the inshore subsistence-oriented fleets constituted almost 30 percent of total landings that year. Shoreline fisheries involve the harvest of over four times the amount of commercial landings of pelagic and bottomfish. While the shoreline fishery is thought to be stable in terms of catch and effort, the accuracy of landings data is uncertain.

Commercial fishing in Samoa is discussed elsewhere and is not the subject of focus here. Last year there were only six *alia* longline operations active in American Samoa, down from 38 or so in 1996. A recent study of the fleet indicated that albacore catch rates are low, consistent with overall decline in albacore fisheries throughout the larger region. Declining market prices, occasional shortage of bait, and increasing fuel prices are contributing factors.

Imported foods account for some 83 percent of the American Samoan diet as measured by value. In 1990, 23 percent of reef fish sold in the local markets was imported, but two years later the figure had increased to 78 percent. Because the domestic small-boat fleet has been unable to provide a consistent supply of fresh fish for the local community over the last decade, seafood increasingly has been imported from Western Samoa.

American Samoans tend to have higher standards of living than their cousins to the west, and generally utilize reef marine resources to a lesser degree. However, in the last few years, local reef resources such as *limu* (seaweed), sea urchins, and *alili* (*Turbo* spp.) have appeared in American Samoan markets for the first time.

As previously discussed, *Fa'a Samoa* is the term for the Samoan way of life or how Samoans live, do, and perceive their world. Since the 1840s, the power and influence of churches and clergy has grown and exerted significant influence of foreign religion on Samoan society and culture. The gradual acceptance of Christianity stems partly from the fact that Samoans have long had a creator concept, and partly from an open spirit of hospitality and willingness to adapt (Figure 10.6).

Fishing-related stories and customs are an important part of Samoan culture. Many proverbial expressions originate from fishing experiences, and the Samoan language is rich with names of marine creatures and fishing gears and methods. While missionaries and other sources have led to change among Samoans, it is important to realize that outside influences have not always been degrading and threatening. Western education has had a positive effect in that it led to the convening of the International Samoan Language Commission, the establishment of a local university, the publication of over 30 books on Samoan language and culture, teaching of Samoan culture in high schools and community colleges, and the perpetuation of the Samoan language through federally funded programs. The churches were the primary vessels for preserving and teaching Samoan language and culture prior to public schools. The Samoan bible remains the leading authority of the formal Samoan language.

Samoans continue to follow some traditional fishing practices, and while commercial fishermen are not required to provide portions of their catch to the chiefs and

Figure 10.6. Fagasa villagers use a traditional *launiu* (coconut frond) weir to encircle and catch *atule*.

the community they will occasionally donate catch to festivals of family or community events. Providing fish to relatives and friends waiting at landing sites (*tapuaiga*) is a traditional fishing practice that is still widely observed. Similarly the practice of providing and consuming fish for Sunday brunch is common to this day. The communal seining of the polychaete worm *palolo* (*Eunice viridis*) in October and November each year continues to involve whole villages. Another fishing method that involves the whole village and is still practiced in certain areas is the *lau*. Villagers gather on the inner reefs and use strings of coconut leaves to surround and trap *atule*. As with *palolo*, it was traditionally taboo to sell the *atule* catch, but it is now commonplace to see the fish sold at roadside.

Some fishermen have been using illegal fishing methods around Tutuila for some years despite the fact that disruption of nearshore marine ecosystems for commercial gain is culturally unacceptable. *Ava niukini*, a traditional fish poison extracted from the local plant *futu*, has been used. One group used dynamite in 2005, and a group of Tongan fishermen reportedly used bleach to land reef fish around Tutuila.

A number of problems currently challenge fishermen in American Samoa. These include the following: (1) airline service necessary for the export of fresh fish is often unreliable; (2) no market for incidental catch landed by operators of the larger vessel fleet; and (3) the Fish Aggregating Device program is sporadic. Also problematic, compared to neighboring Independent Samoa, is a lack of fisheries-related development programs. Moreover, professional grant writers are needed to assist the government in applying for various federal grant monies. Finally, the importance of fishery-related industries in American Samoa cannot be overstated and the potential departure of the tuna canneries is an immediate and serious issue. Tightening of immigration laws has rendered many Independent Samoan cannery workers ineligible to work in American Samoa and thus there is an ongoing cannery worker shortage.

There are some positive signs and potentialities for fisheries development in American Samoa. An increase in the number of American Samoan charter boats may help the struggling tourism industry. Although the giant clam aquaculture project is no longer operational, the facilities exist and could be used once funding is available. A variety of NGO and family fish farms are currently operating in American Samoa.

As regards fisheries data and data collection, DMWR conducts an Offshore Creel Survey, sampling participants in the region's commercial, recreational, and subsistence fisheries on two weekdays and one weekend each week. There is a commercial invoice system as well. This requires that outlets buying or selling fish complete an accounting invoice each month. Inshore surveys are conducted irregularly, but several different studies may provide valuable information about recent and historic activity. Data sources also include (a) community-based MPA programs in four local villages; (b) records of various fishing tournaments held since 1974; (c) the federal longline logbook system; (d) monthly reports of fish processed and landed at the canneries; (e) intermittent market surveys and special projects; and (f) the giant clam project.

Efforts to effectively manage fisheries resources in American Samoa are challenged by issues related to adequate enforcement, representation, and jurisdictional authority. Local enforcement of the community-based MPAs may suffer from ambiguous authority, as well as poaching and variable levels of support beyond community boundaries. Several local residents have voiced the need to elect state legislators rather than nomination through a local council. Fishermen and fisheries managers prefer a fisheries-proactive representative to assist in DMWR-mandated duties. The lack of a unified voice of local fishermen is partially due to the absence of a fishing association. Of several formed over the years, all but one has been disbanded. Jurisdictional issues have arisen between DMWR and the USFWS relating to the administration of Rose Atoll. Maritime boundaries between American Samoa and Western Samoa are not formal and have led to jurisdictional issues regarding regulatory enforcement.

Moreover, there are staffing problems at DWMR. A lack of local biologists is likely due to small numbers of American Samoan students studying marine biology, and the need to improve the marine science program at the local community college. While DWMR biologists report conflicting views on the status of some reef fish stocks and the general health of reef ecosystems, all agree that a local stock assessment is sorely needed. Mr. Aiaoto offered a word of caution regarding collection of valid information while working in villages in American Samoa, noting that it is essential to take a culturally suitable approach and to ask appropriate questions.

Monitoring and Forecasting Ecological Change in the Mariana Archipelago

Judith R. Amesbury, Micronesian Archaeological Research Services

The Mariana Archipelago is subject to a wide range of biophysical sources of change, including typhoons, super-typhoons, drought, fires, El Niño Southern Oscillation (ENSO) events, volcanic eruptions, and earthquakes. Humans also exert impacts.

In Saipan, for instance, bombs and vessel gouging pockmark some reef systems. Military structures have been built to the very edge of the shoreline, and lands farther inland show evidence of large-scale development and agricultural activities. Given the long history of biophysical and human effects in the region, longitudinal data and analysis are required to fully understand and effectively monitor terrestrial and marine ecosystems here.

Assessment and monitoring human and physical environmental sources of change would serve to inform and enhance the development of effective fishery ecosystem plans. Research should focus on factors that shape ecological rhythms and sequences. Understanding of the timing of seasonal fish runs would reciprocally contribute to understanding of subsistence activities. Assessing and monitoring extreme events such as typhoons, super-typhoons, and ENSO events for several years would aid in understanding biophysical limits on long-lived species and other dynamic processes in the physical systems. ENSO events have been shown to lead to a wide range of effects, including lower than average sea levels, higher water temperatures, and droughts. Typhoons can lead to extensive terrestrial runoff. While such events can significantly affect coral reef ecosystems, the associated biophysical processes and dynamics are not well understood in the CNMI.

Habitat structures, key determinants of fish assemblages, are highly variable across Guam and the CNMI. When considering protected areas, habitat structure would ideally be evaluated and planning would seek to ensure protection of a range of habitats. Structural factors might include number and size of holes; rugosity; extent and nature of live coral (especially finger coral); extent of coral cover; water quality; presence of fleshy seaweed and sea grass; presence and extent of mangroves; the presence of barriers that fragment habitat; and the measured or potential influence of typhoons.

Nationally prescribed sampling and testing efforts to assess local habitat quality will soon be implemented in Guam. Surface and nearshore coastal water quality testing will be conducted by the Guam EPA according to nationally standardized protocols. Water, sediment, habitat, and plant and animal life will be sampled from the shoreline to the 60-foot contour. Such efforts in the CNMI are ongoing and include EPA-required water quality testing at 46 sites. The CNMI Marine Monitoring team is assessing and monitoring coral communities, benthic communities, and the abundance of invertebrates and fish in different habitats and watersheds throughout the southern islands.

One effective indicator of potential utility for the development of fishery ecosystem plans in the region is seafood consumption. A range of variables could contribute to such an indicator: consumption of seafood versus other food products, rate of seafood consumption, species consumed, and the percent of locally landed food fish. Consumption in the CNMI has declined dramatically since the 1940s when life in the region was relatively insular and islanders depended so heavily on seafood. Some 365 pounds of seafood were consumed per person per year in 1940. This rate of consumption diminished significantly over the following decades. The situation was similar in Guam where, as of 2002, only 57 pounds of seafood were consumed per person annually. Consumption patterns may reflect growth in the cash economy, and a concomitant decrease in reliance on subsistence fisheries and home cooking.

Given the extent of cultural variability in the CNMI, use and consumption patterns vary extensively. A wide variety of nearshore species are pursued and consumed, including sea cucumbers, small crabs, varieties of mollusks, and a wide variety of reef fish. The importance of seafood in the CNMI is suggestive of a need to assess and monitor security and resiliency of foodstuffs, rates of local seafood production versus imports, and general seafood consumption patterns. This relates in part to the potential for planning a suite of fish refuges in the region to provide protection against all types of disturbances, such as wave exposure, natural predators, and people.

One means through which pertinent environmental and ecosystem conditions might be effectively monitored and communicated for the Mariana Archipelago is through an annual report that would relate to the FEP for the region. A model for the structure and content of such a report would need to be developed through the collaborative input of the Council and other fishery management entities in the area.

Managing Guam's Nearshore Fishery and Fishery Impacts

Jesse Rosario, Indigenous Fisheries Expert, University of Guam

Guam is a relatively small island, 35 miles long and no more than about 9 miles wide. Yet its nearshore ecosystems have historically supported both local villagers and people from neighboring islands. But changing social conditions, pressures, and resource management strategies have preceded several marine ecosystem-related problems.

The tourism industry has been growing. Hotel owners have developed strategies to attract more visitors. Many have begun advertising and implementing various leisure activity programs that are tending to conflict with the resource use patterns of local fishermen. The use of jet skis in and around fishing areas has led to some such problems. Moreover, hotel operators are allowed to manage activities and resources 33 feet seaward of the high water mark. As such, many have undertaken various actions with the intent of improving the experience of their patrons. For instance, some have removed algae from certain areas to improve the bathing experience, with implications for the status of the nearshore ecosystems.

Regulatory actions have had a significant effect on fishing and fishery participants around the island. Guam has five MPAs: one is located north of the island, three along the west-central region and one is south of the island. Establishment of the protected zones has led to heightened tension between harvesters who have had to concentrate fishing effort in ever-smaller areas. Political effects include assertions about inequitable treatment of persons violating protected area boundaries. Some fishermen report that establishment of the MPAs and subsequent spatial changes in fishing effort have led to more rapid depletion of certain fish populations than would have occurred otherwise.

It is thought that local fisheries may also change as a result of policies regarding the definition and regulation of subsistence-oriented fishing. Under Public Law 228, the definition of such fishing effort would be limited to household consumption only and would prohibit customary trade of the catch.

Imperiled by the actions of hotel owner–operators is the *manahak*, a traditionally pursued nearshore species. Typically, pursuit of the fish occurs in spring and summer, with distribution of the catch among relatives and neighbors within and between villages. Disruption of algae beds and JetSki use significantly disrupts this fishery.

The Fishery Information Survey and History (FISH) project has been undertaken to characterize Guam fisheries. Data are being collected regarding the characteristics of the fishers, including local residents and fishers from other islands; fishing gear and techniques; targeted species and seasons; and perceived changes in fishing habits and conditions. The survey is expected to contribute to a historical and cultural assessment of fishing and associated challenges on Guam.

Other local fisheries projects include a public awareness campaign to inform residents and visitors about traditionally important marine resources, biological cycles, and existing regulations. A watershed research project is being developed by the University of Guam.

A future goal is to establish a common platform to facilitate interaction and cooperation between Guam and federal agencies, fishery participants, relevant businesses, and other partners. The intent is to enable initiatives to improve the status of marine resources and ecosystems, increase capacity for regulatory enforcement, and encourage responsible fishing and shoreline management practices.

Fisheries and Social Science Data in the CNMI

John Gourley, Micronesian Environmental Services, CNMI

The CNMI is comprised of 14 islands, five of which are inhabited. The smallest, Farallon de Medinilla, is used as a bombing target by the Department of Defense. The five southernmost islands are fairly well developed limestone platforms with outlying barrier reef and/or fringe reef systems. The nine northernmost islands are more volcanic in nature and there are active vents on a few of the islands. Saipan is the largest island of the group. Most (90 percent) of the approximately 70,000 or so residents of CNMI live on Saipan, and most of the economic activity also occurs here. The populations of Tinian and Rota comprise approximately 5 percent of the total population. A few families live on a couple of the northernmost islands.

The population of CNMI is ethnically diverse. Indigenous Chamorro and Carolinian ethnic groups comprise about 24 percent of the population and have traditionally maintained positions of political power. The sitting governor is the first Carolinian to reach this position.

With regard to ethnic dimensions of participation in the labor force, Filipinos tend to work primarily in the service industry. Many are employed in hotels and restaurants around the islands, and in the construction industry. Persons of Chinese ancestry tend to be employed in the garment factories, an industry limited to Saipan. Persons of Korean and Japanese ancestry tend to work in various retail and wholesale business firms.

Persons of other ethnic backgrounds have also made CNMI their home. For instance, persons from Chuuk, Yap, and the Marshall Islands may immigrate under the Compact of Free Association and many such persons now reside in the CNMI.

Economic conditions in the region are challenging for many. Minimum wage is $3.05 per hour, and low wages are common, except in the public sector. Median income is $25,853, roughly half of the reported median income for families in the United States in 2000. This has implications for pressure on the region's marine resources in that consumptive-oriented fishing and shoreline gathering are common across many of the aforementioned groups, and each group tends to take a different approach to the harvest.

There is extensive diversity in marine species across the region. For instance, there are approximately 256 species of corals and over 1,100 species of nearshore fishes here. Species diversity diminishes somewhat in the more northerly volcanic islands, where the development of coral reefs has been less extensive than elsewhere.

A controversial history of marine management is associated with the northern islands in the chain. Some sanctuaries were designated by public law in the absence of public hearings or opportunities for public comment. Bird Island and Forbidden Island are managed to protect single species. There are currently eight MPAs around CNMI; some protect single species (e.g., sea cucumbers or trochus), while others protect important habitat. These vary widely in terms of primary form of protective measure.

Chapter 11
Social and Economic Indicators and Applications for EAFM in the Pacific

Socioecological Indicators for MPA Evaluation

Patrick Christie, University of Washington, School of Marine Affairs

Marine protected areas (MPAs) are popular mechanisms for protecting marine ecosystems and resources. They have been championed by marine biologists in a variety of contexts. But comparatively little is known about human dimensions of MPAs despite human desire to establish MPAs, do establish MPAs, and are affected by the establishment of MPAs. As such, MPAs should not be evaluated solely in terms of biological success. Rather, both biological and social factors and outcomes should be examined and assessed.

Contestation about the placement, nature, or perceived or actual effects of a protected area can quickly cause problems in a given community. Some such problems can be avoided through effective planning and public input. For example, in cases where user groups are known to rely on resources associated with a proposed MPA for their living or for recreational purposes, carefully negotiated interaction with those groups prior to the setting of rules and boundaries may help mitigate potential conflict. Further research on the effectiveness of such efforts and development of new strategies for minimizing community problems is called for.

MPA-related research and monitoring efforts around the world would benefit by following principles for incorporating social needs and objectives. These include the following: (1) consistently monitoring MPA programs using scientific and participatory methods and indicators such as food security, government support and accountability, improved or restored fisheries, sense of pride in local management, etc.; (2) using comparative in-depth qualitative and quantitative research to develop theory and new models regarding human dimensions of MPAs; (3) integrating research across the natural and social sciences; and (4) linking information generated through research and monitoring to real-time management.

Government agencies typically operate with specific information needs and management goals and objectives. But MPA programs are highly complex and often contention-laden, with issues and questions that extend beyond the informational parameters of such agencies. Thus, MPA-related research should be both mandate-responsive and mandate-independent. For example, mandate-responsive research might involve the conduct of an economic valuation for the purpose of determining

Ecosystem-Based Fisheries Management in the Western Pacific, First Edition. Edward Glazier.

appropriate visitor fees to an MPA area that could have an underwater dive viewing option. In the same setting, mandate-independent research might investigate the capacity of that agency for monitoring or enforcing such an option.

MPAs can be biological successes and social failures. There are, therefore, long-term implications to focusing solely on biological considerations. Evaluative work in the Philippines is relevant. The Tree Hill MPA was established through a community-based participation process. Community participation was initially extensive. While establishment of the MPA was shown to precede improved coral cover, fish abundance, and species diversity, social science-oriented indicators of sustainability were not so positive. Evaluation work indicated that some persons were marginalized during the public participation process. It appears that this preceded disengagement from the group, and subsequently noncompliance and poaching. While biological assessment of the Tree Hill site indicated success in the short term, that assessment is conditioned by problems of "buy-in" by the full range of resource users. The long-term success of the program is therefore in jeopardy (Figure 11.1).

Effective indicators in this context tend to be composite variables best determined and measured through interviews and survey work. One particularly important indicator relates to the size of the affected communities and alternative means of income for those who use or used the resources in the prospective or established MPA. The Fish Project at www.oneocean.org provides some social variables for consideration in evaluating and monitoring the success of MPAs. Potential indicators include the following: (1) use of ecological knowledge in the planning process; (2) establishment of a program to inform user groups and the public about the nature and intent of the MPA; (3) steps taken to minimize conflicts related to the cultural backgrounds of the

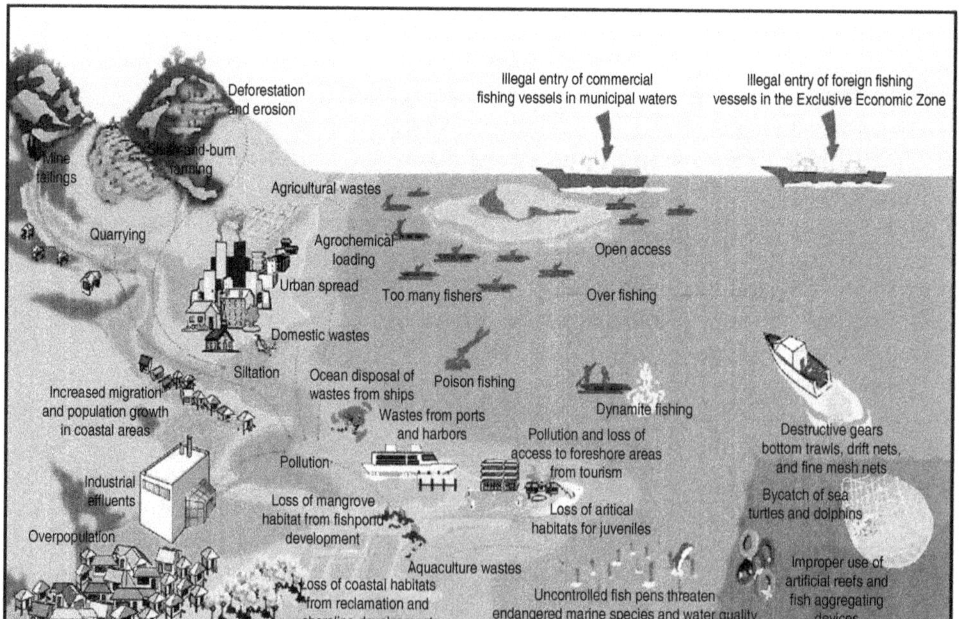

Figure 11.1. Direct and indirect ecosystem relationships (Coastal Resource Management Project). (For a color version of this figure, see Plate 23.)

Figure 11.2. Diverse biological and social goals of the MPA management approach. (For a color version of this figure, see Plate 24.)

involved parties; (4) means for enhancing acceptable relocation of fishing effort or other use of resources; (5) established mechanisms for enforcement; (6) improvements in fish biomass; and (7) improved management of threatened species. Both process-oriented and outcome-oriented indicators are necessary components of assessment and evaluation (Figure 11.2).

[Drs Pollnac and Hennessey discussed the potential for identifying and standardizing social variables and indicators for evaluating and monitoring the success and sustainability of MPAs. Dr Pollnac discussed the importance of specific composite variables for evaluative purposes: (a) satisfactory input of municipalities; (b) preliminary visits by officials; (c) participation of an early core group of stakeholders; (d) alternatives for earning income; (e) monitoring by the community; (f) numbers of initial training meetings or programs; (g) development of an MPA "features" score; and (h) assessment of regulatory compliance. All of these factors were highly correlated with the "performance" of MPAs.

Dr Hennessey suggested that it may be useful to conceive relevant social variables in terms of the way they are interrelated. He also noted the potential utility of incorporating lessons from global MPA networks and experiences when developing indicators and models of MPA performance in the Western Pacific.

Dr Shankar Aswani discussed the importance of assessing spatial and nutritive dimensions of MPAs. For instance, displaced fishing effort can lead to human health consequences in populations dependent on marine resources for purposes of subsistence. This can be indicated in disproportionate effects on the health of women. Moreover, limiting or precluding effort in one area can lead to increased pressure on adjacent biological resources in adjacent areas, with implications for human groups dependent on resources in the latter. Effective assessment of MPAs therefore requires sufficient incorporation of spatial and sociocultural considerations. Given implications for the physical well-being of humans, these may in fact be the most important considerations in the design of MPAs in the Pacific and elsewhere.

Dr Severance indicated the importance of the fact that Dr Christie recognized the colonial history of the Philippines in his interactions with indigenous research participants. The colonial context across the Pacific region is highly relevant to effective research and establishment of management frameworks such as MPAs. Historical processes influence the status and perspectives of prospective research participants across the region. These include the illegal overthrow of the Hawaiian monarchy, treaties in the Samoa Archipelago that were not agreed to by all parties and transfer of colonial power in the Mariana Archipelago. These condition the manner in which local persons may react to new researchers, regulations, and evaluative programs now and in the future.]

Social Science and the WP Council: Know Thy Client and the Devil Is in the Details

Leah Bunce, Senior Director, Conservation International

Social scientists working in the context of marine resource management face challenges beyond those of social scientists working in other realms of inquiry. These include the tendency of resource managers and other scientists to prioritize biophysical research and research findings above social science and social science research findings. Managers also tend to fail to recognize that social research typically involves highly complex issues and subject matter, and that while biophysical researchers may be involved in single projects for months and years, it is often expected that social science projects should be completed much more quickly. Some scientists and managers tend to use terminology which positions social science as "the other" or lesser science, or they simply refer to it in contrast to science. Yet others assume social science simply means outreach and communication. In reality, social science methods are as sophisticated and capable as any other and are employed to examine phenomena that are in reality as or more complex than those addressed by the biophysical sciences. Equal investment of time and energy in the social and biological sciences is clearly called for.

Social science research and monitoring can generate information of profound importance to managers of marine resources around the globe. Particularly viable avenues of inquiry (and determination of indicators) include the following: (1) sociodemographics (gender, ethnicity, literacy, religion, occupation, etc.); (2), participation in activities like fishing or tourism (location, seasonality, type); (3) community infrastructure and ways of life; (4) perceptions and experiences regarding the condition of and threats to marine resources; and (5) mode and manner of governance. Social science can help analyze the effectiveness and effects of MPAs with regard to each of these dimensions, among others.

It is critical that the local social and physical environmental contexts within which these factors are framed are well understood. It is critically important to understand the nature of the marine activities that are taking place, where the efforts of the user groups are focused, relevant dimensions of relationships between the participants, and relationships between the participants and the physical environment.

Spatial dimensions are pivotal. For example, prior to establishing the Dry Tortugas No-Take Reserve, those responsible for establishing the Florida Keys National Marine Sanctuaries undertook spatial examination of the potential biophysical and

Figure 11.3. Community involvement imparts local knowledge. (For a color version of this figure, see Plate 25.)

socioeconomic impacts that could result from the proposed management decisions. The resulting cartographic products illustrated how spatial use patterns in various fisheries could change under a no-take reserve. These were subsequently used in discussions with stakeholders to examine tradeoffs and to help determine the most practical and tractable management actions (Figure 11.3).

Ecological knowledge retained by persons in community settings can offer enormous assistance to scientists and managers involved in ecosystem planning and management. Such knowledge is accessible through social science research methods and can assist in tailoring management strategies to the peculiarities and nuances of the local context. Social science can further contribute to understanding of that context by description and analysis of local customs and traditions, use patterns, and dependence on and values regarding marine resources. Such analysis can demonstrate the value of marine resources in terms that policy-makers and the public appreciate.

Social science can also aid in identifying viable economic and social incentives through which communities may benefit by conserving adjacent natural resources. For instance, such research may identify alternative sources of income, including alternatives that may result from new forms of management. It may also serve to identify and describe key stakeholders and patterns of social relations and tendencies that suggest likely support or opposition to new forms of resource management. Finally, social science can help in identifying human threats to marine ecosystems and situations and sources of potential benefit to the health of those systems.

The Global Socioeconomic Monitoring Initiative, conducted through the NOAA Global Coral Reef Monitoring Network, has involved development of a series of guidebooks, socioeconomic training sessions, and funding for monitoring social factors at specific sites. Dr Orbach noted that these programs are particularly useful because they are at once focused on coral reef issues and customized to the social and cultural conditions specific to different regions of the world. The guidebooks in particular may be useful in developing the FEPs in the Western Pacific region.

Dr Christie cautioned that local people do not always respond well to managers arriving with a guidebook or model in-hand, even if they come with the most collaborative of intentions. He suggested that guides such as "How is Your MPA Doing?" are intended for use by community practitioners and managers rather than social scientists. Questions addressing issues such as where regulations are coming from, how they may be made appropriate for each location, and whether and how they may be incrementally introduced are critically important in community settings, but additional mechanisms may be needed to provide sufficient answers.

Dr Aswani asked how ecosystem social science research might address the issue of changing levels of support for conservation of marine resources generally and resource management programs specifically. Dr Bunce suggested that finding areas of overlap may help assuage waxing and waning support. For instance, Dr Veitayaki has developed programs that combine conservation efforts with programs that provide satisfactory trade-offs for any loss of availability of marine resources. These include programs that enhance potable water resources, enable treatment of sewage, and bring alternative sources of income and opportunity to the community. Dr Bunce suggests that ongoing monitoring of conservation programs, sound partnerships with preexisting social networks of community actors, and stable funding are central elements of successful locally managed conservation programs.]

Addressing Human Factors in Fisheries Development and Regulatory Processes in Fiji: The Mositi Vanuaso Experience

Joeli Veitayaki, University of the South Pacific Marine Studies Program

The vision of an ideal world may serve as a model for understanding the complexities that actually confront those who manage marine resources. In that ideal world, residents of communities adjacent to marine ecosystems would be perennially happy and contented. Natural systems would be sufficiently productive, thereby meeting social and cultural needs. Production and use of natural resources would be facilitated by a variety of income alternatives. Rural development would be carried out smoothly and resource managers would face no obstacles as people transitioned to desired state and ways of living in ways that maintained the sustainability of the natural world and resources around them. Fish populations and fisheries would remain vibrant and readily meet and support commercial, consumptive, and recreational needs and interests without inconvenience to any given sector. The physical and human environments would always be healthy.

The model also makes clear that human beings are pivotal in every meaningful aspect of marine ecosystems and their effective management. Indeed, they define that meaning. Moreover, managing environmental resources is, first and foremost, about managing humans and their activities, and meeting their goals and objectives, including health and happiness.

Capacity building is critical in efforts to influence the behaviors and practices of marine resource user groups—in this case, residents of island villages. It is often the case that the full importance of the ocean and its resources is not grasped by the very people who depend on them. The arrival of new ideas, pressures, and people has in some cases preceded the erosion of traditional patterns of resource management. Assisting people to build understanding and maintain effective care of marine resources while improving local living conditions is best accomplished through community-based initiatives. Changes such as these can be implemented more quickly at the village level than in large-scale settings, and the results will also be visible more quickly. Because collective effort tends to outweigh the sum of individual efforts, partnerships in capacity building are most effective.

Work on Gau Island in Vanuaso District on Fiji exemplifies this approach. People in villages here typically are involved in subsistence practices, with supplementary resources purchased with money earned through occasional participation in wage jobs. These are indigenous Fijians who hold tenure rights to fishing grounds from the high water mark to the outer limits of the reef systems. They also often maintain jurisdiction from the village sites up to the mountains.

A central objective of the project was to improve care of local marine resources on Gau. It was essential to establish rapport with the villages and to win their trust, confidence, and eventually their support. Once this was attained, an attempt was undertaken to collectively recall traditional resource management patterns and customs. The community was also brought together to discuss and evaluate the proposed objectives and means for enhancing treatment of the resources through traditional self-management. It was necessary to gauge the level of commitment of the local populace for undertaking and sustaining the initiative. Once underway, progress was checked on a regular basis, with new concepts and ideas introduced and negotiated along the way (Figure 11.4).

Figure 11.4. View of traditional Fijian settlement. (For a color version of this figure, see Plate 26.)

Focus was applied to the promotion of long-term investment and change. This required a willingness from the community to engage in management activities that might only yield benefits long after their time had passed; several projects started with this point in mind. Another important aspect was the setting up of institutions and champions to propel the work and involvement of all. Publication of the experience was undertaken for similar effect.

Challenges to this endeavor were numerous. On one hand, community development is called for by the people themselves, but consistent dedication and enforcement can be problematic over time. People tend to relax management of resources due to daily activities and constraints. Regular oversight can play an important role here, but with the intent of ensuring that motivation remains endogenous and not externally driven. Enforcement of regulatory measures is also a problem. Existing social control mechanisms may be sufficient, but these may not be able to cope with external elements, such as poaching of resources by outsiders.

It is important to promote critical evaluation of this form of development work. Connection with educational institutions is also important, as a means to transmit the new values and behaviors to younger generations. Connection with and support from the government should be sought as well, potentially opening a way for monitoring changes through an established institutional environment. It is necessary to secure and supervise funding to assist community-based initiatives. Further, given that effective management of resources is economically advantageous in the long term, effort should be made to disseminate experiences as broadly as possible.

The Mositi Vanuaso Project commenced at Vanuaso Tikina on Gau Island. *Mositi* refers to something treasured or deeply valued. Project managers expect to (1) promote participatory decision making in an area where a traditional system of resource management was still being used; and (2) examine the need for and undertake developmental measures that would at once enhance the well-being of villages, while ensuring conservation of the terrestrial and marine environments.

The approach involved a series of workshops in which residents developed objectives and initiatives. Identified was a list of priority development options. This included specific measures, such as (a) improvements in sanitary conditions; (b) use of water catchments and piping and distribution of water; (c) promotion of animal husbandry; (d) reduction in use of pesticides; (e) farming on hill slopes; (f) limiting unnecessary burning; and (g) combating deforestation and embarking on reforestation measures. Also identified were priorities and approaches that were more general, including (a) definition of guidelines for environment-friendly land use; (b) a quest for alternative sources of livelihood or income; and (c) the undertaking of marine resource management and protection of locally valued valuable coastal habitats.

Several positive changes resulted from the enactment of these initiatives. Awareness of existing social and environmental problems increased, perception of responsibility toward the surrounding environment elevated, and basic infrastructure and services in the villages improved.

In sum, the project empowered the community to develop basic infrastructure and services while simultaneously protecting the local environment and its long-term capacity to provide goods and services to those charged with its care. In this respect, the project served to enhance the long-term well-being of villagers in keeping with the ultimate goal of well-integrated and healthy human and biophysical systems.

Sociopolitical Aspects of Developing and Operating a Large-Scale Marine Ecosystems Management Program

Leanne Fernandes, Australian Government: Great Barrier Reef Marine Park Authority

The rezoning efforts of the Great Barrier Reef Marine Park were intended to increase protection of biodiversity throughout the Great Barrier Reef system. The Great Barrier Reef spans thousands of miles along the northeast coast of Australia. It supports numerous fisheries that generate hundreds of millions of dollars in revenue annually, and it is the focus of a several billion dollar tourism industry. Indigenous and colonial-immigrant populations inhabit the adjacent, largely rural land areas.

The great barrier reef marine park authority (GBRMPA) process involved development of the Representative Areas Program. This was established to facilitate a network of no-take areas to protect representative examples of the range of habitats, communities, and species across the Great Barrier Reef. As an ecosystem-based planning measure, it differed from earlier management strategies in that it (1) retained an analytical focus on whole ecosystems and the entire Great Barrier Reef as an ecosystem of itself; (2) addressed relationships between plants, animals, and habitats throughout the system, not just coral reefs; (3) reduced bias toward what is easy to assess and manage and addressed more difficult areas and issues; (4) was not ad hoc or limited to restricted sampling sites, but rather addressed specific data gaps so as to enhance comprehensive treatment; and (5) changed the extent and manner of the use of ecosystem relevant data.

Effective means for ensuring representation of the interests of stakeholders were absolutely critical to the success of the Great Barrier Reef rezoning process. A wide range of user groups were consulted. This included a wide variety of interest groups, aboriginal groups, fishery participants, and other ocean and reef users across a broad region of Australia. GBRMPA representatives held over 200 public meetings before the first formalized Community Participation Phase of the project. Some 800 meetings and other community informational meetings were held thereafter. Workshops with key stakeholders were held throughout. Of particular note in this process was the importance of establishing rapport with key persons in the communities and groups of interest. Such persons were highly influential of others in the community and could therefore spell the success or failure of a given objective.

GBRMPA staff began the process by conducting surveys to assist in developing a communications strategy. Focus groups were held to test the way in which messages about the beneficial effects of rezoning could most effectively be presented. For example, most persons in the region already realized the economic or recreational importance of the Great Barrier Reef.

However, many did not realize the complexity of the ecosystem, nor that the reef encompassed only six percent of the entire marine park area. Thus, communication strategies were formulated to inform the public about the extent of the park, the interconnectivity of reef-associated habitats, and the importance of effectively managing human activities that affected the marine and adjacent terrestrial environments.

Communications also described the Representative Areas Program and issues associated with species and habitat diversity, and reviewed existing zoning. Comments were solicited regarding prospective zoning approaches. Communication materials were tailored to meet the cultural attributes of the various groups, and these were updated throughout the process. Part of the communications strategy involved media campaigns and nationally known figures to promote meetings and solicit public input.

Attitudes and awareness about the park and rezoning program were monitored on a continual basis. Further, government and community representatives were updated throughout the process. This proved critical for meeting cross-jurisdictional challenges.

A Natural Science Steering Committee and a Socioeconomic and Cultural Steering Committee assembled to provide guidance over the course of the project. Members of these committees drew upon relevant research and scientific theory and findings to determine when data was sufficient for defining 70-plus bioregions and to determine appropriate levels of protection for given areas. The Socioeconomic and Cultural Steering Committee also developed principles to confirm assessment of no-take areas and to guide decision making in a manner appropriate to the needs and interest of adjacent human communities.

Scientific advisors determined early in the process that few new biophysical or socioeconomic cultural data were needed to meet program objectives. Staff worked with Queensland fisheries managers to determine how to make best use of extant fisheries information to achieve the biological objectives and to minimize deleterious effects on fishery participants. Such collaborative efforts were fundamental to the rezoning process, particularly when logbook and other data were insufficient for understanding the social context of the fisheries, including the potential effects of displacing participants. Also useful in this regard were census data.

The formalized Community Participation Phase of the project involved solicitation of public comments on a draft rezoning plan. The plan derived from a combination of biophysical and socioeconomic data and analyses. A team of social scientists helped design comment forms and analyze elicited data so as to enable rezoning staff to fully understand the range of public perspectives on the plan. A geographic information systems (GIS) team extracted spatial data from the comments as well. Some 21,000 public comments were ultimately elicited and analyzed, and each rezoning effort reflected public input. Closing surveys indicated that between 80 and 90 percent of the stakeholders and the general public approved of the GBRMPA rezoning process.

The Ahupua'a Model and Its Relationship to Contemporary Government

Leimana DaMate, Association of Hawaiian Civic Clubs, Expert on the Ahupua'a

Efforts to restore ahupua'a are seen by many as valid means for connecting past and future and for protecting and sustaining Hawaii's fragile and unique ecosystems through traditional use and conservation practices. Efforts to develop formal policy for reestablishing ahupua'a have been undertaken jointly by the Pacific Islands Resource Management Institute, the Association of Hawaiian Civic Clubs, and the Office of Hawaiian Affairs.

The remote location of Hawaii has not hindered development of society in the region. Rich ocean ecosystems are enabled by climatic conditions that were and are both favorable and unique. For instance, recent discoveries indicate that the Hawaiian Islands interfere with the easterly trade winds, triggering an "island effect," wherein wind speeds increased between the islands but are significantly weaker on the lee sides. This generates a narrow eastward-flowing warm-water ocean current. Because the current is warmer than surrounding waters, it can generate convection cells in the atmosphere which, in turn, help to sustain the current. It is thought that current may have assisted Polynesians voyaging in the region.

Voyaging Polynesians are said to have arrived in this region between about 100 and 400 A.D. Ahupua'a were gradually developed as an adaptive social process of ho'olaulima (cooperation), wherein early residents interacted closely to produce food and necessary items within three main land zones: ocean, agricultural areas, and upland forests. Spiritual beliefs about a holistic relationship between 'āina (land), moana (ocean), and kanaka (people) helped sustain the system. In concrete terms, ahupua'a refers to a division of the land, usually extending from sea to mountain top. The name derives from the marking of the division boundaries with a heap of stones (ahu) surmounted by an image of a pig (pua'a). Fishes such as the aholehole, 'ama'ama, kumu, pualu or humuhumunukunukuapua'a could also be used as offerings to mark the bounds.

Various terms were used to designate components of the division. The sea portion included the kahakai (beach shore), kulakai (sea plain region), moana (open ocean), or hohonukai (deep ocean) in the makai portion. The landward portion included the i'lima (planting area), pahe'e (grass area), apa'a (dry area), wao kanaka (living zone), the wao nahele (large forest line), and wao akua (small trees). The mountain portions included kuamauna (rounded swell of the mountain) and the kuahiwi (uppermost zone), among others. The social structure of a working system was well defined, with distinct jobs, rights, obligations, and responsibilities in each of the zones and subzones.

Many of the management principles underlying the historic ahupua'a system are now aspects of existing coastal zone and ocean management policies. But reincorporation of the concepts and structure of the system is challenging in that myriad federal, state, and county agencies now enact an intricate system of laws, policies, and programs that do not necessarily recognize the historic boundaries and social processes of the Native Hawaiians. For example, the following is an example of federal laws governing use of the shoreline in Hawaii: Rivers and Harbors Act (1899), National Historic Preservation Act (1966), National Flood Insurance Act (1968) and the Flood Disaster Protection Act (1973), National Environmental Policy Act (1969), Clean Air Act (1970), Coastal Zone Management Act (1972), Endangered Species Act (1973), Clean Water Act (1977), and Magnuson-Stevens Fishery Management and Conservation Act (1996, amended). Relevant state laws are also numerous, and include Chapter 46, HRS, which establishes the counties and their zoning powers; Chapter 205, HRS, which establishes State Land Use Districts; Chapter 205A, HRS, which implements the Federal CZM Act; Chapter 226, the Hawaii State Planning Act; Chapter 343, HRS, which implements the Federal Environmental Policy Act; and a range of other administrative rules.

Nevertheless, effort is being applied to work within existing structures on several fronts: through the Governor's Hawaii Ocean and Coastal Council, through the Ocean Resource Management Plan, through community and local government planning processes, and through grassroots-level action.

Recommendations for reestablishing ahupua'a include the following: (1) incorporating traditional 'aha or decision-making councils into community-based planning and resource management programs and processes; (2) addressing ahupua'a objectives in existing administrative rule and policy making; (3) identifying fiscal resources for moving forward with ahupua'a objectives; (4) creating watershed partnerships;

(5) incorporating Native Hawaiian resource use and management practices into current marine education programs; and (6) incorporating traditional ecological knowledge and ahupua'a principles in prospective coastal development efforts, coastal recreation programs, and resource management programs. Principles relevant to resource management include the following: (a) traditional understanding of fish spawning cycles; (b) use of the Hawaiian Moon Calendar; (c) conserving wetland resources; and (d) consultation with kūpuna (knowledgeable elders). Development of comprehensive ahupua'a maps would be most useful.

In sum, the ahupua'a and related principles were once central aspects of Native Hawaiian society. Their reinstitution in the contemporary context has the potential to enhance conservation and effective management of natural resources. This will require sustained effort to reintroduce important historic principles and social processes within a complex array of existing county, state, and federal laws and agencies. A joint effort toward that end has been undertaken by the Pacific Islands Resource Management Institute, the Association of Hawaiian Civic Clubs, and the Office of Hawaiian Affairs and is now in motion.

Chapter 12
Synthesis: Toward Incorporating Social Science in EAFM

As was made clear during the course of the workshop, the social sciences may be applied in many and various ways to further understanding of human interaction with marine and terrestrial ecosystems, to improve governance aspects of resource management, and to assess the effects of management strategies on people and the marine environment and its resources. In the context of the Pacific islands, social science applications may allow expanded understanding of such interactions and effects in settings where marine resources have long been and remain particularly important for many social, cultural, and economic reasons.

As summarized in the previous pages, a wide range of topics pertinent to ecosystem-based management of marine resources was addressed during the course of the social science workshop. These topics included the following: (1) marine fisheries, fisheries management, and related human and biophysical factors in the Western Pacific; (2) the need for, and utility of, social science in the context of ecosystem-based management in this region and elsewhere; (3) institutional constraints and opportunities for incorporating social science into ecosystem-based management; (4) relevant information needs, useful types of data, and data collection methods; (5) ecosystem-relevant human behavior and resource modeling; (6) indicators for assessing regulatory effects and the performance of management strategies; and (7) scope and scale of social science applications to ecosystem-based management. This chapter section synthesizes workshop discussion regarding these topics and draws on the background context developed earlier in the chapter to discuss elements of a general approach for applying social science to ecosystem-based management across the region.

Drawing on Ancient Concepts and Practices

The long history of human migration throughout Oceania is directly related to accumulation of detailed knowledge and successful pursuit of marine resources. Navigators led intrepid voyagers to distant horizons knowing they had the skills to sustain themselves while seeking land. Once islands were located, colonization and expansion of human populations were based in large part on knowledge of marine ecosystems and resources, and forms of social organization that enabled distribution and consumption of rich sources of dietary protein. In some places and cultures, social mechanisms were developed to formally manage marine ecosystems and resources.

Ecosystem-Based Fisheries Management in the Western Pacific, First Edition. Edward Glazier.

Indigenous Pacific islanders now may draw on lengthy histories and ever-evolving knowledge and traditions of interaction with the ocean and with each other to successfully exploit the marine environment. Persons arriving here during more recent centuries also draw upon traditional and experiential knowledge of the ocean and its resources. This is not intended as token acknowledgment of history. Extensive oral tradition and literature may be drawn upon to conceptualize and plan effective ecosystem-based management in the Pacific islands.

The political and policy implications of indigenous knowledge and marine tenure are also highly significant in the context of contemporary ecosystem-based management. The ahupua‘a system once widely used by Native Hawaiians offers a model for a form of resource management that is attentive to ecological relationships between land and sea within geopolitically specified boundaries. Those boundaries were significant in that they served to delimit attention and use of resources in an otherwise open system, thereby increasing the ability of the konohiki (leader) to monitor and regulate local resources and their use per the needs of the resident population. Ahupua‘a were components of larger moku (districts), and thus monitoring and control of resources also occurred at a district, island, and island-wide basis.

Contemporary advocates of the ahupua‘a system suggest that aspects of that arrangement can and should be considered for potential use under any new form of resource management in Hawaii. For instance, some suggest that konohiki and ‘aha (councils) could once again be established to monitor and assist in decision-making processes regarding natural resources used by people in specific areas. While this would require adaptation to (or of) existing political and management entities and processes, it does follow logically that more and better localized monitoring of island ecosystems and the needs and use patterns of residents could enhance management of resources within and across those systems.

Similar forms of localized resource management processes are being tried in various parts of the coastal zone of the United States, including various watershed models and programs. As Paul Bartram noted during the workshop, there is much potential in strategies that draw on traditional forms of resource use and management in the island context as is successfully occurring in a distinct spatial and cultural context on the island of Moloka‘i. We periodically revisit aspects of the ahupua‘a model in the following pages to illustrate opportunities and challenges associated with establishment of the ecosystem approach in the contemporary Western Pacific.

Existing Institutions and Institutional Parameters

This leads to discussion of existing institutional opportunities and constraints for incremental introduction of the new form of marine resource management. The Western Pacific Council has drafted plans for establishing a system of FEPs based on the geography of this vast region. It is possible that at some point in the evolution of an ecosystem approach an ahupua‘a or similar model would be formally reinstituted in Hawaii. In fact, elements of the Council system have long incorporated concepts inherent in that approach. These include community-based management strategies and projects designed to increase the degree of participation of indigenous persons in management of marine resources. For instance, the Council has developed and is implementing a community development program (CDP) and a community demonstration project program (CDPP). The CDP calls for increased representation of indigenous islanders in regional fisheries management and the CDPP is a funding program that promotes traditional indigenous fishing practices.

The Council is cognizant of the goodness of fit of an ecosystem approach to fisheries management in the island context and it has been proactive in establishing such approaches in the region. Yet, quite obviously, the Western Pacific Council is not alone in its management responsibilities here. This is clearly significant in the ecosystem context in that biophysical relationships occur regardless of sociopolitical bounds, while management interactions occur with direct regard to jurisdictional boundaries. The state of Hawaii has also undertaken programs to enhance community participation in marine resource management efforts, and while its jurisdiction extends to the landward limit of the exclusive economic zone (EEZ), pelagic and other fish species haven't been told about the border and there is no border patrol! For instance, the West Hawaii Fishery Council was established to enable representation of aquarium fish collectors and persons pursuing reef and other fishes for consumptive purposes in the same areas (see Tissot 1999).

Meanwhile, members of the resource user groups also often aren't highly aware of geo-political boundaries and may pursue fish and other marine resources inside and outside of state and federal waters on any given fishing day. As such, management of marine ecosystems and user groups may be most effective where management measures address the realities of the system rather than its imposed political bounds.

Resource managers are well aware of this, and interjurisdictional efforts are not new in the region. But inasmuch as ecosystem approaches increase attention to biophysical systems that do not correspond with jurisdictional boundaries, further interaction and cooperation between agencies and entities may be required.

Lee Anderson and Tim Hennessey discussed problems potentially resulting from changes in the way resource management agencies will operate under the new system of management. They related that, in some cases, actors with skills and capacities that are tailored to existing management regimes will, or have been, forced to quickly adapt to new conditions. Resentment and resistance are not uncommon. Susan Hanna made a strong case for entering into the new management regime with awareness of the possibility for such outcomes, and with readiness to develop unifying goals and objectives across diverse interests and scales of power, control, context, and responsibility. Dr Hanna also made clear the potential utility of indicators for assessing the performance and institutional challenges of new programs over the course of time.

The paradigm shift to an ecosystem approach to management may potentially lead to further institutional complexity and related challenges in all coastal regions of the United States, but perhaps especially so in this unique region of multiple jurisdictions. Although archipelagic-based fishery ecosystem plans (FEPs) may serve to reduce administrative complexities over a vast area, Council and NOAA Fisheries representatives have recognized the potential for new challenges and are undertaking a measured approach to prospective policy changes. Given the immensity of this region, extensive diversity in sociodemographic and sociopolitical context, and the increasing influence of international decisions regarding the migratory species that are so important throughout the region, an incremental and adaptive approach may well be the best way to proceed.

Pursuing Ecosystem Goals and Objectives through Application of Social Science

As stated in the Council FEPs (e.g., Western Pacific Regional Fishery Management Council 2005, p. 5–6), pursuit of an ecosystem approach to management in the region relates in part to NOAA Fisheries' EPAP goal to maintain the overall health and sustainability of marine systems and resources, and to "establish a framework under which the Council will improve its abilities to realize the goals of the Magnuson-Stevens Act through the incorporation of ecosystem principles and science." As depicted in Table 12.1 below, the objectives for meeting that goal have been clearly stated (Western Pacific Regional Fishery Management Council 2005, p. 6). But as discussed by Dr Anderson and others, there is good reason for Council representatives and representatives of other management agencies active in the region to review ways that those general objectives might best be achieved. That is, given the scope of the objectives and potential challenges associated with meeting them, setting of priorities and formulation of specific management measures may be most useful. As indicated in the table, those measures ideally will be formulated based in part on the potential contribution of the applied social sciences.

Research and Monitoring of Direct Ecosystem Relationships

Extensive attention was given during the course of the workshop to data collection methods and modeling techniques. Relatively less attention was focused on development and use of social or economic indicators. Irrespective of depth of coverage, each of these factors relates to measurement, assessment, or monitoring of direct connections between resource user groups and marine ecosystems. These might be termed first-order relationships. Persons harvesting marine resources in the offshore and nearshore waters or shoreline components of marine ecosystems may be readily conceived as important biological components of those systems. Indeed, given that humans are so well equipped to target and capture top predators, we may appropriately be seen as occupying a primary position in the trophic hierarchy of certain ecosystems. As such, the manner in which humans interact with marine ecosystems is an obviously critical consideration in the management of marine resources.

A number of speakers provided insight into methods for understanding those interactions in detail, and in fact an entire workshop might have been devoted to this and related issues. Both Leah Bunce and Bryan Oles described the full range of methodological approaches used individually or in combination to understand, assess, and potentially improve human-marine environmental interactions. The recently published work of Aswani and Lauer (2006) is relevant in this regard, as is the ongoing social network modeling work reported by Jeffrey Johnson, and the MPA assessment work reported by Patrick Christie and Richard Pollnac. Dr Aswani's attention to spatial aspects of sea tenure and the traditional knowledge, pursuit, use, and management of marine resources in the Pacific island context may provide a model for generating in-depth understanding of human-ecosystem interactions in and across specific island areas around the Western Pacific. Dr Johnson's work also is valuable in this regard in that it bridges the interests of biophysical and social scientists by offering viable models for predicting the direct, indirect, and bi-directional effects of and on humans

Table 12.1. Council FEP objectives and prospective role of social science.

Objective	Prospective role of social science
(1) Maintain biologically diverse and productive marine ecosystems and foster the long-term sustainable use of marine resources in an ecologically and culturally sensitive manner through the use of a science-based ecosystem approach to resource management	Determine culturally appropriate protocols for undertaking ecosystem-based management under variable social, cultural, and economic conditions and scenarios in each archipelago
(2) Provide flexible and adaptive management systems that can rapidly address new scientific information and changes in environmental conditions or human use patterns	Determine and document human use patterns and associated socioeconomic factors for each fishery in each archipelago; monitor changes in such patterns and conditions and assess associations with biophysical changes
(3) Improve public and government awareness and understanding of the marine environment in order to reduce unsustainable human impacts and foster support for responsible stewardship	Assess public and government awareness of environmental understanding within and across groups and institutions; identify means for improving venues for communication; identify, characterize, report, and monitor problematic forms of individual and collective interaction with or indirect influence on marine and associated terrestrial ecosystems
(4) Encourage and provide for the sustained and substantive participation of local communities in the exploration, development, conservation, and management of marine resources	Identify potential opportunities for and constraints on sustained community-level participation in these dimensions of marine fisheries; account for inter- and intracultural variability in receptivity to involvement
(5) Minimize fishery bycatch/waste to the extent practicable	Identify fisheries in which bycatch is significant; determine the nature of problematic fishing methods; determine whether individual or broad cultural processes or economic incentives are involved; identify alternative methods or fisheries suitable for offending user groups
(6) Manage and comanage protected species, protected habitats, and protected areas	Identify practical and culturally appropriate means for co-management of such species, habitats, and areas; assess the potential for reintroduction of historic forms of resource management
(7) Promote safety of human life at sea	Identify and assess behavioral factors that contribute to at-sea hazards and identify affordable, amenable, and practical means for reducing these for the various fleets

(*Continued*)

Table 12.1. Council FEP objectives and prospective role of social science. (*Continued*)

Objective	Prospective role of social science
(**8**) Encourage and support appropriate compliance and enforcement with all applicable local and federal fishery regulations	Assess economic and sociocultural factors that may be associated with regulatory violations and identify ways in which regulations and/or user group behavior might be modified to improve compliance. Identify social and cultural settings in which extant customs/sanctions obviate regulations and enforcement and advance these or elements thereof as possible models
(**9**) Increase collaboration with domestic and foreign regional fishery management and other governmental and nongovernmental organizations, communities, and the public at large to successfully manage marine ecosystems	Use economic, sociological, anthropological, and other social science theory and methods to assess the potential for effective collaboration; use such approaches to identify means for improving the effectiveness of such collaboration
(**10**) Improve the quantity and quality of available information to support marine ecosystem management	Identify information needs and perceived shortcomings of extant data from the perspective of managers and decisionmakers working in the region; examine TEK and its potential for improving management of ecosystems

as components of ocean food webs. Johnson's systematic work with traditional ecological knowledge is also highly useful. Dr Christie has presented a strong case for development of indicators capable of assessing success of management regimes (in this case, MPAs) in both biological and social terms. Similarly, Dr Pollnac has developed modeling techniques to assess human-biophysical interactions in the context of ecosystem-based management, with emphasis on measurement of human happiness and well-being, and management measures that would enhance those often-elusive states.

As discussed by Sam Pooley, Dave Hamm, Stewart Allen, Peter Wiley, and others, the ongoing programmatic research of NOAA Fisheries, NOS, HDAR, and other agencies active in the Western Pacific has yielded extensive information of potential use for assessing direct and indirect relationships between resource user groups and marine ecosystems. Council information needs regarding pressure on resources and related aspects of human-marine ecosystem interactions may be met in part through specific topical and spatial analyses of such extant data.

Such data may also be used, potentially in conjunction with data deriving from other research, to assist in developing valid indicators for understanding both the effects of human activities on biophysical systems and the effects of changing biophysical conditions on resource user groups. As indicated by Dr Aswani during the workshop, the latter constitutes a highly significant form of understanding in that, in some island settings, changes in the availability of marine resources can be matters of life and death.

A Note on Social and Economic Indicators

One objective of the workshop was to identify the "best suite of ecosystem indicators related to the human and institutional ecology of marine ecosystems in the Western Pacific and its subregions." While there was background discussion of the potential value of social and economic indicators for assessing interactive relationships and effects between humans and biophysical systems, and their use in other settings, specific indicators of potential utility in the present context were not identified. Based on the rationale that certain "common denominator" indicators could be useful for a range of needs (as discussed by Marc Miller), it may have been productive to facilitate elicitation of perspectives on such indicators during the course of the workshop. But in fact, the eventuality of background discussion may have been appropriate for several reasons.

First, useful social and economic indicators may rightfully be seen as following from specific ecosystem-specific management measures which, in the case of the nascent ecosystem approach, have not yet been fully determined for each archipelago across the region. As we recommend in subsequent sections of this chapter, a venue should be developed to aid in identifying measures that would be most effective for satisfying the Council's FEP objectives and overarching goals. Discussion of specific appropriate indicators could follow in the same venue. Second, as noted above, selection of valid social indicators would ideally derive in part from understanding of extant data and focused consideration of the social and biophysical contexts in question. As such, the above-mentioned venue would ideally be attended by persons highly knowledgeable of that information and those contexts. Finally, an extensive literature on social indicators is available to inform such discussion (e.g., Minerals Management Service 1996; Boyd and Charles 2006).

Ideally, analysis of lessons learned from social indicators research and applications in natural resource settings in other regions would be considered in advance of selection and use of indicators by the Council or NOAA Fisheries in the Western Pacific. One perspective stated at the workshop was that, given developments in fisheries economics, it may be relatively less difficult to determine valid economic indicators of utility for ecosystem-based management and related analyses than it would be for noneconomic social indicators.

An aspect of all social indicator research that bears mention at this juncture is that indicators are viable only insofar as the putative relationship with that which is being indicated is amenable to empirical testing. This is an obvious point at first glance, but in fact, in seeking to understand complex social processes there is always potential for drawing conclusions from spurious associations. At the same time, it should be kept in mind that indicators must be capable of gauging and monitoring the effects of a range of events or processes in addition to those potentially associated with the management measures or event or process in question, and analysts should be prepared to work through a variety of prospective causal and associative relationships.

We emphasize that none of these points is intended to diminish the potential utility of social and economic indicators in the region. Indeed, as numerous workshop participants made clear, valid indicators may be particularly useful as means for assessing and monitoring human-environmental interactions, and as a basis for adjusting resource use policy under the new mode of management.

Research and Monitoring of Indirect Ecosystem Relationships and Effects

Discussion of indicators is also relevant to assessment and longitudinal monitoring of human and physical environmental processes that are *indirectly* related to ecosystem management. We note at the outset that (a) the term "indirect" is used here in the analytical sense and that indirect relationships may be as critical as direct relationships; and (b) direct and indirect relationships are often difficult to parse. For instance, given fluctuating market conditions, operational costs, and other factors affecting participants in the harvest sector, job opportunities outside the fishing industry during off-seasons can in some cases be as critical for the ongoing functioning of a fishing fleet as can the availability of resources during the fishing season(s). Thus, while an indicator such as rate of employment outside the fishing sectors may appear extraneous to the interests of fishery managers, such information may in fact provide a valid if indirect indication of the functional capacity of a given fleet over time.

In fact, many variables and processes that may be seen as indirectly related to extraction of marine resources warrant ongoing assessment and monitoring. For example, during the course of her discussion about Guam, Judith Amesbury elucidated the importance of assessing and monitoring the effects of macro-scale climatic events such as volcanic disturbances and periods of drought associated with ENSO events. These can incur dramatic changes in marine ecosystems. Similarly, as noted by Jesse Rosario, in his discussion about contemporary conditions and challenges for fishers on Guam, the actions of persons with vested interests in coastal tourism can lead to a range of effects which indirectly affect the fleets and shoreline fishers, and by extension those who depend on the resources for consumptive and cultural purposes.

Finally, as discussed by Fini Aitaoto, federal actions in domains other than fishing can also have a dramatic effect on the conduct of local fisheries, as in the case of immigration laws which have precluded Independent Samoans from working in canneries in America Samoa. Given that the cannery managers reportedly are having trouble finding sufficient numbers of employees, this situation has indirect but clearly problematic implications for American Samoa fishers seeking to market their products to canneries in their home country.

Much workshop discussion also naturally focused on potential and actual historical indirect effects of biophysical management measures on fishers, fleets, and communities. This area of consideration is well-covered in the Social Impact Assessment literature and will not be belabored here. A couple of important issues bear reiteration by mention, however. These include displacement of fishing effort as a result of establishing MPAs (as discussed by Mr Rosario and others) and associated implications, including loss of harvest, lost income, and lost cultural opportunities.

Factors associated with fishing communities also bear mentioning. Susan Abbott-Jamieson discussed a wide range of variables and factors that are being monitored by NOAA Fisheries staff around the country (further elaborated in the regional context by Stewart Allen). These have been chosen for monitoring by virtue of their potential for enabling valid assessment of collective engagement in or dependence on marine fisheries, and with the ultimate intent of reducing potentially deleterious indirect effects of regulations on "fishing communities" and the fishing-specific and secondary industries and activities associated with such communities. All such potential effects will need to be considered as aspects of an ecosystem approach that by definition is geared toward understanding and addressing a greater range of human and environmental relationships than has heretofore been considered. As noted by John Petterson, this has implications for understanding and monitoring a range of non-regulatory factors impinging on fishing fleets and communities, from broad macro-social and economic processes to specific environmental events such as hurricanes (see Impact Assessment, Inc. 2006).

This incurs discussion of issues related to the preferred, mandated, and ideal scope and scale of research conducted for purposes of assessing or monitoring human dimensions of ecosystem-based management. Tom Fish related concerns in this regard, noting that truly equitable treatment of all elements of marine ecosystems, inherent linkages with terrestrial ecosystems, and inherent linkages with human societies would require immense outlay of time and energy and hence, some subjective decisions must be made to pragmatically limit or prioritize the foci of resource managers. Such priorities and foci may include macro-level social and economic processes indirectly but significantly affecting fishing fleets, marine ecosystems, and adjacent communities. Numerous presenters discussed or alluded to such processes.

For instance, John Gourley elucidated the implications increasing cultural diversity for the status of marine resources in the Commonwealth of the Northern Mariana Islands. Because many in-migrating groups have arrived here with well-developed methods and preferences for pursuing and using marine resources, macro-level demographic changes may be one of the most important considerations for managers attending to natural resource and ecosystem issues in this distant island region.

Fisheries social science may potentially be applied to further community development objectives of the Western Pacific Council and other entities. Although social and economic conditions in community settings throughout the region are influenced by a wide range of factors not directly related to fishing, engagement of residents in fishing-related industries and activities may benefit such communities in a variety of ways. These include fishing-related opportunities for employment, recreation, and avoidance of detrimental situations and activities. This eventuality was discussed by Council staff economist Marcia Hamilton, furthered by Dr Miller, and exemplified in the Fiji-based community development work of Joeli Veitayaki, all of whom recognized the potential contribution of social science applications in identifying places, situations, conditions, and processes that could involve communities and individuals in the abundance of positive ocean opportunities available throughout the Western Pacific region. While social science cannot be equated with community development *per se*, its application may further understanding of the community context, local receptivity to or need for development programs, and the potential or actual social and economic costs and benefits of such programs.

Choices and Priorities

This section of the chapter has revisited some but not all of the important ecosystem-relevant human dimensions issues discussed during the workshop. Clearly, a wide range of questions and possibilities confront the Council and NOAA Fisheries and other institutions progressing toward adoption of ecosystem approaches to fishery management in the region. Moreover, while some measure of regional specificity was achieved during the workshop, much of the discussion was relatively general in nature and a greater range of specific factors and questions associated with the new paradigm will undoubtedly emerge over the course of time.

Given this ultimately vast array of considerations, setting of priorities may enhance the efforts of managers in the region to begin the incremental and adaptive undertaking of incorporating social science principles, methods, data, analysis, retrospective and predictive modeling, and related considerations in real-time ecosystem approaches to management. In the case of the Western Pacific Council and NOAA Fisheries, these priorities necessarily will relate to respective development of the FEPs and related EIS, and associated discussion and formulation of management measures in the upcoming final workshop and other prospective venues. We therefore articulate the following discussion of prospective approaches and means for setting priorities with those tools for planning and assessment.

Elements of a Social Science Approach to EAFM in the Western Pacific Region

We note once again that our intention is to move beyond ecosystem concepts that are primarily biophysically based. The term "ecosystem" must be understood as encompassing both human and nonhuman elements. In all of the discussion and recommendations below, attention should be given to thorough social scientific description and explanation of the following principal components of marine ecosystems:

1. The biophysical ecology.
2. The human ecology, which has two distinct components:
   a. The human ecology of the constituents, by which we mean the people whose behavior affects, or is affected by, a defined biophysical ecology, or who are otherwise concerned with the state of that biophysical ecology.
   b. The ecology of the governance institutions which have authority or responsibility for formal rules of human behavior with respect to the defined biophysical ecology.

These ecological components—the biophysical, human constituent, and institutional—together comprise the "ecosystems" relevant to fisheries management considerations in the Western Pacific.

Perhaps the most outstanding feature for consideration in applying the social sciences to an ecosystem approach to management is the unparalleled extent of variation in social conditions across the archipelagos. In the case of the Hawaii Archipelago, social, economic, cultural, and demographic conditions vary radically even across a given island, and certainly across the island chain. Factors relating to a shift in governance and management will necessarily vary accordingly.

For instance, population density is quite low on rural Moloka'i and the majority of residents there are Native Hawaiian. Plans for formal reestablishment of an ahupua'a system or ahupua'a-like system of resource use and management on that island would involve different issues and strategies than on an island like O'ahu, where population density is very high, where cultural conditions are relatively diverse, and where urban and rural areas are characteristically quite different in many ways. A shift toward a more traditional form of resource management and governance would likely be more widely received on Moloka'i, while such a strategy would involve a more complex set of considerations on O'ahu. This subsection of the chapter advances a generalized approach intended to assist the Council in addressing such variation while planning for specific applications of social science during the course of adoption and long-term administration of ecosystem-based management in the Western Pacific.

Addressing Variation with an Adaptive Approach

Each archipelago addressed in the Council's draft FEPs is distinct in terms of its sociocultural, socioeconomic, and demographic conditions, in terms of its mode and culture of governance, in environmental terms, and in terms of the types and extent of fishing and other pursuits and uses of marine resources. As noted by Craig Severance during the workshop, local capacity to conduct social research and monitor social and environmental conditions also varies by region. It is essential that all such variation be addressed in planning processes related to the Council's FEP objectives in each archipelago.

For instance, it was made clear during the workshop that understanding of regionally specific social and cultural factors can enhance effective resource management in cross-cultural settings. Thus, the Council's third FEP objective to "improve public and government awareness and understanding of the marine environment in order to reduce unsustainable human impacts and foster support for responsible stewardship" would best be pursued in American Samoa by developing adequate understanding of issues such as those discussed by Fini Aitaoto regarding illegal fishing activities in the offshore waters of Tutuila. Adequate understanding and documentation of variability in how such violators might traditionally be dealt with under *Fa'a Samoa* could assist the Council in planning management measures to address its sustainability and stewardship objectives in the region. In this case, depending on local political considerations (which may well vary by village), contingencies for intervention may or may not be necessary; an effective and adaptive management measure would reflect understanding of the cultural, social, and political context. Similarly, understanding of economic conditions and motivations associated with pursuit and use of pelagic resources among fishery participants on Guam could contribute to an empirical basis from which to collaboratively negotiate an adequate portion of prospective regional quotas on such species—in keeping with Council's FEP Objective Nine.

These are two of many possible examples illustrating the potential value of social science research and monitoring vis-à-vis the Council's FEP objectives. Ideally, specific management measures developed to meet FEP objectives in each unique archipelago would involve use of such information. Although extensive fisheries, census, and other forms of archival information may be used in this regard, increased attention to direct and indirect human-environmental interactions under the ecosystem approach will ultimately require more and more detailed data and analyses. As such, we recommend that an approach be developed for identifying and compiling existing information relevant to human aspects of ecosystem-based management, and for gathering and making available new relevant data regarding core management issues and challenges, and pertinent economic, sociocultural, political, and demographic factors and conditions from across the region. The approach and data framework will need to be flexible and adaptive in keeping with the changing dynamics of marine ecosystems and human interaction with those systems, and with the evolving nature of the ecosystem management approach itself.

Steps for Incorporating Social Science in EAFM in the Region

Based on review of workshop proceedings as described in this chapter, we recommend a series of steps for establishing priorities and incorporating social science methods, models, and principles in the development and long-term administration of management measures under the new management approach. In reiteration, this new approach must articulate with a definition of "ecosystem" that emphasizes the pivotal importance of human beings in marine biophysical systems. These recommended steps include the following.

Establish a Venue for Choosing Priorities and Specific Management Measures

The workshop described in this chapter generated an array of general and specific approaches that may be used to meet the Council's objectives for ecosystem-based management in the Western Pacific. Various challenges, lessons, and cautionary notes were also discussed. All of this material may be useful for the Council as it moves toward full adoption of the ecosystem approach, and this chapter may be consulted for general guidance during that process. More specifically, the social science workshop and these proceedings will contribute to the upcoming ecosystem policy workshop.

But a venue or venues for more specific guidance may also be warranted. Each of the Council's 10 FEP objectives incorporates a complex set of issues and social science considerations, and that complexity is magnified by the fact that the Council's kuleana (responsibility) extends across a vast ocean area and multiple archipelagos with very different characteristics. Establishment of a venue for Council, NOAA Fisheries, and regional social scientists to work toward (a) prioritization of FEP objectives vis-à-vis social science applications, and (b) identification of specific management measures and related information needs to meet those objectives, may serve to resolve some of that complexity. This could potentially occur in or through an existing Council process, but insofar as the shift to the new approach is to be adaptive and incremental, such a venue would ideally be recurring.

Design Research to Meet Prioritized Objectives and Information Needs

Once prospective management measures are identified in association with the prioritized objectives, expertise would ideally be applied to formulate specific plans for conducting social research in the region—as needed to assess the possible effects of implementing those measures. Given that extant data may contribute both to the design of the research and to the necessary analyses, the first and indispensable step in the process would be compilation and organization of relevant data by archipelago. This also would require general expertise and ability to anticipate the kinds and extent of information that would be needed for purposes of modeling and analysis, including knowledge of local and regional data issues and sources. Some field reconnaissance may be useful in identifying salient issues and otherwise hard to identify data and data sources.

Implement a Research and Monitoring Strategy

In cases where existing data is insufficient for assessing the prospective management measures, a strategy for sponsoring and conducting the necessary research and analyses would need to be implemented. Given inevitable limitations on time and fiscal resources, such research would ideally be conducted in conjunction with or under the sponsorship of existing research programs. Alternatively, or additionally, other sources of funding would be identified. As was discussed by Dr Severance and others during the course of the workshop, expertise in fisheries social science can be hard to come by in certain areas of the archipelagos, and thus building long-term local capacity for conducting social science may yield benefits in future years.

With regard to assessment and monitoring of the effects of new management measures, numerous workshop discussants championed the benefits of GIS. The broad utility of spatially oriented social description and analysis is undeniable in the realm of marine fisheries management, and especially given the requisite geospatial parameters of ecosystem-based management. GIS can also assist in relating and monitoring aspects of residence, business, and recreation as these pertain to use of, and human influences on, marine ecosystems and associated resources.

Finally, monitoring of human-environmental interactions in this context would very likely benefit from identification and use of valid social and economic indicators. As noted above, generalized indicators (e.g., local rate of unemployment or availability of alternative forms of employment) may be developed for use in a wide range of settings. More specific indicators could be developed subsequent to identification of specific management measures (e.g., number of active commercial licenses associated with a specific fishery). Again, determination of both generalized and specific indicators would ideally be based on review of the natural resource-relevant social indicators literature and on focused, facilitated discussion between social scientists (and perhaps biophysical scientists) and marine resource managers working in the region, or possessing relevant experience garnered elsewhere.

Depending on the nature and scope of the prospective management measure or policy, it may be prudent to involve stakeholders in early and ongoing discussions about the potential costs and benefits of those measures or policies. The benefits of doing so were elucidated by Dr Fernandes during her discussion of the public comment and participation process associated with establishment of the Great Barrier Reef Marine Park, and by others during the workshop. Identifying actual or potential constraints early in the process typically affords time and enhanced potential for meeting related challenges. A broad literature regarding public involvement in natural resource decision-making processes is widely available and may be consulted.

In any event, the Council is well-versed in the process of engaging the commentary and participation of the general public and specific stakeholder groups in its decision-making processes. Community-based management and community participation are, in fact, stated elements of the FEP planning process (Western Pacific Regional Fishery Management Council 2005, p. 27–28).

Develop and Implement Liaison and Performance Evaluation Programs

Finally, we assert the potential utility of establishing means by which resource user groups may readily interact and communicate on a regular, noncontentious, and interactive basis with management entities in the region. One potential means would be through fisheries liaison personnel versed in cross-cultural communication. This may be particularly useful in this culturally diverse region. Such liaisons could also aid in developing and implementing evaluation programs designed to assess the performance of ecosystem-based management throughout the region(s), and the manner of its reception in host communities.

References

Aswani, S. and Lauer, M. (2006) Incorporating fishermen's local knowledge and behavior into Geographical Information Systems (GIS) for designing marine protected areas in Oceania. *Human Organization*, 65(1), 81–102.

Boyd, H. and Charles, A. (2006) Creating community-based indicators to monitor sustainability of local fisheries. *Ocean and Coastal Management*, 49, 237–258.

Impact Assessment, Inc. (2006) Preliminary assessment of the impacts of hurricane Katrina on Gulf of Mexico coastal fishing communities. Submitted to the US Department of Commerce National Oceanic and Atmospheric Administration, National Marine Fisheries Service Southeast Regional Office, St. Petersburg.

Minerals Management Service (1996) Social Indicators Monitoring Study Peer Review Workshop Proceedings. MMS OCS Study 96-0053. US Department of the Interior, Alaska OCS Region, Anchorage.

Tissot, B. (1999) Adaptive management of aquarium fish collecting in Hawaii. SPC Live Reef Fish Information Bulletin No. 6. December.

Western Pacific Regional Fishery Management Council (2005) *Fishery Ecosystem Plan for the Hawaii Archipelago*. Honolulu.

Chapter 13
Summary Conclusions

This chapter has been compiled to describe the Ecosystem Social Science Workshop held by the Western Pacific Regional Fishery Management Council during January of 2006. This was the second of a three-part workshop program intended to aid the Council in adopting fishery ecosystem plans in the region. A biophysical ecosystem workshop was held in April 2005. A final workshop will synthesize the results of the previous workshops with the intent of developing a framework for regional ecosystem policy and governance.

The social science workshop described herein addressed human dimensions of ecosystem approaches to fishery resource management. The workshop emphasized the three major components of marine systems—the biophysical, the human constituent, and the institutional. A wide range of perspectives were presented on related topics and issues, including the following:

- Marine fisheries, fisheries management, and related human and biophysical factors in the Western Pacific.
- The need for and utility of social science in the context of an ecosystem approach to management in this region and elsewhere.
- Institutional constraints and opportunities for incorporating social science into ecosystem-based management.
- Relevant information needs, useful types of data, and data collection methods.
- Ecosystem-relevant human behavior and resource modeling.
- Indicators for assessing regulatory effects and the performance of management strategies.
- Scope and scale of social science applications to an ecosystem approach to management.

Workshop presentations and discussions were both general and specific in scope, and regional experts were on hand to help ground the discussions with their own perspectives on the realities of island life in the Pacific, and on the various fishery management challenges and solutions that have been encountered and applied in the region.

Summary Points of Particular Relevance to Council FEP Objectives

An extensive assortment of valuable insights, lessons, and pertinent background information about ecosystems, ecosystem social science, and the context of fisheries in the Western Pacific may be derived from the workshop and from these proceedings. Interested persons may consult the body of this chapter for such information. But some

Ecosystem-Based Fisheries Management in the Western Pacific, First Edition. Edward Glazier.

areas of discussion are particularly relevant to the information needs and objectives of the Council as it moves toward full adoption of its Fishery Ecosystem Plans. These lend themselves to summarization and are provided here as a means for bringing the long prior discussion to a conclusion:

1. Definitions and parameters vary and continue to evolve, but there is general consensus that the ecosystem approach to fisheries management is novel in its attention to whole marine systems including relationships among the biophysical, human, and institutional components that comprise those systems.
2. Human beings, groups, and institutions are critically important elements of marine ecosystems, and given their place in the trophic hierarchy, human behaviors, beliefs, and values should be given primary consideration.
3. The Council's approach to ecosystem-based management to date involves adaptive management and emphasis on indigenous forms of resource management; both may be particularly amenable in the Pacific islands context.
4. Indigenous Pacific islanders draw on lengthy histories and ever-evolving knowledge and traditions of interaction with ocean ecosystems and with each other to successfully use that environment. Persons arriving here during more recent centuries also draw upon traditional and experiential knowledge. Both groups may provide valid information and perspectives on viable models for planning and administration of ecosystem-based management in the region.
5. The nascent paradigm shift to ecosystem-based management may potentially lead to further institutional complexity in this unique region of multiple jurisdictions. Given the size of the region, extensive diversity in sociodemographic and sociopolitical context, and the increasing influence of international decisions regarding migratory species, an incremental and adaptive approach may be the best way to proceed.
6. The Council has developed 10 objectives for its Fishery Ecosystem Plans. Given the scope of the objectives and potential challenges associated with meeting them, setting priorities and formulating specific management measures may prove most useful for effectively meeting Council goals. Those measures ideally will be formulated based on the many potential contributions of the applied social sciences.
7. Each archipelago in the region is distinct in terms of sociocultural, socioeconomic, and demographic conditions; mode and culture of governance; environmental conditions; and types and extent of fishing and other pursuits and uses of marine resources. This variation may be effectively addressed for purposes of meeting FEP objectives through appropriate application of social science methods and analysis, including those methods that facilitate public participation in resource management decision-making processes.
8. An array of data collection methods and analytical techniques has been developed to aid in understanding and communicating both the effects of human activities on biophysical systems and the effects of changing biophysical conditions on resource user groups.
9. Selection of social science methods and analytical techniques should be closely tailored to the information needs and objectives at hand, and to particular environmental and societal aspects of each archipelago.
10. Valid social and economic indicators are particularly useful for assessing and monitoring direct and indirect human–environmental interactions, and as a basis

for adjusting resource use policy under the new mode of management. Indicators should articulate with a wide range of climatic, macroeconomic, sociodemographic, regulatory, and community-related factors. In this case, such indicators will need to be developed based on (a) their potential utility for meeting Council objectives, (b) extant data and the social and biophysical contexts in question, and (c) relevant indicators literature.

11. A social science approach to ecosystem-based management in the region should be developed to enhance Council efforts to meet its FEP objectives and to administer the new form of management over the long term. The approach would include a series of related elements, as follow:
    a. A venue or venues for choosing high-priority FEP objectives, specific management measures for meeting those objectives, and valid social and economic indicators.
    b. Design of research to meet prioritized objectives and related information needs.
    c. Implementation of a research strategy to gather and analyze requisite information, and an indicators-based archipelagic monitoring system through which to gauge and analytically parse social change potentially associated with Council actions.
    d. Implementation of a liaison and performance and evaluation program to ensure the validity and effectiveness of the social science approach to ecosystem-based management in the region.
12. Social science cannot be equated with community development *per se*, but application of social science may further understanding of community context, local receptivity to or need for development programs, and the potential or actual social and economic costs and benefits of such programs. Social science may therefore be used to help identify ways in which communities and individuals may participate in the abundance of positive ocean opportunities available throughout the Western Pacific region.
13. Given that a number of fisheries or fisheries-relevant social science research and monitoring programs have been undertaken in the United States and abroad in recent years, the Council FEP social science approach would ideally articulate with these, both drawing upon and contributing to the base of knowledge regarding human interaction with the marine environment and the many related aspects of human behavior discussed during the course of the workshop.

Concluding Discussion

Based on the input of national and regional experts convened for the Ecosystem Social Science Workshop, we have presented valid social science approaches to ecosystem-based management. These may be of potential utility to the Council as it moves toward full adoption of its FEPs across the region. The workshop and documentation have enabled development of background information necessary for initiating refinement of such approaches for real-time application in the Western Pacific. Further work with fisheries managers, compilation and review of archival data, and field reconnaissance will enable full inventory of relevant existing information, identification of salient and ongoing management issues and related information needs, and development of detailed research agendas and designs for specific island areas.

As for biophysical approaches to ecosystem-based management, viable social science approaches must enable understanding of whole systems and relationships between their respective components, including those of user and interest groups, seafood distributors and consumers, and even fisheries researchers and managers and the institutions within which they operate. In the spirit of holistic ecosystem principles and concepts, social science approaches must and can also bear empirically grounded information of predictive utility for management of biophysical components of marine systems.

There is much human and environmental variability within and across the island groups that comprise the vast Western Pacific region. Social science approaches must address such variation and translate findings in a manner that is optimally useful for resource managers seeking to make fair and equitable decisions in an increasingly complex and contested sociopolitical environment. Regional variation notwithstanding, pursuit and consumption of seafood and related cultural processes are constant and critically important aspects of life throughout the archipelagos. As such, there is vital need for understanding and longitudinal monitoring of the full range of factors that may impinge on these activities and processes, including the potential effects of conservation interests and ecosystem-based management.

Ecosystem concepts and principles were developed and applied in adaptive fashion in this region long ago. Indeed, learned ways of efficient interaction with marine and terrestrial ecosystems led to the proliferation of island societies throughout Oceania. Initial periods of trial and error gradually led to the ordering of society in a manner that in certain places and times enabled equilibrium between available marine resources and the demands of human groups depending on them for purposes of survival. By virtue of attention to and accumulation of knowledge regarding the natural world that surrounded them, and through various mechanisms of social control, Pacific islanders were ultimately successful in overcoming various ecological challenges, including those initiated by their ancestors.

The context has changed dramatically over the millennia, and many of the challenges we now face are global in scale. Yet it may be that knowledge of connections within and across island societies and ecosystems, and proven means for managing the activities of those who use and depend on marine resources for so many reasons, remain the most viable points of departure for addressing marine resource challenges in the Pacific in the decades to come.

Section 3
The Ecosystem Policy Workshop

Edward Glazier

Chapter 14
Introduction to Ecosystem Policy

The Western Pacific Council convened the last event in its three-part series of fishery ecosystem planning workshops during the first week of 2007. This chapter summarizes the final workshop and provides essential context for understanding the unique nature of fisheries in the Western Pacific and the manner in which the Council has initiated an ecosystem approach that is well suited to the island and open ocean settings that are characteristic of the region.

Background

The underlying goal of fishery management undertaken in the US exclusive economic zone (EEZ) is to develop and adjust policies that ensure the sustainable use of living marine resources over time. This requires ongoing assessment of marine resources and analytical control of associated environmental conditions. But the internally complex nature of the resources and their dynamic interactions with a wide range of biophysical factors and forces renders this a challenging goal even in the absence of fishery interactions—the variable of particular interest in fisheries disciplines.

Fisheries science and management can indeed be fairly characterized as challenging endeavors, and the latter has generated mixed results. While one perspective holds that the world's fisheries are headed toward failure, others assert that many fisheries have, in fact, been managed successfully and that strategies for appropriate control of pressure on the resources will enable positive future scenarios. Some hold fast to the utility of the scientific method and well-informed management decisions. Because fisheries science and management unarguably address highly complex and dynamic processes, strategies for achieving sustainability inevitably require approaches that are adaptive to changing environmental conditions, to variable pressure on the resources, to new conceptual paradigms and advances in modeling, to new empirical data and analyses, and to ongoing uncertainties.

Management approaches ideally are also adaptive to the unique aspects of the region in question. Species and their dynamic interaction with the marine environment, types of nearshore and deep sea fisheries, regional marketing conditions, the social demography and needs of island residents, and local and regional systems of governance are but some of the factors that at once vary extensively by region and constitute important considerations for effective fisheries management. In some cases, a given region is characteristically complex, and a suite of management approaches are called for; in others, focus on a predominant fishery or species can enable highly effective management.

Ecosystem-Based Fisheries Management in the Western Pacific, First Edition. Edward Glazier.

An example of the latter is reduction of fishing pressure through a sustainable yield-based limited entry program for commercial salmon fisheries in Bristol Bay, Alaska. The strategy has ultimately been successful in biological terms (Hilborn 2006), because it fits the unique nature of the species and fisheries in question. That is, yield potential is, in this case, readily addressed by on-site monitors who can quickly regulate highly focused fishing pressure in response to escapement and fishery performance data. The process is augmented through overall annual limits on the size of the commercial fleet and numbers of recreational and subsistence permits, with allocation issues resolved via a regional regulatory process that incorporates public input (Brady 2004). Of significance from a biological perspective, population biocomplexity appears to contribute to the sustainability of the fishery despite cyclically unfavorable oceanographic conditions in the adjacent North Pacific Ocean (Hilborn et al. 2003).

Despite the biological successes of management in this case, participants in the Bristol Bay commercial salmon fisheries have long been challenged by depressed salmon prices resulting in part from saturation of world markets with farmed salmon products. In this regard, while the single-species approach appears to have ensured an abundant resource, and allocation decisions are successfully brokered through a process that enables the meaningful input of fishery participants, that approach has largely failed to address the constraints of the macroeconomic context in which the fishery is executed. The overall benefits of biologically successful management in this case are, therefore, uncertain.

For instance, it should be noted that individual and collective capacity to fish for consumptive and cultural purposes in rural Alaska is today often based in part on income derived through jobs in the commercial fishing industry. There are therefore instances in which participation in subsistence fisheries diminishes when the commercial sector is constrained by new regulations or other limiting factors. (Impact Assessment, Inc. 2007). If sustainability is to assume real meaning for the most deeply involved participants, then a more holistic strategy than that encapsulated by a single-species approach may be called for. It can be argued that such a strategy would necessarily prioritize human values, needs, and experiences, and address them as integral and pivotally important elements of marine ecosystems.

An Ecosystem Approach for the Western Pacific

Much knowledge has been gained through the traditional species-based approach to management in the Western Pacific. But the Council has increasingly recognized the suitability of an approach that emphasizes relationships between those resources and the unique physical and human environmental contexts in which they are situated. The Council's past efforts have led to recognition that its region of jurisdiction is unique and well suited to a science-based management approach that is responsive to the dynamics of large, open–ocean marine ecosystems and to complex social and economic connections between islands, islanders, adjacent marine ecosystems and jurisdictions, and associated marine resources.

A Vast and Complex Region

As noted in the previous chapters, the Western Pacific is indeed vast and unique. Numerous and varied nearshore and deep sea fisheries occur here, and unlike fisheries administered by other councils in the United States, fisheries in the Western Pacific are conducted from small islands located many thousands of miles from North America or any other continental land mass. Societies vary widely in terms of historical and contemporary economic, cultural, political, and linguistic attributes, and such variation can be notable both within and across island settings and associated fishing fleets.

Although great strides have been made in scientific understanding of diverse reef, demersal, neritic–pelagic, and pelagic species and fisheries in the vast Western Pacific, many unknowns and uncertainties remain. The scientific process is ongoing. Social scientific inquiry is also ongoing and similarly challenged by the size of the region and the diversity of conditions across the archipelagos and their respective islands.

It should be noted that some of these areas share offshore jurisdictional boundaries with other nations, thereby lending a level of complexity to governance that is unique among the fishery councils in the United States. Areas of shared international boundaries in the region include the following: (1) Palmyra Atoll and Jarvis Island, which are adjacent to the Northern and Southern Line Islands governed by the Republic of Kiribati; (2) Howland and Baker Islands, which are adjacent to the Kiribati-governed Phoenix Islands; (3) American Samoa, which is adjacent to the Independent State of Samoa and Tonga, and to the Cook Islands, Niue, and Tokelau; (4) Wake Island, which is adjacent to possessions of the Republic of the Marshall Islands; (5) Guam, which is adjacent to possessions of the Federated State of Micronesia; and (6) the Northern Marianas islands, which are adjacent to various islands of Japan.

The Western Pacific Council is also responsible for managing migratory and highly migratory pelagic fishery resources across a vast portion of the Pacific. This is increasingly complicated in that numerous groups and conventions now address management of resources across international jurisdictional bounds, including those of the US EEZ. These entities include the Inter-American Tropical Tuna Commission, the Interim Scientific Committee for Tunas and Tuna-like Species in the North Pacific, the Western and Central Pacific Fisheries Commission, the Secretariat of the Pacific Community, the Multilateral Treaty on Fisheries between the Government of Certain Pacific Island States and the Government of the United States, and others.

Addressing Uncertainties with an Adaptive and Incremental Strategy

The Council has adopted an approach to fisheries management that is novel in its attention to whole marine systems and to the physical and biological relationships among the components of those systems. While the ecosystem approach holds promise for addressing the aforementioned complexities of fisheries science and management, some envision uncertainty in how it might be applied and what its benefits might be in the realm of management. The approach can therefore be seen as presenting a conundrum to some scientists and managers, wherein the intricacies of marine systems are widely recognized as important subjects of inquiry of direct relevance to understanding fishery dynamics, but generating adequate understanding of complex and often relatively unknown trophic relationships is seen as daunting, overly taxing of available resources, or too great a departure from already-productive areas of inquiry.

But it should be noted that the Council continues to seek information that is based on the foundation of empirical science, and although the ecosystem approach necessarily involves expansion of attention into a larger realm of scientific inquiry, this is to occur incrementally and adaptively rather than abruptly. In fact, the new approach may be seen as involving an initial period of gradual change wherein the species-based approach is converted to a place-based approach that reorganizes and complements rather than replaces ongoing scientific inquiry and management. This is in keeping with the approach as envisioned by NOAA's Ecosystem Principles Advisory Panel in 1998:

> Ecosystem-based management can be an important complement to existing fishery management approaches. When fishery managers understand the complex ecological and socioeconomic environments in which fish and fisheries exist, they may be able to anticipate the effects that fishery management will have on the ecosystem and the effects that ecosystem change will have on fisheries. (Ecosystem Principles Advisory Panel 1999)

The Suitability of EAFM Policy in the Western Pacific

An EAFM is seen as particularly amenable to the Pacific island context for many reasons. For instance, historic management strategies undertaken here effectively recognized human and biophysical relationships and interactions and therefore provide conceptual models for planning a new approach. Island settings foster common recognition of such relationships and interactions, and an ecosystem strategy organized by archipelago may improve investigation and monitoring of such relationships and interactions at local and archipelagic levels of analysis. This is likely to reduce administrative burdens associated with management of single species pursued by multiple fleets across distant islands and archipelagos.

As discussed in the preceding chapter, several attributes render islands in the Western Pacific particularly suitable for examining ecological processes and, by extension, for applying ecosystem principles to management of marine resources. First, they are small relative to continents, and they tend to present distinct and isolated settings for certain forms of investigation.

Further, marine life congregates at islands (Sibert and Hampton 2003), and as Vitousek (1995, p. 11) asserts, islands afford the "opportunity to understand controls on ecosystem structure and function" and to develop models that "can be applied as the basis for understanding more complex continental systems." Similarly, Kirch and Hunt (1997) assert that understanding long-term feedback effects of ecological change in such settings may yield much insight into similar processes in larger island and continental ecosystems around the world.

The sea and the bounds between land and sea and their respective biophysical systems are readily envisioned from islands (Berkes 1999, p. 69), and marine resources are invariably important in social and economic terms in island settings. But long-term residents of islands typically recognize that living marine resources are finite and sometimes challenging to acquire. Moreover, many goods and services are not available unless they are imported through trade or other economic transaction. Viewed in historical perspective, such limitations have required islanders to develop extensive knowledge of marine resources and the factors that constrain or enable their availability, abundance, and acquisition (see Poepoe et al. 2003). Local and traditional knowledge of marine ecosystems and relationships between their components can thus often be extensive in island settings in the Pacific.

Prelude to the Ecosystem Policy Workshop

The Council's fishery ecosystem plans (FEPs) address complex relationships between populations of organisms, habitats, oceanographic conditions, human communities and societies, and other dimensions of marine ecosystems. The approach involves the input of scientists, policy experts, and persons who fish or in some manner benefit from fishing and, as noted throughout this text, an incremental and adaptive approach is being used to implement the FEPs across the region.

The Ecosystem Policy Workshop is reviewed in depth in the following pages. In brief, workshop participants reviewed the cross-jurisdictional and cross-cultural settings that are characteristic of the region, and they discussed options for enhancing the ecosystem approach in each island group. Participants once again defined marine ecosystems to include humans and their institutions, and they examined the needs and interests of indigenous fishing practitioners and others in this context. Finally, the group discussed needs and opportunities for ecosystem research and long-term monitoring in the Western Pacific.

As reviewed in subsequent sections, participants generated a number of immediately practical results on the final day of the workshop. These included the following: (1) policy options for meeting the Council's goal of empowering communities and working with local governments to develop place-based fishery management plans; (2) viable means for establishing effective long-term consultation with communities through the Council's Regional Ecosystem Advisory Committee (REAC) process; (3) recommendations for documenting traditional ecological knowledge (TEK) through effective and culturally sensitive collaboration with indigenous practitioners; and (4) possible opportunities for acquiring funding and deploying human resources that would enable long-term ecosystem research and monitoring across the region.

Organization of the Chapter

Following this introductory discussion, summaries of notes and transcripts recorded during the Ecosystem Policy Workshop are provided. The materials are organized and presented in chronological sequence.

A series of presentations were given during the morning of the first day of the Ecosystem Policy Workshop. These summarized the previous two workshops and provided the context needed to inform subsequent discussion of the region's fisheries and associated management challenges. Facilitated sessions were held during the late morning and afternoon hours to aid in developing an integrated science framework for meeting the Council's ecosystem information needs and management objectives.

During the morning hours of Day Two of the event, experts discussed the various challenges confronting fishery scientists and managers across the region, and approaches that may assist the Council as it develops and implements the new FEPs. Specific policy issues were addressed through facilitated interaction during the afternoon hours. Topics included the following: (a) challenges associated with introducing a new system of management in a region of many agencies and jurisdictions; (b) involving indigenous practitioners and other persons and groups in the management process; and (c) implementing the ecosystem approach through valid long-term research and monitoring.

Discussions during Day Three of the workshop were focused on review and synthesis of findings and recommendations generated during the previous days of the event. A discussion of how to assist the Council in its Regional Ecosystem Advisory Committee process was held during the morning hours. This was followed by facilitated discussion and concluding prioritization of information and policy recommendations for developing and implementing an integrated approach to EAFM in the Western Pacific.

The summary section of this chapter revisits the outcome of the ecosystem policy workshop in the context of the two previous events, and it reiterates and contextualizes policy options and recommendations generated during each event.

References

Berkes, F. (1999) *Sacred Ecology—Traditional Ecological Knowledge and Resource Management*. Taylor & Francis Group, London.

Brady, J. (2004) An Overview of the Alaska Salmon Management Program. Technical Report: Wild Salmon Center. North Cape Fisheries Consulting, Anchorage.

Ecosystem Principles Advisory Panel (1999) Ecosystems Based Fishery Management: A Report to Congress by the Ecosystem Principles Advisory Panel. US Department of Commerce, National Oceanic and Atmospheric Administration, National Marine Fisheries Service, Silver Spring.

Hilborn, R. (2006) Fisheries success and failure: the case of the Bristol Bay salmon fishery. *Bulletin of Marine Science*, 78, 487–498.

Hilborn, R., Quinn, T. P., Schindler, D. E., et al. (2003) Biocomplexity and fisheries sustainability. *Proceedings of the National Academy of the Sciences (PNAS)*, 100(11), 6564–6568.

Impact Assessment, Inc. (2007) Summary of socioeconomic trends and current conditions in the North Aleutian basin planning area. Prepared for the Information and Research Planning Meeting for the North Aleutian Basin. US Department of the Interior, Minerals Management Service, Alaska OCS Region, La Jolla, CA, and Honolulu, HI.

Kirch, P. V. and Hunt, T.L. (1997) *Historical Ecology in the Pacific Islands—Prehistoric Environmental and Landscape Change*. Yale University Press, New Haven.

Poepoe, K., Bartram, P., Friedlander, A. (2003) The use of traditional knowledge in the contemporary management of a Hawaiian community's marine resources. In: *Putting Fisher's Knowledge to Work*, (eds. N. Haggan, C. Brignall, L. Wood). *Fisheries Centre Research Reports*, 11(1), 328.

Sibert, J. and Hampton, J. (2003) Mobility of tropical tunas and the implications for fisheries management. *Marine Policy*, 27, 87–95.

Vitousek, P. M. (1995) The Hawaiian Islands as a model system for ecosystem studies. *Pacific Science*, 49, 2–16.

Chapter 15
Ecosystem Policy

The final event in the Council's ecosystem workshop series was held during early January 2007. Local, regional, and national policy and topical experts were convened for three days to assist in synthesizing output from the preceding workshops and to develop viable ecosystem policy options for use in the Council's fishery ecosystem planning process.

Objectives and Issues

A critical objective of the event was to discuss the challenges of implementing new marine resource management policies in the diverse social and biophysical settings that are characteristic of the region. The workshop involved deliberation on three basic issues of relevance to planning and implementing an ecosystem approach in the Western Pacific.

Participants discussed the concept of institutional ecology and related governance issues with the intent of identifying policy options for maximizing the potential benefits of an ecosystem approach to management in the cross-jurisdictional and cross-cultural settings that are characteristic of the region. A range of challenges was addressed in this regard, including issues of scale, institutional inertia, interagency coordination and sustained allocation of fiscal resources in support of a new system of management, and pursuit of equity and fairness in resource decision-making processes.

The group also examined policy options for addressing the needs and interests of indigenous fishing practitioners and other resource user groups across the region. Special attention was given to the Hawaiian system of managing resources in and adjacent to ahupua'a or political land divisions within which available resources from mountain to sea were and are produced, managed, and utilized, including resources from the deep sea (e.g., see Kirch 1985, Minerbi 1999). This system provides an example of the potential value of local monitoring and management of marine resources, and regional representation of the needs and interests of the human constituents of marine ecosystems.

Finally, participants discussed options for enhancing the benefits of fishery ecosystem research and monitoring in the region. This discussion was particularly important in that it was intended to assist the Council as it and its constituents increasingly address connections within and between biophysical and social components of the region's marine ecosystems.

During the course of the workshop, participants examined many of the institutional challenges associated with integrating ecosystem principles into fisheries management,

Ecosystem-Based Fisheries Management in the Western Pacific, First Edition. Edward Glazier.

and they ultimately developed various priority recommendations for implementing the approach in the region.

Overview of Findings

Although change in the marine environment is associated with many factors and processes, it was recognized that fishery managers are best equipped to influence humans and the effects of their activities. Thus, the ideal focus of management agencies was seen to be upon humans and their position in and relationship to marine ecosystems, a situation that clearly warrants additional attention to social science applications to fishery management.

Workshop participants agreed that policy makers and managers need to define essential ecosystem terminology to maximize understanding across the biophysical and social sciences and to reduce uncertainty in the definition of management objectives. Similarly, it was determined that ecosystem-based resource management will be enhanced when the conceptual and physical bounds of marine ecosystems are clearly delineated. Participants also agreed that efforts to increase rapport between scientists, managers, and persons pursuing and using marine resources would serve to minimize conflicts, and thereby enhance the chances that the new form of management will succeed.

Organizers of the workshop sought the participation of those who could effectively provide expertise on a variety of topics of particular relevance to fishery ecosystem planning in the Western Pacific. Some participants had been involved in the previous events and were asked to inform further deliberations on ecosystem concepts and marine policy based on their generalized expertise. Others were asked to attend based on specific geographic or topical expertise. As such, a combination of generalized and regional expertise was brought to bear on the issues at hand.

Following a brief round of introductions, the Ecosystem Policy Workshop was initiated in earnest. This and subsequent sections summarize the event and provide context through which to better understand the potential benefits and challenges of the ecosystem approach in the Western Pacific.

References

Kirch, P. V. (1985) *Feathered Gods and Fishhooks: An Introduction to Hawaiian Archaeology and Prehistory*. University of Hawaii Press, Honolulu.

Minerbi, L. (1999) Indigenous management models and protection of the ahupua'a. *Social Process in Hawaii*, 39, 209–225.

Chapter 16
Opening Presentations

A Place-Based Approach to Ecosystem Management

Kitty Simonds, Executive Director, Western Pacific Regional Fishery Management Council

The Western Pacific and its respective islands and archipelagos are unique and most appropriate for implementation of a place-based approach to ecosystem fisheries management. The ecosystem approach will require close relationships with existing government agencies and nongovernmental entities. Moreover, it will be a particularly appropriate means for empowering communities and for ensuring that long-accumulated traditional and local knowledge of marine ecosystems and resources is available for purposes of effective management.

Considerations regarding community involvement are now codified in the recently reauthorized Magnuson-Stevens Fishery Conservation and Management Act (MSFMCA), and a series of Puwalu or conferences was being held in Hawaii to more fully incorporate Native Hawaiian perspectives into the fishery management process in the region. There will be potential for bureaucratic challenges in implementing an ecosystem approach to management in the Western Pacific, which calls for workshop participants to muster their expertise to assist the Council in its efforts.

Western Pacific Context: Associating Ecosystem Principles

Samuel Pooley, Director, NOAA Fisheries; PIFSC

Ecosystem science, while novel in some ways, actually develops directly from the kinds of research that have been undertaken in the region over the past decades. Moreover, it will involve many, if not all, of the same challenging issues as have been addressed by the existing mandates.

Challenges that may arise will most likely be associated with (a) effectively inspiring institutions already deeply engaged in research and management of complex natural resource issues to engage a new paradigm which is not without uncertainties; (b) the costs of administering new programs given ongoing fiscal demands; and (c) coordination of efforts to implement a new system of management in a large and complex multijurisdictional region.

Ecosystem-Based Fisheries Management in the Western Pacific, First Edition. Edward Glazier.

Integrating Social Sciences

Michael Orbach, Duke University Marine Lab

The social sciences have as much or more to contribute to ecosystem science and an ecosystem approach to management as do the biophysical sciences. This requires development and use of parallel language and concepts. For instance, the term and concept of *connectivity* that is increasingly used to describe relationships between physical components of marine systems is also useful for describing the way people interact with the ocean and its resources, with each other while pursuing those resources, and with the institutions that govern those activities. Similarly, the concept and term *resilience* used to describe biophysical responses to sources and vectors of change can be used to describe social responses to sources and vectors of change, including those associated with marine ecosystems.

While defining marine ecosystems in terms of integrated biophysical, human, and institutional components requires attention to a larger and more complex field of inquiry, it can and must be accomplished. The attributes of biophysical systems, populations of user groups, and government institutions can be defined and mapped, and the symbiotic relationships between them can be deciphered and analyzed. Significantly, regulatory institutions cannot directly affect the biophysical environment. Rather, the biophysical environment is indirectly affected through mediation of human behavior.

Two basic questions for workshop participants to consider as the meetings moved forward are as follows: (1) how does an ecosystem approach to fisheries management (EAFM) differ from single species management and (2) what is the potential timeline for implementing the new approach; that is, at what point in the existing regime should implementation of new principles and approaches begin? Regarding the latter, a place-based island or archipelago-centric approach may prove timely. This would ideally emphasize the human and institutional ecology of the subregions, and connections between people and resources in those areas. This principle is at the heart of an ecosystem approach to fishery management and is conceptually opposite from concepts underlying single-species management approaches, since these generally emphasize the biophysical resources and factors and work toward the people, often almost incidentally.

While all should be sensitive to the fact that National Oceanic and Atmospheric Administration (NOAA) and the councils have well-specified authorities and responsibilities, there is latitude for restructuring the system. This could involve development of partnerships with constituents and agencies not generally or heretofore addressed in the existing approach to fishery science and management.

There is a need to bridge differences between the natural and social sciences in moving toward an effective new approach to fishery management in this and other regions. All parties involved in fisheries management should consider ways in which the outputs from the natural and social science workshops are complementary and ways in which they differ, and how the sciences might best be integrated to enable a more holistic approach to ecosystem management across the unique and diverse archipelagic subregions that comprise the Western Pacific.

Overview of Historic US Ecosystem Management

David Fluharty, Professor, University of Washington School of Marine Affairs

As of the late 1980s, the term "ecosystem management" was not yet widely used in scientific literature. Today, however, articles on ecosystem management and its implications abound in a variety of journals and reports. This change began in the late 1980s when a federal report was written to describe the inability of fishery managers to resolve certain issues under the then–current fishery management plans. These related to the need to control for a variety of environmental factors impinging on assessment of fishery resources. Although the authors described the potential merits of an ecosystem approach to solving the issues, administrators did not act on the findings.

In 1996, National Marine Fisheries Service (NMFS) appointed a 20-person panel to study the potential applications of ecosystem principles in US fisheries management. The panel agreed that more effective control of marine fisheries was in order, and that this could be accomplished through better enforcement of regulations, monitoring target and bycatch harvests more carefully, and through calculated control of harvest capacity. It was also deemed that a number of prerequisites would need to be satisfied if the ecosystem approach were to reach a point at which it could respond to some of the problems being encountered in the management of marine fisheries in the United States. The panel defined a mission and developed seven principles that could guide the future of an ecosystem approach to fishery management. These are as follows:

- Ability to predict ecosystem behavior is limited.
- Ecosystems have thresholds and limits affecting ecosystem structure.
- If limits are exceeded, changes can be irreversible.
- Diversity is important to ecosystem functioning [debated].
- Multiple time scales interact in and among ecosystems.
- Components of ecosystems are linked.
- Ecosystem boundaries are open.
- Ecosystems change with time.

The panel struggled with key questions associated with EAFM. These included the following: (a) how to deal with different scales of activities; (b) how to deal with open boundaries; and (c) what kind or level of change is acceptable? The panelists concluded that any policy advice generated through their efforts should be unbiased, particularly as regards evaluation of the effectiveness of a particular change in management. Panelists recommended that NMFS apply a precautionary principle in implementing ecosystem management, and that the agency seek to learn from experience and consider incentives for establishing real change in the way fishing fleets operate. Regarding the latter, panelists agreed on the importance of understanding what motivates people to comply with regulations. It was determined that any changes in the management process must enable equity and fairness—outcomes that are not typically concerns in a top–down form of resource management.

The Panel developed a fishery ecosystem plan (FEP) to provide ideas for coordinating the efforts of regional fisheries managers to move beyond the status quo. Although it was well received by Congress, the plan was effectively shelved because of budgetary and administrative constraints.

Since 2003, however, there has been a proliferation of ecosystem management initiatives around the nation, including the North Pacific Council's Aleutian Islands Fisheries Ecosystem Plan; NOAA's approval of the Chesapeake Bay FEP; funding of the West Coast, Gulf, and New England Councils to begin an FEP process; the Pacific Fishery Management Council's decision to start the FEP process; and, more broadly, NOAA's own interest in ecosystem approach to management. The NOAA initiative to address the Gulf Hypoxic Zone was mentioned as another example of integrated ecosystem assessment.

The US Ocean Policy Report includes recommendations for moving away from single species management. It also recommends doubling the amount of funding for NOAA, creating regional ocean ecosystem councils, and refining fishery management to use an ecosystem approach. The Ocean Policy Council (OPC) was established as a result of those recommendations. Most recently, Vice-Admiral Conrad Lautenbacher announced a new Regional Collaboration strategy. Although it is not yet fully funded, the plan calls for integrating ecosystem assessment, resilient coastal hazards management, and integrating weather and climate approaches.

While some scientists argue that an ecosystem approach is too difficult to effectively define and implement, it may be readily attainable when the focus is on managing human behavior rather than managing the entire ecosystem. The recent emphasis on an ecosystem approach to management may potentially bring about the following changes:

- Marine fisheries will be managed for abundance, not scarcity.
- Fishing capacity and employment will likely diminish.
- Marine fisheries will involve higher levels of income and use of more sophisticated technology.
- Fishing practices leading to extensive impact on habitat will be replaced by alternative techniques.
- There will be greater use of spatially explicit management measures.
- Fisheries restrictions and regulations will serve to meet corollary goals, such as conservation of biodiversity.

EAFM in the Western Pacific—an Example of Joined-up Ocean Governance or a Ball of Confusion

Paul Dalzell, Western Pacific Regional Fishery Management Council

A range of factors and concerns suggestive of the need to implement an EAFM in the Western Pacific include the following: (a) sufficient understanding of the population dynamics of protected species, such as monk seals and green turtles, requires attention to a wide range of physical and human environmental factors; (b) evaluation of the effectiveness of Marine Protected Areas needs to be undertaken, thus requiring a range of valid data; (c) changes in fishing mortality, such as that associated with jack species in Hawaii, requires greater attention to ecosystemic factors; (d) the population dynamics of exotic invasive species, such as the explosive growth of ta'ape, which has a limited local market, need to be better understood in more holistic context; and (e) the changing human demography of the Western Pacific and associated changes in fishing and reef collecting practices require attention to a broad array of factors and variables.

The Council is currently in the process of reorganizing extant management measures and regulations and its overall institutional framework as it shifts from a species-based approach to a place-based archipelagic approach to management. As noted elsewhere in this chapter, FEPs are being developed for each of the archipelagos and for the region's pelagic fisheries. The latter will focus on basin-scale and Pan-Pacific stock issues, thereby involving the least amount of change to the existing system of management.

Ten objectives formulated to guide implementation of an EAFM policy in the region are as follows:

1. To maintain biologically diverse and productive marine ecosystems and to foster the long-term sustainable use of marine resources in an ecologically and culturally sensitive manner through the use of a science-based ecosystem approach to resource management.
2. To provide flexible and adaptive management systems that can rapidly address new scientific information and changes in environmental conditions or human use patterns.
3. To improve public and government awareness and understanding of the marine environment in order to reduce unsustainable human impacts and foster support for responsible stewardship.
4. To encourage and provide for the sustained and substantive participation of local communities in the exploration, development, conservation, and management of marine resources.
5. To minimize fishery bycatch and waste to the extent practicable.
6. To manage and comanage protected species, protected habitats, and protected areas.
7. To promote safety of human life at sea.
8. To encourage and support compliance and enforcement with all applicable local and federal fishery regulations.
9. To increase collaboration with domestic and foreign regional fishery management and other governmental and nongovernmental organizations, communities and the public at large to successfully manage marine ecosystems.
10. To improve the quantity and quality of available information to support marine ecosystem management.

Actions the Council will need to undertake as it moves forward with its ecosystem planning process and with implementation of an ecosystem approach over the course of time were also identified. These included the following:

1. Develop appropriate indicators of ecosystem conditions, including socioeconomic indicators;
2. Develop place-based conceptual food web and population models with various dynamic forcing mechanisms;
3. Implement new data collection and analysis programs to better understand ecosystems;
4. Continue adaptive management, using best available scientific information;
5. Incorporate traditional knowledge into management;

6. Continue and increase participation in international management forums;
7. Participate in meetings and workshops with neighboring nations; and
8. Increase community participation in the Council process.

The Western Pacific Council has created several new advisory committees, panels, and teams to meet the above objectives. These include advisory panels composed of fishery participants and other knowledgeable persons in each archipelago; planning teams for reviewing the FEPs and plan administration; standing committees composed of Council members; a Community Demonstration Projects Advisory Panel; and a Community Development Advisory Panel. Finally, an REAC is being established for each archipelago. These will be composed of persons from various branches of government, nongovernmental organizations, and others involved or interested in the ecosystem approach to fishery management in the region (Figure 16.1).

Given the many layers of involvement in the management process under the new FEPS, there is potential for "bureaucratic paralysis." The number of governing element involved in the effort and sometimes conflicting agendas may make for a challenging arena for effective management, and solutions for this potential problem will need to be addressed in the planning process and as the effort moves forward. Discussion of challenges associated with the prospective institutional complexities of an ecosystem approach to management is an important objective of the current workshop.

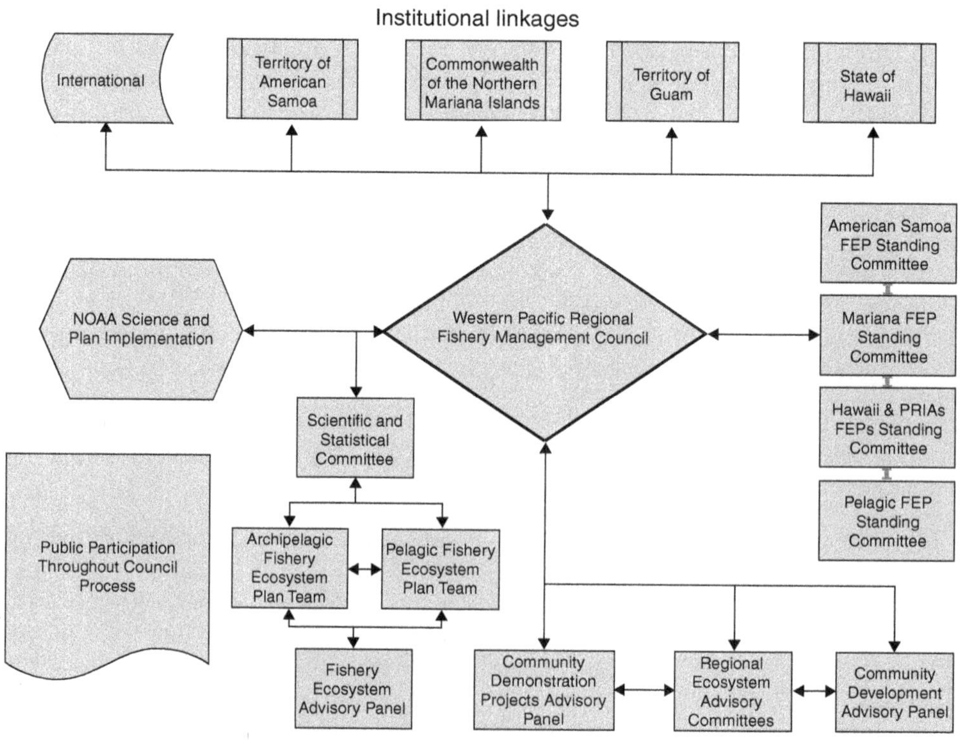

Figure 16.1. Institutional linkages between governing elements under the new FEPs.

Review of the Ecosystem Biophysical Workshop

David Kirby, Senior Scientist, Secretariat of the Pacific Community

Several important shared understandings emerged during the course of workshop discussions concerning the role of ecosystem modeling: (1) ecosystems are complex systems and the quantitative integration of available knowledge will inevitably lead to complex models; (2) details that are difficult to model can nevertheless have potentially significant effects on outcome; (3) some models appear to be capable of addressing great complexity but can yield spurious results and/or questionable precision; (4) complex models may be used to explore strategic trade-offs and risks even if their results are not precise; (5) humans tend to make decisions based on what they understand, and thus there is a need for models that are relatively straightforward; moreover, communication at the interface between science and management and between science and policy must be clear and precise; and (6) language regarding the structural assumptions and limitations of data, indicators, and models must be made patently clear.

Some basic recommendations would enhance implementation of an EAFM in the Western Pacific: (1) Once scientists know what policy options or potential outcomes should be modeled, they should determine what data should be obtained, what indicators should be monitored, and what approaches should be undertaken to support analysis of such options; (2) managers should be able to use such data and accompanying analyses to think through potential policy outcomes and effects in both biophysical and social terms; and (3) hard trade-offs and decisions are inevitable; scientists ideally will work to reduce uncertainties associated with data, models, and indicators.

Operational objectives for the Council's FEPs include the following: (1) conserving and managing the [target] species; (2) minimizing bycatch; (3) managing trade-offs; (4) accounting for feedback effects; (5) establishing appropriate ecosystem boundaries; (6) maintaining ecosystem productivity and balance to ecosystem structure; (7) accounting for climate variability; and (8) using adaptive approaches to management.

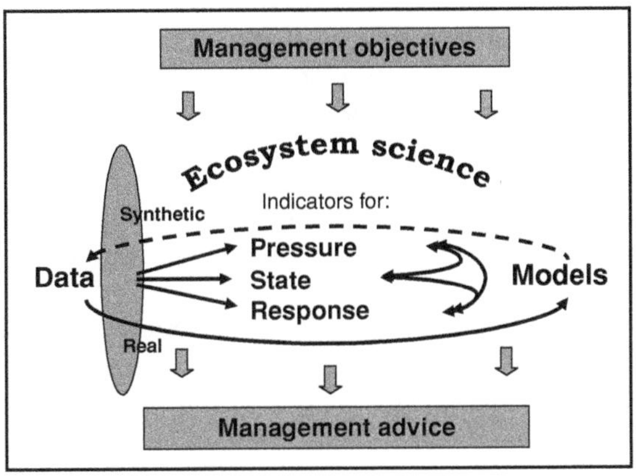

Figure 16.2. Conceptual model for the role of ecosystem science in support of fisheries management.

Figure 16.2 illustrates the role of ecosystem science under an EAFM as envisioned by participants at the biophysical workshop. Under the parameters of this model, the externally derived management objectives are supported by the data, models, and indicators of relevance to the needs of the Council.

Several weaknesses of the meeting include the fact that the meeting itself had no specific policy questions to consider, without which discussion of specific data, indicators, and modeling approaches were necessarily abstract. Moreover, the synthesis of the meeting as described in the proceedings was not so much a summary of what had been previously discussed as an overview of policy-making in the context of an ecosystem approach to management. Ecosystem science could potentially be progressively ignored over the course of time.

On a more positive note, workshop participants demonstrated the scientific basis for an effective EAFM. Provided that management goals are made clear and funding is duly allocated, there is no reason why natural science should not continue to contribute significantly to the management of marine resources across the Western Pacific.

Overview of Hawaii Archipelagic Ecosystem Research Plan (HARP)

Frank Parrish, NOAA Fisheries, PIFSC

Hawaii archipelagic ecosystem research plan (HARP) is a long-term multiagency research plan designed to address ecosystem-relevant information needs in and across the Hawaiian Archipelago. HARP was initiated following the Northwestern Hawaiian Islands Symposium, with the goal of defining new and emerging research priorities and advancing scientific inquiry in support of an ecosystem approach to resource management. The organization is guided by an Executive Management Team, a Steering Committee, and a Drafting Team. The HARP mission is as follows:

> Sustainable conservation and management . . . through improved understanding of the unique physical and biological attributes of the Hawaiian archipelagic marine ecosystems, their interconnected dynamics, and their interactions with human beings.

Among the initial tasks undertaken by the HARP Drafting Team was provision of a plan of action and timeline for developing an ecosystem research plan (see Figure 16.3). During this process, the following research principles were identified:

1. Select testable hypotheses consistent with vision statement;
2. Understand physical, biological, and social processes at an archipelagic scale;
3. Employ comparisons between the Main Hawaiian Islands (MHI) and Northwestern Hawaiian Islands (NWHI);
4. Acknowledge understanding of the human component as essential to long-term ecosystem management; and
5. Conduct research at a scale and intensity that will advance ecosystem modeling and forecasting.

The HARP Drafting Team currently anticipates a 10-year time frame for HARP, though a starting point has yet to be determined. Requisites for any proposed action

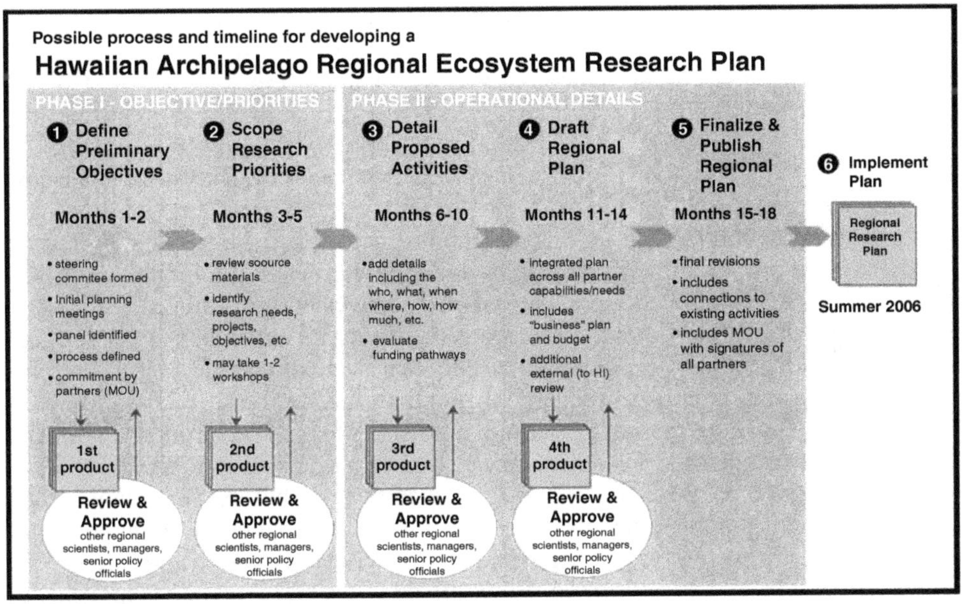

Figure 16.3. Timeline for developing HARP.

include establishment of a panel of international experts that would provide independent review of the plan, suggest revisions, and request modifications to complement existing national and international resource management priorities.

The HARP Drafting Team examined, compared, and analyzed several ecosystem research plans from other regions. In so doing, the team identified six notable process-oriented priorities and themes. These are as follows:

1. Ecosystem indicators and metrics, which include physical, chemical, biotic, and remote sensing indices.
2. Native biodiversity (endemic and vanishing) and invasive species.
3. Connectivity, which includes hydrodynamics of the archipelago, movement studies on adult taxa, transport modeling, and population genetic structure.
4. Monitoring of human interactions and anticipation of human impacts to the marine ecosystem.
5. Sustainability, resilience, and recovery, which focus on pathways of and modifiers to resilience.
6. Modeling and forecasting, which include identifying the variables, resolution, and relevant scale for ecological models; reviewing existing models and conduct of gap analysis; evaluating parameter research and model validation; and developing a capacity for ecosystem forecasting.

The Drafting Team also conducted a series of focus group interviews with local experts to obtain structured input regarding each of the themes listed above. It is notable that biological, physical, and social considerations are integrated into each.

Review of the Ecosystem Social Science Workshop

Susan Hanna, Professor of Marine Economics, Oregon State University Coastal Oregon Marine Experiment Station

The purpose of the Council's Ecosystem Social Science Workshop was to discuss social science requirements for EAFM in the Western Pacific. Humans were defined as integral to rather than merely exogenous factors exerting effects on marine systems. Workshop participants represented a wide variety of disciplines, agencies, and institutions. This enabled a diversity of perspectives and experience relevant to the task of discussing the human dimensions of fishery management in a large and complex region. Although a single unifying theme was lacking at the outset, the participants clearly united across disciplines through agreement on three basic themes: (1) management is about managing people, not fish; (2) incentives are paramount to implementing effective fisheries management; and (3) context is everything.

The first theme recognizes the importance of context in that long-standing fishing practices and management concepts may be usefully integrated into existing management efforts. Traditions and experiential knowledge are at once varied and extensive in the Western Pacific, and these may contribute to adaptive and integrative approaches to management across the region.

Second, management objectives ideally will direct the course of social science research and analysis. In other words, social scientists should not impose their own research interests in this area unless they are relevant to the needs of decision makers. For example, scientists could design culturally appropriate protocols for defining sustainability within a particular context and then again for helping to understand the potential for behavior that would promote such sustainability. Social scientists could be stationed so as to monitor the behavior of resource users and thus enhance the potential for adaptive management. There is also potential for social scientists to assess the need and ideal means for public input, and for addressing the feasibility of cooperative management of marine resources. Community-based management is as institutionally complex and challenging as any other form or dimension of natural resource management.

A third theme recognizes that social scientists must monitor human interactions within ecosystems in terms of the direct flow of goods and services. In order to do that, social scientists will need to develop valid social and economic indicators (the fourth theme emerging from the workshop). Indicators should relate clearly to management objectives, involve distinct measures, build on existing practices, utilize the benefits of local ecological knowledge, and assume some measure of commonality across fisheries.

The fifth point or theme involved indirect ecosystem relationships, such as tourism. These are as important to monitor as are direct relationships, such as those between fishers and fish. Indirect linkages or factors warrant consideration as they bear great potential for affecting biophysical systems or changing the social or economic context within which marine fisheries are undertaken.

There can be tension between advocates of flexible and responsive management and persons charged with making sure established mandates and regulatory processes are closely followed. In other words, while an adaptive or flexible approach is logical and useful in some cases, it can interfere with management requirements (such as sufficient time and due notice for public comment).

Finally, there is a need for building social science knowledge into management. This should occur directly and systematically and can be accomplished by meeting the following objectives: (a) clearly identifying needs for social science research and related data and modeling; (b) setting priorities for incorporating social science data and analyses into existing plans and programs; (c) adaptively responding to emerging challenges via social science research and data; and (d) monitoring and evaluating the outcomes of such work in terms of the degree to which it enhanced efforts to better manage the fishery in question.

Report from the Western Pacific Ecosystem Social Indicators Working Group

Stewart Allen, NOAA Fisheries; PIFSC

The two day Social Indicators workshop, held in Honolulu in 2006, began with a series of preplanning meetings and discussions that followed the Ecosystem Social Science Workshop held in January 2006. The initial discussions were intended to lay the groundwork for an effective meeting later in the year.

The first day of the actual workshop focused on review of the vast literature on social and economic indicators. Participants explored the Global Socioeconomic Monitoring Initiative for Coastal Management (SocMon), the document titled *How's Your MPA Doing?*, and a number of important texts and journal articles reviewing the utility of indicators developed to gauge social change resulting through and from a variety of vectors and sources. The premise of the exercise was that review of indicators literature would likely yield a deeper understanding of the rationale for and suitable use of specific indicators in the context of natural resource management.

Participants developed several scenarios to enable grounded discussion of the utility, applicability, and predictive capability of specific social and economic indicators. A template was developed to describe the name, function, and relationship of various indicators. This facilitated discussion of the indicators vis-à-vis Council management actions, issues of measurement, and data collection and analysis. The process of conceptualizing the indicators in this systematic way was highly useful.

Several ways in which social indicators can contribute to fisheries management in this and other regions were determined. For instance, indicators could be used to monitor impacts predicted to occur in an Environmental Impact Statement, and to identify populations and resource use trends and conditions relevant to Council actions over the long term.

Working through the template, participants focused on the FEP objectives and measurements capable of gauging progress toward achieving Council objectives. To help simplify the task, participants separated the broad FEP objectives into parts. For example, by looking at just one part of the following phrase, "maintain biologically diverse and productive marine ecosystems and foster long-term sustainable use marine resources in an ecologically and culturally sensitive manner," the group was able to produce nearly 40 potentially useful social and economic indicators, including the extent of use of traditional ecological knowledge in decision making.

Following is a list of 10 principles or guidelines determined by workshop participants to facilitate an effective approach to development and use of valid indicators. These include the following:

1. Indicators need to be developed and in some cases measured with involvement of a broader community to ensure they are meaningful, comprehensible, and useful for other purposes (such as building social capital).
2. Indicators constitute an important component of a comprehensive social science research plan for implementing and monitoring the Council's FEPs; indicators and indicator-related research findings should also be part of the social science information made available to decision makers and the public.
3. A principal method for developing and measuring several of the social science indicators should be a broadly distributed household survey; this would be conducted in addition to research directed toward specific sectors (as is currently the norm).
4. A valid science of ecosystems must view human beings as pivotally important in marine systems and as important elements of trophic systems; humans are *not* exogenous elements of otherwise "natural" systems.
5. Just as each indicator will have characteristics that render it useful for analysts and managers, the full set of social science indicators eventually adopted should be matched to a balanced set of methods, scales, and types of information requirements (both primary and secondary).
6. Social and economic indicators should be developed and considered in the context of biophysical indicators.
7. Ecosystem management and ecosystem indicators should provide avenues for bringing constituencies and jurisdictions together, particularly since comanagement and community management approaches are likely to be necessary for effective ecosystem management.
8. Indicators can be "value-free" or "value-laden" depending on what they are measuring and how they are phrased; both scientists and managers need to be sensitive to this issue.
9. Causality can be difficult to determine through use of indicators; if important, the level of accuracy needed should be made explicit.
10. It is critical to pilot test the indicators and emplace a data management system through which to care for and readily analyze the data (a time-consuming but necessary undertaking).

Group Discussion

A variety of issues were reviewed as follow-up to the presentations. A recurrent discussion addressed the appropriate scope of ecosystem fisheries policy. A number of participants asked whether the mission of scientists and managers in that case was to undertake (a) ecosystem management or (b) an ecosystem approach to fishery management. Although there was general agreement that the mission was ecosystem-based fishery management, there was extensive discussion about terrestrial linkages to marine systems, especially in the context of islands, and whether such linkages fell within the realm of responsibility of fishery managers. The group reached some consensus on that issue, as it was determined that relevant processes on land (such as pollution in upland and estuarine areas) were indeed important considerations within a holistic or ecosystemic approach to management of marine fisheries.

A similar discussion addressed the issue of public trust and the scope of public interest in the management of marine resources in the Western Pacific. Again, it was determined that the scope of interest exceeded those who involved in direct pursuit and use of the resources. The general public also had a stake in the future course of management, and various human activities not related to fishing had the potential to affect marine resources and fisheries. It was felt that expanding the scope of management considerations to include the general public and the specific interest groups was in keeping with the ecosystem approach which, once again, addressed a larger set of relationships than had heretofore been addressed in the recent history of fishery management. The Council's Regional Ecosystem Advisory Committee (REAC) process was discussed as a means for increasing the scope of input and involvement in the management of marine resources across the archipelagos.

Facilitated Discussion: Interactions between Scientists and Policy-makers

Following the presentations and associated interaction among participants during the morning and early afternoon hours, John Kirkpatrick of Belt Collins Hawaii, Ltd. facilitated group discussion regarding the interface of science and policy in the context of an ecosystem approach to fishery management. The discussion ultimately focused on the appropriate role of scientists in the process of resource management and policy-making.

Some points of common ground were identified between biophysical and social scientists present at the workshop. For instance, both groups accepted that science involves a reasonable level of adherence to the principles of objectivity, replication, and refutability. Both regarded sufficient data and valid indicators as essential for good science.

Finally, while there was general agreement that any form of science should be distinguished in some manner or to some extent from the subjective process of making management decisions, participants tended to acknowledge the need for ongoing dialog between scientists and managers. This was effectively stated by Dr Hanna as follows:

> Different people play different roles at different times, and when you are designated as one of the policy-makers or as one of the science advisors, you fulfill one role or the other. However, it is very helpful for both policy-makers and scientists to be versed in the language of the other, to engage in the kind of dialogue that allows each to be informed by the other. But that is very different from a science advisory body taking on the responsibilities of making policy decisions, or decision-making bodies taking on the responsibility of muddling through science.

Participants were asked to consider what means could be used to bridge biophysical science, social science, and traditional knowledge as the ecosystem approach to the management process moves forward. There was initial consensus that integration and effective use of what is at times disparate and at times complementary information is one of the important roles of the Council as it implements the ecosystem approach across the region. But the actual mechanics of this process were seen as likely to be somewhat variable and to some extent uncertain. It was stated that at times integration of scientific data falls under the purview of Council plan teams, and at others it occurs during development of environmental impact statement (EIS) or similar large-scale documentation and assessment efforts. It was also asserted that a reserve of data and knowledge is available at the regional science center and via Council staff.

Extensive attention was applied to the potential role of geographic information system (GIS) as a potentially highly effective means for integrating ecosystem-relevant data from each of the sciences. Some participants asserted that much of the data necessary for implementing an EAFM already exists, but that an effective tool is needed to integrate such information so that it can be widely and readily used. It was asserted that geography is the discipline best equipped for integrating highly variable data and applying it in a place-based management regime such as that being developed by the Council, and GIS was advanced as the ideal tool for so doing. The potential value of GIS in this context was asserted by a number of persons, including John Petterson of Impact Assessment, Inc., who stated the position as follows:

> If we're starting to map landings, effort, CPUE, and other variables in terms of geography, then we're moving toward an integrated data system. We are now able to depict whether the fishery is being conducted in the inshore waters or offshore zone, three miles, 32 miles, benthic or pelagic, etcetera. Once we start mapping these sorts of variables against the many factors and processes being addressed by the biologists and the ecologists—that is the nexus of interest. That is the link between the world of the biophysical sciences and the world of the social sciences. That is the point of integration. We need to look at the variety of data points and how they link up in the ecosystem . . . the data are linking us together.

Finally, there was discussion about the nature and roles of traditional knowledge, and appropriate ways in which such information may be integrated into the process of managing marine resources. It was made clear at the outset that while such information is often invaluable in its own right and an excellent complement to data generated by formal scientific research, it cannot and is not intended to replace data generated through such research. Examples of programs designed to incorporate local and traditional knowledge into formalized research were discussed at some length. Craig Severance framed relevant points of the discussion in relation to the Council's own nascent REAC process:

> . . . I would like to think that all scientists, including indigenous scientists, would be curious enough to want to learn how other people think and how people from other scientific disciplines think, because if we're going to work together on this, we have to understand where everyone is coming from. If the Council is going to take this process out following from what I view as a pretty successful exercise with the local Hawaiian community, and they're going to take it to the Samoan Council and to the Chamorro community and to the Carolinian community, and to the other communities in other parts of the region that are multi-ethnic in nature, they're going to have to do it in a way that's legitimate. Which means that respect for local knowledge is essential . . . An appropriate and effective means of integrating traditional ecological knowledge and formal scientific knowledge will also be called for . . .

Chapter 17
Policy for Indigenous Resource User Groups

The Ecosystem Policy Workshop involved extensive examination of challenges associated with effectively addressing the needs and interests of indigenous fishing practitioners and other fishery participants in the various archipelagic subregions of the Western Pacific. A series of presentations were interspersed with interactive and facilitated discussions.

Presenters discussed a variety of challenges associated with management of marine resource challenges in the Commonwealth of the Northern Marianas Islands (CNMI) and Guam. These reportedly include the following: (a) depletion of nearshore fisheries; (b) lack of enforcement; (c) commercial and residential overdevelopment; (d) diminishing fishing grounds; and (e) regional economic emphasis on tourism with concomitant deleterious effects on marine fisheries. Management problems in American Samoa reportedly include management inertia and contested maritime boundaries and fishing areas, among others. A variety of historical and contemporary challenges were described in relation to marine fisheries in the Hawaiian Islands.

Workshop participants anticipated that asking local agencies across the Western Pacific to work together on implementation of the Council FEPs would not be an easy task. For instance, it was thought that competing interests and agendas were likely to present various challenges. Workshop participants recognized the need to understand cultural aspects of management in the regions of interest, and it was suggested that incentives might be designed to enhance the efforts of the Council.

Participants made several additional recommendations for enabling a streamlined approach to an EAFM across the archipelagos. For instance, it was felt that management problems on Guam and CNMI might best be approached by appointing a liaison to help initiate dialogue between these somewhat culturally distinct island areas. Other recommendations include efforts to (a) improve enforcement of nearshore fishing regulations on the populated islands; (b) integrate local knowledge with formal science-based approaches to management; and (c) increase local involvement in the management process.

The issue of empowering island communities was discussed at length during the second day of the workshop. It was recognized that there is much local and traditional knowledge in the islands through which to better understand resources, ecosystems, and human use thereof. There was also discussion of the desirability of enabling meaningful local and regional participation in management of marine resources rather than imposing rules and regulations from the outside.

Finally, a range of issues associated with interagency cooperation were discussed at some length. Participants agreed that successful implementation of the FEPs will require the support of local agencies and institutions, and that attaining such support in a context of multiple interests and agendas may well be challenging. It was felt that

Ecosystem-Based Fisheries Management in the Western Pacific, First Edition. Edward Glazier.

there might be a role for social scientists in developing a better understanding of the sociocultural context of fishery management in the archipelagos, and that this could help streamline data gathering and incorporation of regional needs and interests via the Council's REAC system and other forms of local involvement in the management process.

Ecological Risk Assessment for Species Caught in WCPO Tuna Fisheries: Inherent Risk as Determined by Productivity–Susceptibility Analysis

David Kirby, Senior Scientist, Secretariat of the Pacific Community

Ecological risk assessment research and modeling work was recently undertaken for the Scientific Committee of the Western and Central Pacific Fisheries Commission. The effort attended to a need for information regarding fisheries-related interactions between target, at-risk, and incidental species typical of the pelagic fisheries in the region.

The productivity–susceptibility model employed used data regarding biological characteristics of the species of interest to estimate the degree to which they are interacting with a given fishery in the absence of precise catch or mortality estimates. Effects on a target species bear differential implications for other parts of the ecosystem, such as nontarget species. The model thus illustrated certain connectivity dimensions of pelagic fishery ecosystems. The work is particularly germane to ecosystem-based fishery management given recent legislation intended to address incidental catch of certain species in various parts of the region.

Observer data was employed to identify the species landed in pelagic fisheries across the region. Biological characteristics and indicators of productivity were assessed for each. Indicators included age-based metrics, maximum length, length at maturity, and reproductive strategy. Data regarding the condition of the fish when captured and whether it was kept or released were also used. Data were aggregated across the region for the period 2001 through 2005. Based on these factors, composite indices of risk and susceptibility were derived for a range of open ocean species.

Bird–fishery interactions usually result in avian mortality. Effective management measures would, therefore, ideally focus on preventing such encounters. Turtle encounters are not usually fatal and, as such, effective management measures would most effectively include postencounter treatment, including hook extraction and rest and recovery before release. Finally, encounters with sharks are not usually fatal. If live-landed sharks were promptly released, fishing-related shark mortality could decrease by some 30 percent.

While the modeling effort involved some subjective determination in terms of weighting the indices, the model does enable an objective means for ranking extent of susceptibility of a range of species to fishery interactions. The analysis can be applied to nonfin species as well, providing the existence of the requisite data.

Management and Policy Challenges in CNMI

John Gourley, Micronesian Environmental Services

The Northern Marianas are comprised of 14 islands: nine of which are volcanic in nature and five of which are comprised of raised limestone. The three inhabited islands—Saipan, Tinian, and Rota—are located in the southernmost part of the archipelago. The population is comprised of several large multiethnic communities. Indigenous residents are outnumbered by workers who have arrived from other regions; slightly less than half of the total population is indigenous Chamarro or Carolinian. About 90 percent of the total population of the region resides in Saipan. There are numerous cultural differences between the people of the Northern Marianas and Guam, even though the indigenous populations share the same ethnic background (Chamorro). Challenging economic conditions, a diversity of habitats and reef ecosystems, and the vast size of the archipelago reportedly lead to a range of natural resource management challenges.

The depletion of nearshore fish stocks is one of the most pressing fisheries issues of the day. In order to address this and other resource management concerns, policy makers in the CNMI region recently set forth several objectives. These include efforts to (a) improve enforcement of nearshore regulations for the populated islands; (b) increase public involvement in the process of managing resources across the region in a general sense; and (c) specifically enhance opportunities for public involvement in science-based approaches to research, monitoring, and the balanced development of resource management programs.

While regulatory processes have an effect on the manner in which fisheries in the region are conducted, a variety of additional factors indirectly influence them. For instance, the Farallon de Medinilla region is an important fishing ground for residents, but weather conditions limit access to that area for about six months of the year. Cost factors also present various constraints. For example, lack of available capital has thus far deterred local fishermen from purchasing large ice makers, which would have a dramatic effect on the capacity of local fleets to meet demands for fresh seafood. Fishermen currently must carry ice to the grounds at Farallon de Medinilla or go without.

There are two marine protected areas in the CNMI: the Managaha Marine Conservation Area (MMCA) and a protected area in Saipan. The implementation process for the MMCA is one that is working. The Department of Fish and Wildlife collaborated with local researchers to set up public outreach and education forums, to determine effective boundaries, and to conduct postimplementation monitoring surveys.

The 2005 Micronesian Challenge required governing bodies across Micronesian to commit to the preservation of at least 30 percent of nearshore marine areas and 20 percent of forested areas in each of the countries and territories across the archipelagos. While these percentages seem arbitrary rather than based on the tenets of science, many in the CNMI are taking the Micronesian Challenge very seriously and see it as a vehicle for advancing the rationale for establishment of marine reserves. Thus far, officials from Guam, CMNI, the Republic of the Marshall Islands (RMI), the Republic of Palau (ROP), and the Federated States of Micronesia (FSM) have signed on to the commitment.

Government officials of the CNMI are also discussing options for establishing marine reserves around the western seamounts. These are under Council jurisdiction. There is a need for improved understanding of jurisdictional responsibilities and the potential socioeconomic effects of establishing reserves in the EEZ surrounding the

CNMI. There appears to be significant momentum behind the Micronesian Challenge, with various nongovernmental groups supporting the effort.

Workshop participants should consider ways to successfully generate some commonality and solidarity between resource user groups on Guam and in the CNMI. Some benefits may follow from employing the ecosystem approach to bring representatives from these typically distinct political entities together for meaningful discussion. The Council might designate some key person to mediate between representatives from the two regions so as to facilitate dialogue needed to enact a successful common ground public forum.

It is often the case that expatriate professionals make comments at public hearings while long-term local residents do not. This is not a matter of indifference, but rather a cultural issue. Many local residents are not comfortable offering comments at public hearings. This means that comments often reflect the opinions and agendas of only a small percentage of interested persons, and a more encompassing approach may be warranted.

Management and Policy Challenges in Guam

Judith Amesbury, Micronesian Archaeological Research Services

Longstanding differences between Chamorro residents living on Guam and Chamorro residents in the CNMI are reportedly a political artifact of World War II, when Chamorros in the CNMI were used by the Japanese as scouts and interpreters on Guam. Efforts undertaken in 1969 to unite the two regions into a single political entity ultimately failed, in part due to historical animosity. It may be difficult to effectively address these differences in the event that persons from each of the two subregions are encouraged to interact in implementing the ecosystem approach across the region.

The US military recaptured Guam in 1944. Today, military interests are said to be pervasive and likely to increase dramatically over the next six years as approximately 8,000 military personnel relocate to the island with family members and various other persons needed to support an increase of this size. It is anticipated that the influx will lead to a 15 percent increase in population, with implications for the capacity of existing physical and service infrastructure and potential environmental effects associated with expansion of roads, construction of new sewage facilities, pressure on sources of potable water, and so forth.

The general nature of relations between the people of Guam and the US military has been alternately welcoming and negative. In the 1990s, some residents initiated a movement to remove the military and acquire lands taken by the federal government. But the downturn in the Asian economy reportedly led to a rethinking of this process on the part of some, in that military activities are central to the regional economy.

Relations between the government and local fishermen are also said to be periodically problematic. Establishment of several marine preserves around Guam were not welcomed by some, and some fishermen refute that the Division of Aquatic and Wildlife Resources enabled fair representation in its public comment process. Some fishermen may have not become aware of the hearings. As such, the protocol for disseminating public hearing notices may need to be reexamined.

Local fishermen are also said to be expressing claims of inequity since some areas are closed to extractive uses but allow other activities to continue. Some resident fishers are said to object to disturbances caused by recreational jet skiers. Although there is a ban on jet skiing in East Agana Bay during certain fish runs, some contend there is inadequate enforcement. Some assert that a resultant need to fish in distant or unfamiliar waters, and that safety and the availability of fish in such areas are issues of increasing concern.

Another concern involves apparent redundancy in coastal resource planning. Jurisdictional uncertainties sometimes lead to situations in which pressing or latent problems do not get addressed. Moreover, because tourism is critically important in the region, there has been a tendency toward overdevelopment, and hotel properties reportedly are tending to limit access to historically important fishing grounds.

But, the region is not without beneficial policies and situations. The Guam Fishermen's Co-op, the island's small boat commercial fishing cooperative, has reportedly established policies that combine conservation with good business. For example, the Co-op hopes to reduce pressure on reef fish by teaching pelagic fishing methods, using the vessel purchased through the Council's Community Demonstration Project Program. The Co-op also addresses issues associated with the quality and safety of seafood on Guam.

Empowering communities across the region leads to potential benefits throughout the various fisheries affected by management procedures. Some sort of shared management process is needed to ensure that the needs, interests, and values of residents are not overlooked in the years to come.

Management and Policy Challenges in American Samoa

Fini Aitaoto, Council On-Site Coordinator for American Samoa

Some of the important fishery issues, governance factors, and cultural attributes of relevance to implementation of the ecosystem approach to fishery management in American Samoa include the following: (1) there may be some level of administrative and jurisdictional inertia and/or confusion regarding a new approach to management in the region; (2) maritime boundaries and fishing areas between American Samoa and Independent Samoa are to some extent contested, and this may complicate any management efforts; (3) the large amount of fresh seafood imported from Independent Samoa is indicative of diminishing production in American Samoa; this suggests a need for deeper understanding of current trends and conditions in American Samoa fisheries; and (4) there is ongoing contestation between the US Department of Fish and Wildlife and Territorial government agencies regarding refuge policies at Rose Atoll; this may complicate collection and/or analysis of data regarding fisheries and related issues specific to the area.

The traditional *matai* system of governance was discussed in some depth, and it was asserted that this might present some challenges should the new system of resource management require significant changes to existing policies. Conflicts of authority can occur on occasion, and some measure of cultural sensitivity is warranted. For example, territorial government officials have, on occasion, attempted to restrict land use in areas where tenure or ownership is claimed by persons with ascribed status under the cultural and political parameters of traditional village life. In some cases, this can lead to an attempted overriding of previously exercised authority, with a range of problematic long-term implications.

The traditional mode of governance presents clear benefits in that chiefly authority can serve to put a swift and absolute stop to detrimental activities or behaviors associated with use of natural resources. Meanwhile, government agencies can merely issue a restriction, which may or may not yield the desired results. American Samoans and their established mechanisms of social control must be approached with respect and diplomacy if the matai system is to yield effects that are complementary to the objectives of fishery managers with responsibilities in the region.

The FEP should include formalized and culturally sensitive protocols and procedures, which would aid in implementation of the new approach to fishery management in the region. This would serve to help scientists collect fisheries data more efficiently, minimize duplication of effort, and provide a timeline for delivery of data and reports.

Report on the Puwalu Series

Leimana DaMate, Association of Hawaiian Civic Clubs

The overarching intent of the three Puwalu (conferences) held in Hawaii during 2006 was to convene Native Hawaiian cultural practitioners from each district on each island, and other interested individuals and groups, to address issues of relevance to contemporary management of marine resources in the Main Hawaiian Islands.

The first meeting in the series emphasized the gathering of cultural experts and practitioners and the need for resolution to move forward with incorporating indigenous values and knowledge into the management process. The second meeting emphasized the input of educators, and the need to revitalize and validate traditional systems of knowledge in formal and informal educational settings around the state. The third meeting was focused on development of working policies needed to advocate for the perpetuation of Native Hawaiian values, practices, and interests.

The theme of the series—"truth"—is symbolized in the illustration of Kūʻula, the Hawaiian god rising from the ocean holding a wana (sea urchin) in his hand (Figure 17.1). Each meeting built on the knowledge, resolutions, and directives of the one before it. One objective of the meetings was to protect and restore ecosystem integrity by promoting the restoration of traditional resource use and management practices implemented in the ahupuaʻa.

Prior to the first meeting in October 2006, conference organizers identified the principal groups with interests in the management of marine fisheries in the islands, and they identified persons who would likely be able to represent the respective groups. The groups included representatives from (a) fishing communities, including Native Hawaiian communities; (b) commercial, recreational, and subsistence fishers; (c) federal, state, and county government agencies; and (d) nongovernmental organizations with an interest in marine fisheries and related issues.

Representatives from a range of public and private sector agencies and organizations subsequently participated at some point in the event. Participating entities included the Association of Hawaiian Civic Clubs; the Department of Land and Natural Resources; the Office of Hawaiian Affairs; the Hawaii Tourism Authority; the Office of Planning—Coastal Zone Management; the Department of Business, Economic Development and Tourism; Kamehameha Schools; and the Office of Hawaiian Affairs, among others.

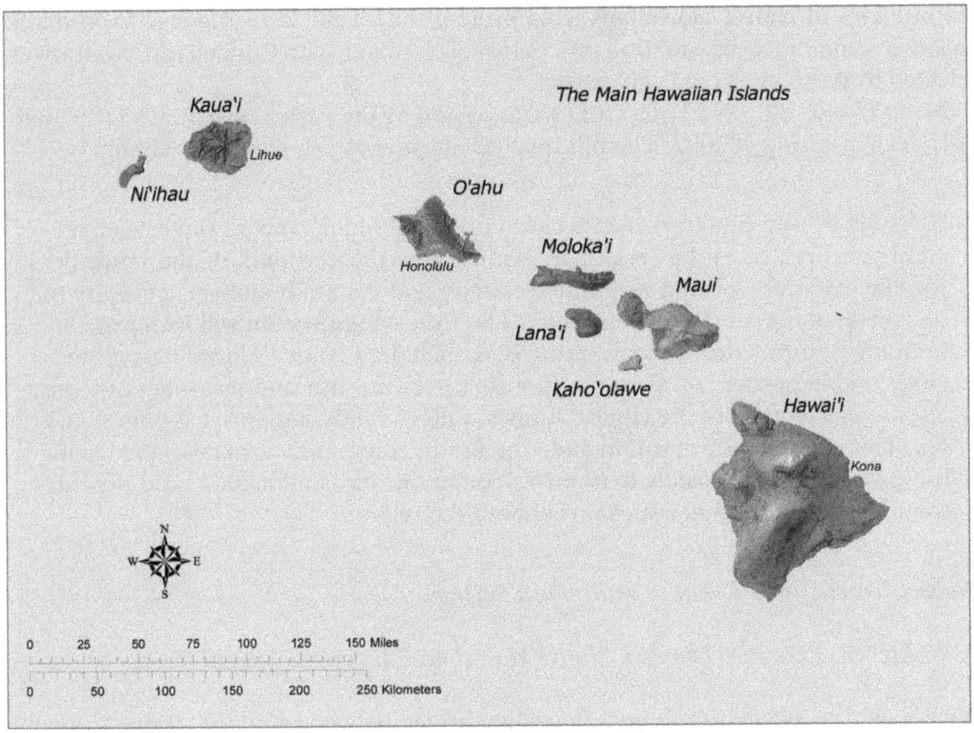

Figure 17.1. Kūʻula rising from the ocean. (For a color version of this figure, see Plate 27.)

The Puwalu organizers convened traditional lawaiʻa (fishing) and mahiʻai (agricultural) practitioners from each of the 37 moku or traditional land districts in the islands. The meetings were successful because participants chose their own representatives from each moku.

The first meeting was characterized as the most challenging as it required organizers to establish trust among a wide range of participants. Participants were united by their shared concern for protecting and sustaining their natural resources. A resolution was passed that would "unite Native Hawaiians to move forward, to live, to grow, to gather together, to stand firm and to restore and perpetuate the Hawaiian way of life."

The second Puwalu involved the participation of educators from each of the private, public, immersion, and charter schools in the state of Hawaii. The educators learned that natural resources, seasons of harvest, and types of fishing and gathering practices vary by island. Participants also realized that a protocol would need to be adopted for asking kūpuna to share traditional knowledge, as the sacred nature of that knowledge can limit discussion of traditions with outsiders. Plans are under way to integrate knowledge derived through traditional place-based Hawaiian science into educational programs throughout the state.

Participants of the third workshop have worked to develop legislation incorporating the concepts and objectives of ʻAha Moku, which are regional councils representing the interests of Native Hawaiians from around the island. The intent is to formally enable a community consultation process for each island, with Council representatives selected by participants in each district.

Note: House Bill 1948 H.D.2 S.D.1 was passed by the 24th Legislature of the State of Hawaii in spring of 2007. The bill specifies its purpose, in part, as enabling:

> A system of best practices that is based upon the indigenous resource management practices of moku (regional) boundaries, the contours of the land, the specific resources located within those areas, and the methodology necessary to sustain resources and the community. The ʻAha Moku system will foster understanding and practical use of knowledge, including Native Hawaiian methodology and expertise, to assure responsible stewardship and awareness of the interconnectedness of the clouds, forests, valleys, lands, streams, fishponds, and sea. This council system will include the use of community expertise and establish programs and projects to improve communication, education, and provide training on stewardship issues throughout the region.

Bridging Practice and Policy in Indigenous Science

Colin Kippen, Executive Director, Native Hawaiian Education Council

What would be needed to build a figurative bridge between traditional practitioners who depend upon marine resources and decision makers who typically do not? Such a linkage will be an essential requirement for successful implementation of ecosystem-based fisheries management in a region such as the Western Pacific.

The strategy of enhancing communication between practitioners and policy makers, as is being undertaken by the organizers of the Puwalu series, is fully appropriate to the issues at hand. Government officials tend to be once removed from the natural environment with which practitioners are so familiar. Thus, bringing the concerns of those practitioners and their descendants to a forum such as the Puwalu series is critically important to the future of Native Hawaiian and other indigenous societies in the Western Pacific. The process may be compared to an inverted Christmas tree, whereby the people closest to the resources are brought into the system, and unlike the "traditional Christmas tree where the person at the top of the tree represents all of the branches below and one person speaks for all, it is actually an inverted model which breaks down all expectations."

The Puwalu series was successful in facilitating communication in that all interested parties were provided with information needed to make the best possible decisions. The strength of the ʻAha Moku system is its capacity to encourage effective communication between members of island communities that are often quite different in nature in both historic and contemporary terms. Although this can involve much time and effort during initial periods of consultation, the process can move very quickly once participants reach consensus on a given issue. While agreement with a given decision will not always be universal, there is widespread respect among participants for the process through which decisions are being made.

The ʻAha Moku process is capable of supporting the cross-generational interests of cultural practitioners. The process is envisioned as a long-term venue for ensuring that the concerns of the kūpuna or revered elders are heard and addressed. These include concerns about the future, and the importance of passing knowledge of the environment and traditional practices to future generations of Hawaiians. Education is the foundation of the bridge between the past and future, and the culturally appropriate consultation structure of the Puwalu process has helped the kūpuna understand the need for a proactive attitude in transmitting their knowledge to others. The system will enable practices and knowledge developed over thousands of years of Hawaiian history to be effectively communicated to future generations.

Pan-Pacific Issues and Challenges

Paul Dalzell, Senior Scientist, Western Pacific Regional Fishery Management Council

Approximately two-thirds of global tuna production derives from this vast region, with roughly half of the tonnage landed by fleets active in the Western and Central Pacific. Several agencies regulate and manage tuna fisheries in the Pacific. These include the Western and Central Pacific Fishery Commission (WCPFC), the Inter-American Tropical Tuna Commission (IATTC), and the Forum Fisheries Agency (FFA), which is based in the Solomon Islands. The independent nations of the Pacific comprise the FFA.

Palau, the Federated States of Micronesia, the Marshall Islands, Kiribati, the Solomons, Papua New Guinea, and Nauru comprise an important subgrouping within the FFA. These are collectively titled the Parties to the Nauru Agreement (PNA). The PNA receives the largest share of revenues from a treaty between the United States and the independent Pacific island nations that enables access to the region by the US purse seine fleet. The PNA derives the majority of its revenue from the aku or skipjack fishery as the resource. By way of contrast, the Central South Pacific does not have a centralized skipjack resource and its fleets are more deeply involved in longline fisheries.

The Secretariat of the Pacific Community (SPC) is an apolitical organization with departments that provide scientific and technical advice on marine fisheries issues to member nations and territories. The SPC Oceanic Fisheries Program has been responsible for conducting most of the stock assessments and monitoring programs in the region for the past quarter century. Although the Program primarily addresses stocks and fleets in the Western, Central, and South Pacific, it also monitors longline activities in the Eastern Pacific.

A range of issues and concerns relevant to the context surrounding pelagic fisheries management in the Pacific include the following: (a) highly productive longline fishing; (b) generalized overfishing of certain resources; (c) live reef fish enterprises in the Pacific, mostly China based; (d) the "China Syndrome," in which many marine resources are exported to China—particularly beche demer, trochus, green snail, pearl oysters, and reef fish; (e) rapid population growth in many island nations, with between two and five percent annual growth not uncommon; and (f) loss of skilled labor to industrialized nations around the Pacific Rim.

Longline fishing has increased dramatically over the course of time in the Western and Central Pacific Ocean. The fleet has increased in size from about 2,000 vessels to around 4,500 vessels in the last few years. This mode of harvest has become increasingly popular in recent years in American Samoa, Samoa, Fiji, Tonga, and Papua New Guinea. Meanwhile, the size of the purse seine fleet in the Eastern Pacific and other regions, which had begun to decline, has recently increased in size.

Extensive fishing pressure on yellowfin and bigeye tuna stocks in the Pacific is a particularly acute challenge for managers. Pressure stems in large part from the activities of the purse seine fishery, and from use of various gear around fish aggregating devices (FADs). FAD fishing and generalized use of purse seine gear are tending to affect juvenile components of bigeye and yellowfin stocks, and bigeye populations are particularly stressed by longlining. Substantial changes in allowable take were, therefore, recommended by the Science Committee at the third meeting of the WCPFC.

In light of this information and the stated interest of many Pacific island nations in development programs that include expansion of marine fisheries, the challenges will be to at once develop fisheries and conserve tunas. One unintended consequence of fisheries expansion in the region is that failure to meet quotas may lead some Pacific island nations to lease unused quota(s) to distant water fleets of nations outside the region.

The WCPFC is attempting to reduce pressure on tuna populations by instituting a changeover from limiting the number of purse seine vessels to limiting the number of overall vessel days that can be fished by the various fleets and nations. The conservation effects of the new regulatory changes in the region are not yet known.

Group Discussion

The group engaged in some initial discussion of the risk–susceptibility model presented by Dr Kirby. Dr John Sibert of PFRP noted that the model was somewhat inconsistent with what was known about certain species, such as skipjack tuna, which appeared to be proliferating despite extensive fishing pressure. He noted that there were uncertainties in the rationale underlying the model in that "rather than adding to the list of species that are in trouble, we should be looking for processes that are counterintuitive, processes that indicate something meaningful about the ecosystem." Dr Lee Anderson of the University of Delaware School of Marine Affairs, addressed the model in terms of its potential for indicating specific trade-offs; that is, areas and issues that managers would need to address in terms of the effects of fishery interactions on nontarget species under the expanded parameters of an ecosystem approach to management.

The Micronesian Challenge was also a topic of discussion. Dr Marc Miller of the University of Washington's School of Marine Affairs spoke to the multiple objectives of marine protected areas. He noted that while the areas could be established to enhance fish stocks, they were also established for reasons that had little to do with marine fisheries or which, in fact, precluded fishing entirely. Dr Miller raised a range of issues associated with the way in which marine reserves were portrayed to the public, and the suitability of the strategy for some areas but not others.

Dr Sam Pooley, Director of the NOAA Fisheries Pacific Islands Fisheries Science Center, discussed the governance dimension of an ecosystem approach to fishery management in that regard, noting the difficulties in arriving at a governance structure that can address the plurality of localized value systems, on the one hand, and globalizing processes and pressures, on the other hand. He asserted that clarity and transparency of objectives, such as those underlying the Micronesian Challenge, were requisites for effectively addressing the local implications of changing management strategies.

Extensive discussion was also applied to resource use conflicts associated with the recent increase in ethnic diversity in the CNMI. It was argued that fishing and gathering practices unique to newly arriving populations of residents were exerting new pressures on marine resources in the region, and that social research might effectively improve understanding of the nature and scope of that pressure and related issues of concern.

Susan Hanna lauded the virtues of the planning process underlying the Puwalu initiative. She noted that unlike many community-based management approaches, the effort in that case was undertaken systematically and that it bore potential lessons for similar work undertaken elsewhere around the world. There was also extended discussion of the unique nature of representation under the ‘aha moku mode of governance.

Sam Pooley compared the system to a "community of governors" through which competing jurisdictions could represent their respective interests without losing individual autonomy. Strong assertions were made by a number of participants regarding the overall benefits of maintaining the integrity of traditional ecological knowledge and other elements of indigenous culture through mechanisms of governance such as those discussed during the course of that portion of the workshop.

Finally, a range of issues were discussed in relation to Mr Dalzell's presentation on management issues related to pelagic species across the Pacific. The fisheries were particularly important since some 85 percent of the total catch in the Western Pacific region derives landings of pelagic species. Noteworthy discussion included review of factors that render an ecosystem approach useful for understanding and managing the fisheries. These included (a) the political and managerial need to address bycatch and incidental take of protected species; (b) the significance of fishery–resource interactions at seamounts and other ocean features; (c) a range of macrolevel environmental factors such as El Nino Southern Oscillation and its effects on patterns of tuna movement and migration; (d) population dynamics of tunas vis-à-vis fishery interactions; and (e) the many political factors associated with management of migratory or highly mobile species in an area of multiple jurisdiction and a complexity of management arrangements.

Facilitated Discussion: Management Scenarios, Challenges, and Solutions

John Kirkpatrick facilitated discussion on a range of ecosystem policy issues during the afternoon hours of the second day of the workshop. The original purpose of the session was to elicit the deliberative thought of participants on the following policy-related issues:

Policy Issue 1: Institutional Ecology

- Can participating agencies be expected to incorporate a new decision-making framework into their respective management agendas? How might this be enabled in a way that would minimize human impediments to potentially beneficial results? That is, what intra-agency and interagency challenges might be expected and how might these be overcome?
- How might sociocultural and institutional variability complicate development and implementation of an EAFM policy in the diverse archipelagic subregions that comprise the Western Pacific?

Policy Issue 2: Indigenous Practitioners and other Resource Users

- How might indigenous and traditional interests, approaches to ecosystems, and management of marine resources be incorporated into the contemporary ecosystem-based fishery management process across the region?
- What are the implications of an ecosystem approach to resource management for the full range of resource users, and how might such persons respond to attendant policies and regulations? How might the challenges be equitably addressed?

Policy Issue 3: Facilitating Ecosystem Policy Process with Valid Research and Monitoring

- What scientific research and monitoring efforts need to be undertaken to enable an effective ecosystem approach to policy and management in a region as diverse as the Western Pacific? What challenges might be expected in this regard and how might these be met?
- To what extent can such efforts be coordinated to yield integrated biophysical and social science research products and monitoring programs of utility for managers and policy makers here? How can such integration be maximized?
- Are existing human and fiscal resources sufficient to enable effective ecosystem research and monitoring work in the region? If not, what is needed and what new resources might be identified?

In order to focus discussion on these issues, participants were asked to select an area in which the ecosystem approach to fishery management will actually be applied through the Council FEP process. Given its diversity of cultures, environmental conditions, and fishery issues and factors, the CNMI was chosen for focused heuristic discussion.

Participants were subsequently asked to identify: (1) the major ecological zones of the region; (2) the most significant marine resources; and (3) user groups of significance to an ecosystem approach to management efforts in the region. The group collectively identified the major ecological zones as including seamounts, pelagic zones, deep slopes for the deepwater bottomfish complex, fringing reef areas, barrier reefs, lagoons, and sea grass beds. Resources identified from the lagoon outward included reef fish, invertebrates, algae, mollusk, coral, sea grass, the water column, pelagic fish, bottomfish, substrates, crustaceans, turtles, sharks, cetaceans, and birds. User groups included commercial, recreational, and consumptive-oriented fishers; Chamorros, Carolinians, and other Micronesians; immigrants; snorkelers; seafood consumers; persons using the resources for visual, spiritual, emotional, or other non-consumptive purposes; cruise ship owners and operators; military vessels; scientists, researchers, educators, and students; and children and future generations.

In terms of governance, several agencies were identified as sharing responsibility for management, rule making, and enforcement of the region's marine ecosystems and associated resources. These agencies (from lagoon to deep slope waters) include the Division of Environmental Quality (DEQ), Coastal Resources Management (CRM), the US Environmental Protection Agency (EPA), the Army Corps of Engineers (ACOE), the US Coast Guard (USCG), the WPRFMC, NMFS, USFWS, the Department of Public Works (DPW), National Park Service (NPS), and the Department of Interior (DOI).

Participants in the exercise were then asked to envision a process through which all of the agencies involved in governance of deep slope activities around the CNMI would be convened to deliberate on ecosystem-based management of resources in that ecological zone. The participants identified the following particularly salient concerns and challenges deemed likely to be encountered during the process:

1. Appropriate delineation and application of jurisdiction was seen as particularly important given: (a) limited resources available to conduct ecosystem research and monitoring and (b) interest in avoiding duplicative management and enforcement efforts. Issues of scale and application of understanding at a scale appropriate for addressing interaction between ecosystem components and processes were seen particularly important. As one participant noted: "Everybody has jurisdiction, but few exercise it on an ecosystemic scale."
2. The legal capacity and reach of authority under existing systems of governance vis-à-vis the extent of application of governance powers in reality was raised as an important issue in this exercise. In other words, it was asserted there is a fundamental disconnect between what governing authorities can do and actions that are actually undertaken. For instance, a marine ecosystem may be heavily impacted by siltation of reefs caused by shoreside development. Agencies may have the power to act in a way that would ultimately regulate such activities and satisfy resource management mandates, but typically they do not. In short, agency involvement is often both narrowly defined and practiced.
3. Because agency representatives involved in management of marine resources in the CNMI were thought likely to be pursuing a variety of mandates and agendas (similar to personnel in resource agencies elsewhere in the United States), participants felt that full support from the Governor of the CNMI would streamline the gathering of support from the cognizant agencies. It was felt that a leader with full authority in support of the principles of an ecosystem approach to management could offer integrating support in his region of responsibility.
4. Concern was expressed about the tendency for biologists to determine policy when policy makers or agency directors do not have the background needed to determine effective means to desirable outcomes. Thus, it was felt that certain social or economic factors or incentives may not be addressed in the information gathering, analytical, or decision-making processes associated with implementing an ecosystem approach to management of marine resources in the region. These concerns were not perceived as specific to the CNMI but were rather emblematic of the situation in settings around the United States.

Participants also identified the zones and resources over which NMFS and the Council exert management authority. These zones include (a) barrier reefs and seamounts that are more than three miles from shore; (b) the deep slope; and (c) the open ocean and associated pelagic species, sharks, turtles, and cetaceans. Significantly, the application of ecosystem principles and attention to connectivity between the biophysical and human components of marine ecosystems expands attention of federal agencies to a larger field of ecological zones, species, and human constituents. As such, the level of interaction between federal, state, and territorial agencies is likely to increase, with the Council continuing to respect local jurisdictions.

Participants in this exercise anticipated that several challenges result from increased interaction with local agencies during the implementation phase of the FEP process.

Again, it was asserted that the competing interests and agendas of local and regional agencies might render streamlined interaction problematic in some cases. It was asserted that effort should be applied to understanding the incentives that drive the actions of agencies and their representatives, and how those relate to interagency dynamics. It was determined that sustained cooperation might require formalized agreement and appropriate incentives such as reciprocal sharing of data.

It was also asserted that in order for the FEP in CNMI (and elsewhere) to be successful, it would also need to be supported by the local polity. Some degree of involvement in the management process by local constituencies will also enhance the chances for success. There was discussion of this issue in terms of the reauthorized Magnuson–Stevens Act, which mandates extensive public involvement and use of traditional ecological knowledge in the process of managing resources in the Western Pacific. As such, it was recommended that the Council consider marine ecosystems in the CNMI through the eyes of village leaders and resource users. This would require equitable opportunity for input from Chamorro, Carolinian, and other cultural practitioners throughout the region.

Dr Kirkpatrick presented workshop participants with an emergency scenario in which the island of Hawaii was isolated by an earthquake. Participants were divided into three groups to discuss a logical course of action needed to sustain the region's fisheries for the sake of survival. The point of this heuristic exercise was to encourage participants to consider the ways in which people are linked to marine ecosystems in the Western Pacific, and appropriate and logical governance strategies for responding to disruptions to such systems.

All three groups developed creative means for responding to the variety of problems that would inevitably result from sudden isolation and dependence on finite natural resources. Significantly, each group employed scientific principles to arrive at solutions for sustaining important terrestrial and marine resources over the course of time. Of particular note in this regard, each group's plan relied extensively upon the knowledge systems of indigenous cultural practitioners and others familiar with marine ecosystems in the region. The issue of regulation of human behavior was also important as it was recognized that a system of governance and allocation of resources would be particularly necessary during times of scarcity.

While the scenario minimized the external complexities of modernity and globalization, it was generally felt that the exercise was useful in that it inspired participants to reflect on decision-making processes associated with the most basic challenges of effective fishery management.

These included challenges associated with (a) maximizing the utility of a finite base of natural resources for the collective good of the polity, (b) arriving at and maintaining system of governance that enables sustainability of natural resources, and (c) allocation of limited resources in a manner that is fair and equitable.

Chapter 18
Options for Community and Agency Interaction

Participants discussed a variety of options for maximizing opportunities for and benefits of fishery ecosystem research and monitoring through community involvement in the management process. As summarized in subsequent sections, the discursive input and recommendations generated by workshop participants will guide the Council as it moves forward with the ecosystem approach across its region of jurisdiction.

In reiteration, the Council is expanding its consultation process to enable additional opportunities for public input in the management process. This involves outreach to user groups via the REAC process, and ongoing implementation of the Community Development Program and Community Demonstration Project Program. The latter continue to be administered with the intent of providing technical and fiscal resources to indigenous communities around the region.

Workshop participants reviewed several challenges likely to be encountered as the Council engages in an ecosystem approach to fisheries management (EAFM). Means for interacting with the variety of agencies and communities in the region were discussed once again. It was determined that the Council would ideally identify ways to provide leadership while enabling the sharing of knowledge and responsibility for the stewardship of natural resources. There was also deliberation on how best to evaluate the performance of an ecosystem approach to management. It was stated that without clear performance objectives, progress cannot be readily measured.

As described in the final sections of this chapter, several high-priority recommendations for the Council FEP process were generated. These are associated with what were widely perceived as critical needs for (a) clearly defined roles, rules, responsibilities, and terms of reference for ecosystem-based management in the various archipelagic subregions; (b) mapping and monitoring of physical and human environmental impacts following from or in some manner addressed by the new management approach; (c) developing partnerships for research and monitoring; and (d) building social capital and developing trust in and empowering communities around the region.

Extensive discussion was applied to the overarching issue of whether implementation of the Council's place-based ecosystem plans should be incremental in nature, or whether this should proceed via a wholesale changeover from the existing process of managing single species across the region. There was also extensive deliberation on the nature of the ecosystem planning process and how the ecosystem mode of management would in reality differ from the existing mode of managing the region's fisheries. It was eventually determined that the Council will indeed proceed incrementally and adaptively and that this strategy would likely preclude a range of challenges potentially resulting from an overly rapid shift to the new paradigm and approach.

Ecosystem-Based Fisheries Management in the Western Pacific, First Edition. Edward Glazier.

A useful summary of the unique elements of the ecosystem planning process and the nascent EAFM to management in the region was offered by Stewart Allen of NOAA Fisheries Pacific Islands Fisheries Science Center. Dr Allen discussed the Council's new approach and its unique attributes as follows:

> The Council is adopting an archipelagic approach, which involves reliance on a new institutional mechanism, the Regional Ecosystem Advisory Committees, which is a kind of expansion of kuleana (realm of responsibility). It involves greater emphasis on community co-management and on the applications of Hawaiian science. It involves greater reliance on biophysical ecosystem models and concepts. It involves greater reliance on biophysical and social indicators. [In terms of governance], it's about collaboratively extending—it doesn't consider existing authorities as some sacred boundary beyond which we will not cross. So to me, what we've been moving toward is a willingness to go beyond NMFS and Council traditional authorities and responsibilities, where appropriate, and to recognize other influences on the ecosystems that we manage a portion of through our fisheries management processes, and do that through increased collaboration with other jurisdictions and entities. To me, we've already identified the suite of ways in which the Council wants to move beyond our existing processes.

The Regional Ecosystem Advisory Committees

Paul Dalzell, Senior Scientist, Western Pacific Regional Fishery Management Council

REACs have been established for each of the archipelagos. The advisory bodies are comprised of invited Council members with expertise in marine fisheries and related issues, and representatives from federal, state, and local government agencies, businesses, and nongovernmental organizations with responsibilities or interests in land-based and nonfishing activities potentially affecting the marine environment.

The REACs will provide a mechanism for the Council and member agencies to share information about relevant programs and activities, and to coordinate management efforts to better address factors impinging on the status of marine ecosystems within and beyond the jurisdiction of the Council. The Committees are also intended as a mechanism for the public to provide and gather information about area-specific issues affecting fisheries and related aspects of community life.

A newly established Marine Education Program will also be part of the new approach. This program will promote the integration of traditional knowledge and marine science into educational programs around the region.

Council staff is visiting the archipelagos in 2007 to discuss the EAFM with community representatives. The shift to an ecosystem approach to management and implementation of the FEPs will be incremental in nature.

[Note: Initial REAC meetings were held in the Commonwealth of the Northern Mariana Islands (CNMI) and on Guam during February 2007 and in American Samoa and Hawaii during April 2007. Each of the REACs expressed overarching concerns about loss of traditional ecological knowledge and deterioration of cultural practices directly and indirectly related to marine resources and ecosystems. A principal issue of concern on Guam was loss of access to shoreline areas associated with beachfront development and marine reserves. Concerns regarding loss of traditional Chamorro life ways were expressed by REAC representatives in the CNMI. Such concerns were also expressed by representatives in Hawaii, although there was favorable discussion regarding perpetuation of traditional fishing practices and related patterns of culture in certain areas around the Hawaiian Islands. Additional concerns around the entire region include enforcement problems, pollution, and detrimental aspects of tourism (see Western Pacific Regional Fishery Management Council 2007).]

Community Consultation and Interaction

Jarad Makaiau, Habitat Coordinator, WPRFMC

With mandated foci on management and regulatory processes, governing bodies sometimes disregard the fact that it is local residents who are most closely involved in the use of marine resources. Such persons care for and depend upon those resources for income, sustenance, and a range of social and cultural purposes and outcomes. Management and regulatory processes function to condition the behavior of persons who on occasion may not attend to traditional or normative behavior regarding proper use of marine resources.

The Council's goal in implementing the FEPs is to assist and empower individuals and communities dependent on healthy and productive marine ecosystems. Because not every individual and group can be immediately assisted, however, efforts are necessarily focused on addressing the most salient issues and challenges. That is, resources are first applied to persons and groups with the most pressing and feasible goals. Communities will bring their issues to the REAC, which will then ask the Council if it is an appropriate and feasibly addressed issue. If so, the Council will apply resources to the issue with the intent of enabling the community in question to assume some measure of collaborative responsibility for effectively meeting the challenge.

The Council's Community Development Program can serve to provide communities with technical support to address problems such as pollution or habitat degradation. The Community Demonstration Project Program can also help solve such problems. The newly authorized Marine Education and Training Program authorizes the Western Pacific and North Pacific Councils to provide funding and technical expertise to promote the incorporation of traditional knowledge into the management process. In short, the nonregulatory component of the FEPs provides for application of various resources to meet a range of fishery-related concerns and challenges in communities throughout the region.

The Community Demonstration Project Program could be used to fund collaborative research between NOAA Fisheries scientists and participants in the region's fisheries. Program funding is provided on a year-to-year basis.

A team will be appointed to address Council-approved REAC recommendations. The Committees will ideally meet at least three times a year in their home locations. Regarding pelagic fisheries conducted in the region's exclusive economic zone (EEZ), the Council may work through the REAC to assist the community in addressing a suitable strategy for addressing management issues in a given offshore area. Formalized scientific information and/or traditional ecological knowledge needed to inform the prospective management strategies will be determined on a case-by-case basis. The voluntary fisheries data collection initiative was intended to gain basic understanding of small boat fisheries around the offshore banks near Guam. The

Council's overarching goal is to empower communities and provide them with some of the tools needed to collaboratively assist in the management of their fishery resources and marine ecosystems.

Group Discussion

A number of issues discussed by Mr Dalzell and Mr Makaiau were reviewed by participants during the morning hours of the final day of the workshop. It was recognized that the Council will gradually and incrementally implement management measures to further operationalize the FEPs. But some participants asserted that even under an incremental approach a long-term vision and related management objectives would be needed to maintain continuity of the ecosystem approach across the region. Moreover, it was asserted that (a) without clear objectives relating to that vision, it would be difficult to evaluate the overall effectiveness of the new approach and (b) the regulatory component of the ecosystem approach would ideally involve a regular review process to identify lessons learned during its incremental implementation. Thus, the approach could be revised as necessary for more effective implementation across the archipelagos.

Council staff members related that the Western Pacific Council had collectively arrived at a vision for the future of the ecosystem approach and that objectives had been developed to satisfy that vision. Those relate primarily to the process for deepening relationships with island communities over the course of time, and to immediate and practical plans for initiating that process. Efforts were currently being undertaken to successfully initiate the REACs, which were intended to improve the Council's understanding of the biophysical and human dimensions of the region's marine ecosystems and thereby introduce a more effective and empowering management regime.

It was agreed that the REAC process could and would allow the Council to consider and address issues extending beyond those it had traditionally considered, such as terrestrially generated pollution and other factors affecting comprehensively envisioned marine ecosystems. It was determined that (a) that would occur largely through enhanced and expanded advisory roles vis-à-vis other governing bodies in the region, (b) direct regulatory changes could only be effected in the EEZ, and (c) it might be useful to interact with local agency representatives during the early phases of the effort to explain the intent and nature of the advisory role of the Council in that regard.

Significantly, the Council staff reported that increased attention would also be given to the well-being of communities of islanders who were, to some extent, involved in and/or dependent on marine ecosystems in the region. Understanding of problems and needs in such communities would be communicated through the REACs, and with Council support, certain issues could be addressed through the actions of the Community Demonstration Projects Advisory Panel and Community Development Advisory Panel. Thus, by increasing opportunities for communication and administration of programs designed to assist communities in solving salient problems, the new approach could, for example, heighten the potential for commercial fishery participants to more successfully market their seafood products via a cooperative, or to inform agency representatives cognizant of a pressing shoreline access issue and potential options for addressing the problem. One workshop participant noted the potential value of social indicators for assessing the level of success of the approach in the communities. Finally, staff also noted that the Community Development Program might enable the Council to assist in addressing challenges that were not directly related to fishing but that affected communities where fishing was critically important.

One workshop participant repeatedly asserted the potential utility of GIS applications under the ecosystem approach to management. He noted that the spatial purview of the involved agencies and organizations could be mapped, as could spatial aspects of their respective mandates and regulations. Various physical and human ecological factors of relevance to marine fisheries and fishery management could also be mapped. Using such layers in conjunction, it would be possible to depict the interface of jurisdictions, mandates, ocean and climate factors and processes, spatial dimensions of populations of marine organisms and habitats, resource use patterns, and other factors—with great potential for integrating new and existing data and analyses for purposes of biophysical and social assessment and long-term monitoring under the new ecosystem approach to management. In short, total ecological relationships, conditions, and trends might be depicted. It was asserted that a GIS used in that manner had great potential for effectively monitoring fishery interactions and controlling for environmental factors over time, and for assessing changes related to the highly adaptive nature of commercial fishing fleets.

Some attention was applied to issues regarding participation and representation in the REAC process. It was asserted that social science research might be useful for generating in-depth understanding of social and cultural conditions and factors in the communities, thereby potentially optimizing the participation of knowledgeable persons not likely to become involved given various social or cultural constraints. Social science research methods might also be useful for gaining an understanding of interagency dynamics, potential challenges to effective interaction, and potential solutions for working collaboratively to solve pressing issues.

A number of participants discussed the need for a Council ecosystem science plan for both biophysical and social sciences. It was recommended that that should include an interdisciplinary approach to ecosystem-based management in the region.

Finally, the group discussed comanagement and community empowerment. It was made clear that the intent of either process should be to augment rather than replace formal scientific information and advice. It was argued that involving fishery participants in the management process could serve to better incorporate the knowledge and perspectives of persons most directly involved in use of marine resources and/or those who were most informed of factors affecting or influencing the status of the region's marine ecosystems broadly conceived.

Problematic aspects of achieving such a system and ensuring favorable outcomes given the complexities of multiple jurisdictions, agencies, and island communities in the vast Western Pacific became a topic of focused discussion that carried the participants through the remaining hours of the final day of the workshop. We review that discussion in some depth by way of concluding summary and recommendations in the following chapter.

Reference

Western Pacific Regional Fishery Management Council (2007) Communities voice concern about loss of traditional practices. In: *Pacific Island Fishery News*. Newsletter of the Western Pacific Regional Fishery Management Council. Spring, Honolulu.

Chapter 19
Conclusions and Recommendations

The Ecosystem Policy workshop provided a forum through which participants contributed their knowledge and expertise to ecosystem-based fishery management in the Western Pacific. The meeting was not about formulating ecosystem policy *per se*. Rather, it was intended to enable deliberation on important ecosystem management issues and provide insight and lessons from persons working in marine fisheries or relevant fields in other parts of the region, nation, and world. This chapter reviews key elements of that insight, revisits some of those lessons, and generates summary conclusions and recommendations intended to assist the Council as it incrementally and adaptively moves forward with ecosystem principles in the years to come.

Overview

It should be kept in mind that the term *policy* derives from the Greek *politeia* or *politēs*, meaning citizen. Policies regard principles or procedures for guiding *people*. The emergent workshop emphases on human communities, political aspects of cross-jurisdictional management of marine resources, and expansion of management considerations into the realm of human ecology, were thus logically appropriate for a venue addressing matters of policy.

This in no way diminishes the critical importance of the biophysical sciences, the information generated through such investigation, or the dire need for understanding of the physical properties and processes of marine ecosystems. Rather, matters of marine policy force recognition that all means of acquiring knowledge of marine systems, and that knowledge itself, ultimately relate to human objectives. In the case of fishery management under existing federal mandates, objectives involve optimized sustainable human use of the ocean's living resources.

The Council has determined that ecosystem principles will likely optimize management of marine resources across its region of jurisdiction. In moving toward implementation of those principles, the Council has drafted FEPs, established REACs and related processes and entities, and convened national and region experts to deliberate on biophysical, social, and policy dimensions of an ecosystem approach. NOAA Fisheries has concomitantly undertaken an Environmental Impact Assessment process to examine the potential biophysical and social effects and implications of the approach.

The Ecosystem Policy workshop examined a wide range of issues relevant to EAFM. Participants discussed governance issues and policy options that would ideally maximize the benefits of the approach in the cross-jurisdictional and cross-cultural settings that characterize the region. The group also examined options for addressing the

Ecosystem-Based Fisheries Management in the Western Pacific, First Edition. Edward Glazier.

needs and interests of indigenous fishing practitioners and other resource user groups across the region. Finally, workshop participants discussed options and opportunities for fishery ecosystem research and monitoring in the Western Pacific. The following material reviews select elements of the discussions, framed in terms of the potential benefits of the new strategy.

Potential Benefits of the EAFM

The potential for an ecosystem approach to heighten scientific attention to connections between biophysical and human components of marine ecosystems was an important underlying theme of the workshop. As noted at the outset of this chapter, islands are in various ways amenable to scientific inquiry using ecosystem principles, and the approach is useful in concept and practice for controlling environmental factors affecting fish stocks and marine fisheries.

From a managerial perspective, the shift from a single-species approach to a place-based approach is seen as beneficial in that it will reduce the administrative complexities of studying and managing species and fisheries across rather than within areas that are highly varied in terms of their environmental and political attributes. That is, the FEPs consolidate and reorganize management provisions so that each archipelago and its marine resources are addressed as a distinct management unit. As previously noted, given tendencies of movement and migration, and related international jurisdictional factors, pelagic species are being managed separately.

The ecosystem approach is being advanced for its potential to enhance understanding of relationships between the marine environment and user groups, and to provide the latter with expanded opportunities for contributing to the management process. This involves increased attention to issues of social and sociopolitical connectivity and necessitates expanded relationships between the Council, fishery participants, communities, and governing entities across the region. Of note, the new process has some potential for blurring spatial or other distinctions historically imposed on biophysical systems whose components and processes are not readily bounded.

In enabling expanded connections with island communities, the new approach will enable scientists and managers to benefit from traditional and local knowledge of marine resources, marine ecosystems, and longstanding fishing and shoreline food-collecting practices. This is potentially highly significant in a culturally diverse region such as the Western Pacific. Each archipelago is home to indigenous peoples who have, to greater and lesser degrees, accumulated and transmitted centuries of knowledge regarding marine and terrestrial components of island ecosystems.

The approach is also significant in terms of its potential for identifying and potentially mitigating overuse of marine resources or potentially deleterious fishing or food-collecting practices undertaken by long-term residents or newly arriving immigrant groups. Enhancing or expanding connections with human communities may serve to identify and mitigate a range of factors impinging on the health of marine ecosystems. Finally, the approach may provide mechanisms for community development initiatives both related and unrelated to marine fisheries.

It was widely agreed that the ecosystem approach to fishery management would involve expansion of scientific attention to a larger field of physical, environmental, social, and political processes, factors, and issues. A wholesale shift in existing stock assessments, attention to biomass issues, and species-based scientific inquiry is by no means indicated. Rather, the intent is to incrementally and adaptively shift to an approach that more fully attends to ecosystemic processes such as those associated with food webs, predator/prey relationships, endangered species interactions, and so forth. While such expansion will require increasing levels of funding, policy workshop participants identified a number of venues potentially supporting ecosystem-based biophysical and social research and monitoring, as described further along in this concluding section.

It should be noted that the sustainability and productivity goals of revamped ecosystem research and monitoring are in keeping with the missions of the fishery councils and NOAA Fisheries. Moreover, potential opportunities for public participation in ecosystem research and monitoring programs around the region are in line with the Council's stated interest in empowering communities. It was determined that such opportunities could be facilitated through the Council's Community Demonstration Project Program and other programs.

Summary Recommendations for Maximizing the Benefits of the New Approach

This section draws on the findings and recommendations of the biophysical and social science ecosystem workshops, and the final ecosystem policy workshop, to present a series of options and recommendations to the Council as it moves forward with its FEPs and related processes for better understanding and addressing the context, information needs, and potential benefits and liabilities of the new mode of management.

Clearly, as is described in previous sections, participants expressed many doubts, concerns, caveats, and conditions regarding the future course of planning for an implementing the ecosystem approach in this and other regions. Participants identified and elucidated a variety of factors likely to challenge successful implementation of the new strategy. These included, among others: (a) institutional inertia; (b) uncertainty in terms of the capacity of science and existing data to sufficiently address highly complex ecosystem processes; (c) limited funding available for expanding physical and social scientific inquiry in support of the new approach; (d) lack of well-defined ecosystem-specific management objectives; (e) increased bureaucratic complexity; and (f) lack of well-defined incentives for agencies to collaborate in implementing a new strategy.

But participants also worked diligently to arrive at solutions to these and additional challenges. Following is workshop-generated guidance for meeting such challenges if and as they arise in future planning and implementation phases of the new management regime.

Biophysical Workshop Recommendations Reiterated

As discussed previously (Western Pacific Council 2006), participants in the first Council ecosystem workshop generated six basic points of policy advice for the Council to consider as it continues with its ecosystem planning process. These are as follows:

- As a default, a precautionary approach should be employed in implementing the ecosystem approach to management in the region, but precautions should relate not only to the status of marine resources but also to those who depend on the sea for food, cultural practices, employment, and other aspects of life.
- The fishing industry and managers should endeavor to be proactive in changing the burden of proof regarding the impacts of fishing, with industry taking an active participatory role in research, monitoring, resource conservation, and sustainability.
- "Insurance" or spatial and other latitude will be useful in adaptive development and implementation of ecosystem-related policy.
- Lessons should be drawn from other regions and used adaptively in the Western Pacific.
- Proper incentives should be used to aid in the achievement of management goals.
- Fairness and equity should be duly considered in the ecosystem approach to management in the region.

Additional policy-relevant recommendations provided in Western Pacific Council (2006) include the following: (1) clearly define and articulate management/policy issues and questions along lines of urgency and identified needs; (2) assign a centralized resource entity with sufficient seniority and appropriate financial and human resources to establish and maintain a centralized data reference and contact point (the "who, what, where, and how" of data); (3) review and evaluate all currently available data and data collection schemes (biological, social, economic, etc.) and initiate and maintain data "mining" and recovery activities; (4) undertake initial assessment and analysis of available data to identify strengths, weaknesses, and gaps in available information; (5) identify and initiate adaptive management experiments at an ecosystem scale; (6) ensure data collection and models/analyses for EAFM are coordinated with and driven by clearly identified management needs and issues; (7) encourage keeping all models/analyses at the most "simple" level, that is, avoid the temptation to build large, exceedingly complex models; (8) ensure adequate support and resources for clearly identified ecosystem-scale research, monitoring, and analysis; and (9) evaluate a suite of indicators (fishery based as well as new and emerging ecosystem based) in an evolving and adaptive process.

Social Science Ecosystem Workshop Recommendations Reiterated

A number of recommendations and policy advice also emerged from the Council's Ecosystem Social Science Workshop. Of overarching importance was the necessity for envisioning both the biophysical ecology of marine ecosystems and the human ecology of those systems, wherein the latter involves (a) the human ecology of constituent groups and (b) the ecology of governing institutions.

Options and recommendations for incorporating social science into ecosystem management in the region included the following: (1) establish venues for identifying priority issues and objectives to address with social research and monitoring; (2) design social research to meet prioritized objectives and related information needs; (3) implement a social, economic, and sociodemographic research strategy for the archipelagos; (4) develop and implement liaison and ecosystem social science performance evaluation programs; (5) employ an incremental/adaptive strategy coupled with appropriate incentives; (6) identify valid social and economic indicators to assess and monitor direct and indirect human–environmental interactions and to adjust resource use policies; (7) apply social science to assess the potential for public input and community development programs; and (8) effectively relate new research programs to ongoing programs.

Of note from the perspective of the social science workshop, application of precautionary strategies to management should not be limited to biophysical aspects of ecosystems only. Rather, the approach may also be envisioned as a means for safeguarding the well-being of resource user groups and fishing-associated communities until better biophysical data and analyses can be generated.

Policy Workshop Recommendations for Enhancing Participation

Among the most essential points of advice generated during the final workshop were those relating to the process of engaging communities and agencies in the Council's REAC process. It was perceived that the effort could involve a variety of challenges, but that these were by no means insurmountable. A range of options were reviewed for their potential to ensure programmatic success.

For instance, it was recommended that the Council should clearly determine its terms of reference and engagement prior to initiating formal relationships with agencies and individuals via the REAC process. This would involve determination of the rationale for involving groups and their representatives and identification of the expectations for each. These factors and the overall objectives and intent of the process would then be clearly communicated to REAC participants, thereby generating a context of transparency and trust.

Workshop participants felt strongly that Council representatives should clearly communicate to REAC participants that the intent of the effort is not to expand its jurisdiction, but rather to enhance opportunities for empowering communities and solving fishery-specific and other challenges. It was agreed that building trust would take time, but that it is essential to the long-term success of the program. It was suggested that the Council could enhance and streamline participation by identifying a problem of interest to multiple agencies and groups, and to which available human and fiscal resources would be readily applied. That is, it was felt that a common threat or need is an effective incentive for galvanizing cooperation and encouraging the collective interest to transcend competitive interests. Conversely, it was recommended that potentially contentious issues, such as allocation of marine resources, be avoided.

One participant discussed the engaging effects of data sharing. This was advanced as a relatively easy and cost-effective means for gaining trust, building community capacity, and furthering the goals of ecosystem-relevant science and monitoring. Discussion focused on the possibility of engaging fishers and other resource users in the research and monitoring process, with oversight and technical advice provided by agency representatives and scientists who are capable of addressing the various cross-cultural challenges likely to be encountered around the islands. It was asserted that by involving fishery participants in the science and management process, the community in question will inevitably become more empowered and its residents will gain a sense of trust in the management process.

One potential challenge relates to the concomitant desirability for locally staffed long-term monitoring and the uncertain probability of finding local persons able to sustain involvement over the course of time. However, intense local dedication, as exemplified in the ‘Aha Moku process being implemented in the Hawaiian Archipelago, may well serve to minimize such challenges. Long-term dedication and vested interest coupled with the multigenerational scope and educational emphasis of the ‘Aha Moku process render it a useful model with potential for guiding community involvement in research, monitoring, and management of marine resources elsewhere in the Pacific.

An effective element of the ‘Aha Moku process is its attention to variation in the historical background and contemporary interests of indigenous communities, and its capacity for effective representation at a regional level. The process benefits from the manner in which the Council has been able to provide leadership and enable authority to be enjoined by the participants.

Issues of geographic scope and scale were considered by workshop participants in formulating a recommendation to the Council to avoid an initial undertaking that is overly encompassing. While it was agreed that the Council would need to address some issues at a broad scale, it was asserted that it might be beneficial to experiment at a more localized place level and to address dynamic relationships between residents and the marine and community environments at that modest scale. A related recommendation to build upon existing strategies or groups was seen as potentially useful by some participants, but potentially complicating by others.

While there was general appreciation for the potential utility of traditional ecological knowledge in contributing meaningfully to the ecosystem approach to management, it was felt by some that such knowledge, of itself, might be insufficient for understanding certain ecosystem processes. Formalized science may also be insufficient for addressing certain largely unknown environmental processes, including global climate change and its implications for the status of marine ecosystems and resources. Knowledge uncertainties notwithstanding, participants tended to express respect for and advocate use of the full suite of knowledge gathering tools available to resource managers in the region. In the event formalized science and traditional ecological knowledge are not in agreement, the disparity should be taken as a learning opportunity and a rationale for formulating and testing hypotheses.

Trial and error was recognized as fundamental to an adaptive approach to management. It was felt that mistakes are inevitable, and that the point of scientific method was to learn from failed trials. This applies to application of knowledge deriving from formal science and from traditional fishing practices.

The REAC process is amenable to assessment and evaluation. Valid indicators that a successful institutional ecology has been established through the process include the following: (a) level of sustained participation; (b) ongoing internal commitment of resources; (c) willingness to engage in formal partnerships; and (d) positive feedback from the community that the process is yielding successful results. Indicators of community well-being, such as nutritional status and educational success, may also be useful for assessing the benefits of the program.

Policy Workshop Recommendations for Identifying Fiscal and Human Resources

Workshop participants also worked to identify options for resources needed to conduct an expanded program of ecosystem research and monitoring across the region. For instance, it was determined that human resources could include properly administered undergraduate internship programs through which students could gain experience and exposure to traditional knowledge while contributing to scientific research and monitoring efforts.

As described above, it was also suggested that human resources available in a given community could be applied to formal scientific research and monitoring. This could be organized so that the data collection process is conducted with technical assistance and data management provided by dedicated staff from local agencies and with sufficient ongoing interaction between those agencies and persons actually using the information for analysis of marine fisheries and marine ecosystems in the region. Again, this would ideally involve a reciprocal arrangement wherein the data and analyses are shared between all parties, with due attention to issues of proprietary and confidential information.

Interaction with local fishery participants and cultural practitioners in this context would require use of appropriate cultural protocol and procedures that attend to the cultural sensitivities of sharing and disseminating such information. Council on-site island coordinators, consulting cultural practitioners, and social scientists familiar with the cultural and linguistic subtleties of island societies in the Western Pacific would ideally play central roles in the interface between the Council and prospective sources of local and traditional knowledge relevant to understanding marine ecosystems and managing marine fisheries across the archipelagos, including those of the deep sea.

A range of options for funding an expanding realm of ecosystem-relevant scientific inquiry and monitoring were discussed by participants at the policy workshop. These included Work Force Training Act funds, a variety of nontraditional sources of federal funds, and funds and research partnerships with nongovernmental organizations, including philanthropic institutions. It was noted that in the absence of novel funding options, difficult tradeoffs were likely to affect implementation of certain research and monitoring priorities.

Of significance to this discussion and to the Council's interest in empowering communities and expanding connections with communities around the region, new stipulations in the reauthorized Magnuson–Stevens Act allow for provision of monies to support fishery demonstration projects "that foster and promote traditional indigenous fishing practices," and qualifying Community Development Programs in Alaska and the Western Pacific. Readers are encouraged to consult the Magnuson–Stevens Fishery Conservation and Management Act, Public Law 94–265, as amended by the Magnuson–Stevens Fishery Conservation and Management Reauthorization Act, Public Law 109–479. The Act was available at the time of this writing at the following address: http://www.fws.gov/fisheries/fwco/pdfs/Magnuson-StevensACT.pdf.

Conclusions and Final Recommendations

The ecosystem workshop series convened by the Council has enabled collection and dissemination of the na'auao (knowledge or wisdom) of scientists, managers, fishery participants, cultural practitioners, and policy experts from around the nation and region. The preceding pages have reviewed and synthesized select elements of that base of knowledge to provide the Council with a concise source of information of utility for managers, staff, and scientists seeking to apply ecosystem principles around the region. Readers are reminded that a broad range of additional summary information is available in the full reports on the biophysical and social science workshops held in 2005 and 2006, respectively. The policy advice and recommendations generated during the course of the previous workshops remain viable and have thus been provided in the concluding section of this book.

The final workshop that is the focus of this chapter led to generation of additional insight into a number of issues of pertinence to the formation of ecosystem-related policy in the Western Pacific and other regions around the nation. Participants generated a number of practical results for the Western Pacific Council. These included options, recommendations, and insights particularly useful for (a) meeting the Council's goal of empowering communities and working with local governments to undertake place-based ecosystem fishery management; (b) establishing effective long-term consultation with communities through the Council's REAC process; (c) documenting TEK through effective and culturally sensitive collaboration with indigenous practitioners; and (d) identifying possible venues for funding and human resources needed to enable long-term ecosystem research and monitoring across the region.

As noted above, the focused attention of the final workshop on issues related to community empowerment and the potentialities of traditional knowledge and expanded local involvement in the management process was in keeping with new stipulations in the reauthorized Magnuson–Stevens Act. That attention gave pragmatic advice to a process that recognizes the value of local and indigenous systems of knowledge, and the potential benefits of expanding connections between the Council and its constituent communities and adjacent governing bodies. Witherell (2004) underscores the importance of these processes in a manner appropriate to the diversity of island communities in the Western Pacific. This is noted in the *Proceedings of the Conference on Fisheries Management in the United States* sponsored by the Regional Fishery Management Councils, the Fisheries Commissions, and NOAA Fisheries in 2003 and 2005. Witherell (2004) asserts:

> The differences between various fishing communities are important to document: the history, geography, relative level of dependence on the fishing industry, values and norms, and long-term goals of the community are all significant factors in determining an appropriate action. The involvement of stakeholders through collaborative research is one way to incorporate community needs and knowledge into the decision-making process … Use of local, cultural, and traditional knowledge is critically important to achieving a better understanding of the impacts of a proposed regulatory change and potentially innovative solutions. In this way, including local traditional knowledge is beneficial to both the resource and the participants in the process, and can stimulate community-driven initiatives that are well supported and successful…. Because fishing communities are so variable, it is inherently understood that communities both within and among regions will require different protection measures. One way to respond to this need is by incorporating local and traditional knowledge into the analytical and public policymaking process, in part by providing for a stronger community presence in the Council's advisory panels and committees.

A multitude of issues, policy options, and recommendations have been advanced through the Council's ecosystem workshop series. An integrating theme from each workshop was the desirability and value of valid science-based data and analyses with which resource managers can formulate policies that positively affect the future of marine ecosystems and associated resources and user groups across the region. Such information can take many forms, including but not limited to, traditional ecological knowledge of marine resources, public input on salient issues and challenges, biological stock assessments, risk and susceptibility analyses, analysis of the effects of oceanic regime shifts on marine fisheries, economic modeling of fleet costs and revenues, food web analysis, geographic depiction of total ecosystems, sociodemographic profiling, and analysis of the institutional ecology of fishery management.

Returning to the original premise of this book, the endeavors of fisheries-relevant science are as or more complex and comprehensive as those of any formalized field of inquiry, and an ecosystem approach may serve to yield deeper understanding of those complexities.

Yet, as indicated throughout the workshop series, an ecosystem approach to management further expands the information requirements of the various fishery sciences and disciplines. It therefore necessitates an even wider range of formalized research methods, analytical approaches, areas of inquiry, and modes of interaction with persons who are highly knowledgeable of and/or directly involved with the ocean, its resources, and the range of factors that can impinge on the sustainability and productivity of marine ecosystems and fishing societies.

Given this expanding realm of inquiry, and the desirability of using the full suite of science-based means for improving fishery management through application of ecosystem principles, a final recommendation of the workshop series relates to the need for a formalized ecosystem research and monitoring plan for contributing directly to the Council's fishery management goals and objectives across the region. Such a plan would address the need to:

- inventory existing relevant biophysical, social science, and traditional ecological knowledge data and research programs and projects;
- identify management objectives specific to implementation of the ecosystem approach across the archipelagos;
- identify sources of funding for new research, analysis, ongoing monitoring, and programmatic evaluation;
- articulate ongoing and new research, data management, and data analysis strategies with specific management objectives; and
- develop means for disseminating the analyses in a manner that would best support implementation of the Council's ecosystem approach and associated projects and programs.

Finally, such a plan should be designed to contribute to the full range of ecosystem management goals of the Council, with particular attention to the unique and highly varied attributes of human communities and their integral relationship with marine ecosystems and resources across the Western Pacific.

References

Western Pacific Regional Fishery Management Council (2007) Communities voice concern about loss of traditional practices. In: *Pacific Island Fishery News*. Newsletter of the Western Pacific Regional Fishery Management Council. Spring, Honolulu.

Witherell, D. (2004) *Managing Our Nation's Fisheries: Past, Present, and Future*. North Pacific Fishery Management Council under NOAA Award No. NA03NMF4410302.

Appendix A
Speaker Biographies

Susan Abbott-Jamieson is an anthropologist and senior social scientist at NOAA Fisheries Headquarters, Office of Science and Technology, Silver Spring, Maryland. She joined NOAA Fisheries following a 25-year career as a faculty member in the Department of Anthropology at the University of Kentucky. She is currently guiding the development of the NOAA Fisheries social sciences portion of the Economics and Social Sciences Program. This is part of the agency's effort to improve its ability to meet its mission-related social science research requirements.

Fini Aitaoto is a licensed grant writer and federal grants instructor. He was the former Acting Director and MIS Manager for the Department of Marine and Wildlife Resources for more than 20 years. He has served on all of the WPRFMC Plan Teams and is the Council's American Samoa On-Site Coordinator. He is former Executive Director for two NGOs and is a Samoan High Chief.

Stewart Allen is senior social scientist with NOAA Fisheries' Pacific Fisheries Science Center and leads the Center's Human Dimensions Research Program. Dr Allen is also a member of the WPRFMC Science and Statistical Committee. He has worked as a social scientist in a variety of natural resource settings since 1980 and has extensive research and analytical experience with human–ecosystem interactions.

Judith R. Amesbury is an archaeologist with Micronesian Archaeological Research Services, Guam. She received her education at the University of Arizona. Before moving to the Pacific, she worked on Native American and Spanish sites in Arizona, as well as the Neanderthal cave site of Tabun in Israel. In Hawaii, Amesbury worked for the Bishop Museum and the State Historic Preservation Division. She has now been conducting archaeological research in Guam and the CNMI for more than 20 years. Her area of expertise is analysis of archaeological faunal remains, which has led to an interest in long-term fishery data, indigenous fishing, and fishing communities.

Lee G. Anderson earned a PhD in Economics from the University of Washington in 1970. He is Professor of Economics and Marine Studies at the University of Delaware. He has written or edited six books and over 60 scientific papers on fisheries economics and the economics of fisheries management. He is a past member and chairperson of the Mid-Atlantic Fishery Management Council, and past President of the International Institute of Fisheries Economics and Trade. Dr Anderson is currently President-Elect of the North American Association of Fisheries Economists and a member of the Ocean Studies Board. His current work deals with simulation models,

Ecosystem-Based Fisheries Management in the Western Pacific, First Edition. Edward Glazier.

design, and implementation of ITQ programs, the economics of fishing in time and space, and marine reserves.

George "Bud" Antonelis has been the Chief of the Protected Species Division, Pacific Islands Fisheries Science Center, National Marine Fisheries Service, Honolulu, Hawaii, since 1996. He oversees research on population assessment and monitoring, foraging ecology, and health and disease of Hawaiian monk seals, cetaceans, and sea turtles in the Pacific Island region. Prior to his work in Hawaii, he was the Northern Fur Seal Task Leader at the National Marine Mammal Laboratory, National Marine Fisheries Service, Seattle, Washington, where he conducted studies in Alaska for nine years on fur seal population dynamics, foraging ecology, and migration patterns. He also spent 12 years of his early career as a Wildlife Biologist conducting ecological studies on five different pinniped populations at San Miguel Island, California. Dr Antonelis graduated with a BS and an MS from San Diego State University, and obtained his PhD from the University of Nagasaki.

Jerald Ault is Professor of Marine Biology and Fisheries at the University of Miami's Rosenstiel School of Marine and Atmospheric Science. Dr Ault is an internationally recognized leader in quantitative fisheries science for his theoretical and applied research on tropical coral reef fish and tarpon–bonefish population dynamics, ecosystem modeling, and resource management decision-making under uncertainty. A particularly unique aspect of his work is the development of large-scale spatial coupled biophysical ecosystem simulation models to assess the response and risks of multispecies fisheries to exploitation and environmental changes. He is well published regarding his research on the statistical assessment, modeling, and management of marine fishery ecosystems, and has been featured on the *CBS Evening News with Dan Rather*, *NBC Nightly News with Brian Williams*, *Good Morning America*, *National Geographic Channel, CNN, Animal Planet*, *PBS Waterways, Voice of America TV* and regional TV news; and, in the *Miami Herald*, *LA Times*, *Chicago Tribune*, *Reuters International,* and *New York Times*.

Paul K. Bartram has over 20 years of experience in marine and coastal resources use, assessment, and management throughout the Pacific basin. Mr Bartram is a member of Hui Malama o Mo'omomi, a community organization that is revitalizing and applying traditional Hawaiian knowledge to coastal fisheries conservation on the island of Moloka'i, Hawaii. He managed *'Imi 'Ike* ("search for knowledge"), a Native Hawaiian Education project that incorporated traditional Hawaiian learning approaches into public school curricula on Moloka'i. Bartram also serves as an adviser to community-based fishermen's organizations in Guam and American Samoa, and he regularly consults for the Western Pacific Regional Fishery Management Council and other fishery organizations in the region.

Russell Brainard, an oceanographer with the National Oceanic and Atmospheric Administration (NOAA) for over 25 years, serves as co-Principal Investigator of the Census of Coral Reef Ecosystems (CReefs) project of the International Census of Marine Life and is Chief of the Coral Reef Ecosystem Division (CRED) of NOAA's Pacific Islands Fisheries Science Center in Honolulu. The CRED leads an integrated, multidisciplinary, ecosystem-based program of research, benthic habitat mapping, and long-term ecological monitoring of the United States–affiliated Pacific Islands to promote conservation and management. This work involves conducting biennial Pacific Reef Assessment and Monitoring Program cruises to each of the 55 islands and atolls in Hawaii (main Hawaiian Islands and the Northwestern Hawaiian Islands), American Samoa, Guam, the Commonwealth of the Northern Mariana Islands, and the Pacific Remote Island Areas. These Pacific RAMP surveys monitor the fish, corals, other invertebrates, and algae in the context of their varying benthic and oceanographic habitats.

Leah Bunce is Senior Director of the Marine Management Area Science Program within Center for Applied Biodiversity Science of Conservation International. Dr Bunce is a marine social scientist with roots in the natural sciences. She has focused on applying science to conservation throughout her career. She has a BA in Biology from the University of Pennsylvania and a PhD in Environmental Studies from Duke University, where she focused on the sociocultural aspects of marine conservation in developing nations. Dr Bunce joined the National Oceanic and Atmospheric Administration's (NOAA's) International Program Office in 1998 after consulting for the World Bank, Organization of American States, and the Island Resources Foundation. At NOAA, she developed a global program for socioeconomic monitoring for coastal managers, coordinated various international coral reef and marine protected area activities, and served as one of the social science advisors to the agency.

Villy Christensen works with ecosystem modeling and has a background in fisheries research. Dr Christensen utilizes ecosystem modeling as the main tool to understand how human exploitation impacts marine ecosystems. As part of this, he has been central to the development and dissemination of the *Ecopath* approach and software and has focused on trophic dynamics of aquatic resources. He has led a large number of training courses and workshops throughout the world, centered on developing ecosystem approaches to fisheries management. Ecopath modeling is the de facto standard for ecosystem approaches to fisheries management and is being applied throughout the world. Dr Christensen currently focuses on how to communicate science and how to use science as input to the decision-making process. This involves use of advanced gaming technology and visualizations combined with research on the decision-making process.

Patrick Christie received his BS in Zoology in 1987 from the University of Wisconsin–Madison, his MS in Conservation Biology in 1993 from the University of Michigan, and his PhD in Natural Resources and Environment in 1999 from the University of Michigan. He has been Assistant Professor, School of Marine Affairs at the Jackson School of International Studies, University of Washington since 2001. Dr Christie recently concluded a three year research project in the Philippines and Indonesia investigating challenges associated with coastal environmental management over time. His current work includes feasibility studies of expanding World Heritage sites to include marine systems globally and implementing ecosystem-based fisheries management models in the Philippines. Dr Christie conducted graduate research on the Caribbean Coast of Nicaragua, where he studied the potential of participatory research for improving environmental management. He was previously involved in the implementation of community-based MPAs in the Philippines as a Peace Corps Volunteer. He is Associate Editor for the journal, *Coastal Management*.

Paul Dalzell is Senior Scientist and Pelagic Fisheries Coordinator for the Western Pacific Regional Fishery Management Council. He has worked in the Pacific Islands for the past 30 years and has published a range of papers and reports on the biology, stock assessment, and management of coastal and pelagic fisheries across the region. While at the Secretariat of the Pacific Community in New Caledonia, he co-authored a comprehensive review of Pacific Island coastal fisheries. Mr Dalzell oversees a variety of Council programs and projects across the region, including those related to recreational fisheries, commercial bycatch, and protected species.

Leimana DaMate has been working to integrate Native Hawaiian cultural values and practices into governmental and regulatory processes since the mid-1970s. She is currently Ocean Resources Program Director for the Association of Hawaiian Civic Clubs, a national confederacy of 51 Native Hawaiian organizations created in 1918 by Prince Jonah Kuhio. Ms DaMate represents Native Hawaiian interests as a member of Governor Lingle's Ocean and Coastal Council and consults with Native American and Alaska Native groups on various cultural and natural resource issues. She has worked extensively with Hawaiian communities on every island, documenting and researching cultural values and practices as they pertain to conservation land, ocean, and associated ecosystems, and is involved in restoration of ahupua'a lands through various cultural resource management processes.

Gerard DiNardo is the supervisory research fishery biologist with NOAA's PIFSC.

Leanne Fernandes has extensive academic and professional experience in sustainable use of natural resources and has conducted related research and applied work in various locations in the Caribbean, North Sea, the Maldives, and Australia. Her PhD involved development of a multicriteria decision support process for coral reef management, and she holds a master's degree in Resource Economics. Dr Fernandes was Manager of the Great Barrier Reef Marine Park Authority (GBRMPA) Representative Areas Program from 1999 to 2005. This program involved the rezoning of the entire Great Barrier Reef Marine Park. She is now Director of the GBRMPA Community Partnerships Group, formed to continue and build upon the Authority's community engagement work conducted through the rezoning process.

David Fluharty is Associate Professor in the School of Marine Affairs at University of Washington. He is also Wakefield Professor of Ocean and Fishery Sciences. Dr Fluharty received both his BA and his MA from the University of Washington, and his PhD from the University of Michigan. His areas on interest include climate variability and fishery management, international management of fisheries and marine animals, nonrenewable (oil, gas, minerals) natural resource management, training programs for natural resource managers from developing countries, and management institutions.

Michael Fogarty received a doctorate from the University of Rhode Island and came to the Northeast Fisheries Science Center in 1980, where he studies changes in marine ecosystems in response to fishing. Dr Fogarty is Adjunct Associate Scientist at Woods

Hole Oceanographic Institution and NOAA's Northeast Fisheries Science Center. He has served on numerous national and international panels and committees including the Scientific Steering Committee of the US GLOBEC program, which he chaired from 1997 to 2002, the Global Ocean Observation System (GOOS) Steering Committee, and the Coastal Ocean Observation Panel of GOOS.

John Gourley is owner and principal of Micronesian Environmental Services, a firm specializing in environmental regulatory permitting issues associated with terrestrial and marine environments. Previous training grounds include the Virginia Institute of Marine Science, UT Port Aransas Marine Lab, and the US Fish and Wildlife Service. Mr Gourley arrived in Saipan in 1989, working as a fishery biologist for the Division of Fish and Wildlife. He has been active in the CNMI consulting arena for the past 11 years. A former WPRFMC Advisory Panel and Plan Team member, Mr Gourley is associated with the Industry Advisory Council of the Center for Tropical and Subtropical Aquaculture, and the environmental and government affairs committees of the Saipan Chamber of Commerce.

Neil Gribble, from the Queensland Dept of Primary Industries & Fisheries, is a co-investigator on the Seabed Biodiversity Project and is responsible for delivery of prawn species composition and spatial distribution information. Dr Gribble has extensive experience leading projects on fish and crustacean biology, ecological physiology, and ecosystem modeling.

Susan Hanna is professor of marine economics at Oregon State University. Her research and publications are in marine economics and policy, with a focus on economic performance of fishery management, ecosystem-based fishery management, and application of incentive-based tools and institutional design. Dr Hanna serves on the Science Advisory Board of the National Oceanic and Atmospheric Administration and the Independent Science Advisory Board for Columbia River Basin Salmon Recovery. She is a former member of the Science Advisory Panel, US Commission on Ocean Policy; Ocean Studies Board, National Research Council; Scientific and Statistical Committee, Pacific Fishery Management Council; Marine Fisheries Advisory Committee, National Oceanic and Atmospheric Administration; National Research Council Committee to Review Individual Quotas in Fisheries, and NRC Committee on Protection and Management of Pacific Northwest Anadromous Salmonids.

Timothy Hennessey is professor of Marine Affairs and professor of Political Science at the University of Rhode Island. He holds an undergraduate degree from Brown University and a PhD from the University of North Carolina at Chapel Hill. Dr Hennessey has also held senior research appointments at the Woods Hole Oceanographic Institution, Dalhousie University, and the University of British Columbia. He is interested in the design and analysis of governance institutions as these relate to the management of natural resources. He recently completed a National Sea Grant funded study on the governance dimensions of large marine ecosystems with a number of colleagues from URI and The Northeast Science Center in Woods Hole. He completed a large-scale comparative analysis of governance systems in six estuarine watersheds, funded by the National Academy of Public Administration. His most recent publication is *Large Marine Ecosystems: The Human Dimension*.

Russell Ito is a fisheries biologist with PIFSC's Fisheries Monitoring and Analysis Program (FMAP).

Jeffrey Johnson is Senior Scientist at the Institute for Coastal and Marine Resources and serves as Professor in the Departments of Sociology, Anthropology, and Biology and Biostatistics at East Carolina University. He conducted a long-term research project supported by the National Science Foundation comparing group dynamics of the over-wintering crews at the American South Pole Station with those at the Polish, Russian, Chinese, and Indian Antarctic Stations. He is interested in network models of complex biological systems and is currently working with several ecologists to examine problems associated with trophic dynamics in food webs. His most recent work funded by NSF involves the development and testing of cognitive models of Inupiaq understandings of the Kotzebue Sound ecosystem in the Arctic.

Kurt Kawamoto is a fishery biologist and program manager for PIFSC's Fisheries Monitoring and Analysis Program. He is currently involved with barbless circle hook research and dissemination to the fishing community regarding advances in hook technology.

Colin Kippen is Executive Director of the Native Hawaiian Education Council and holds an appointment from the secretary of the Department of the Interior as a member of the Review Committee of the Native American Graves Protection and Repatriation Act. He is an advocate for Native people generally, and Native Hawaiians specifically. Colin is former Senior Counsel to the Senate Committee on Indian Affairs, Deputy Director of the Office of Hawaiian Affairs, judge for various Northwest Indian tribes, and a trial lawyer and prosecutor in Washington State.

David Kirby is a cross-disciplinary ocean scientist with additional interests in natural resource management and ocean governance. He studied at the University of Wales (BSc, MSc) and Leicester University (PhD) and has worked in the United Kingdom, New Zealand, Germany, and presently in New Caledonia, where he is Senior Fisheries Scientist with the Secretariat of the Pacific Community's Oceanic Fisheries Programme. Dr Kirby is interested in the adaptation of fisheries management and of the science base used to support it in order to better address the ecosystem governance goals that follow from UNCLOS.

Patrick Lehodey is Principal Fisheries Scientist in ocean ecology/biology at SPC. He has been working for some years to better understand the relationship between environmental variability and the distribution, abundance, and recruitment of tuna. The difficulties inherent in making direct measurements of both environmental conditions and fish stocks in the central tropical Pacific has led Dr Lehodey to develop a specialized computer model (SEAPODYM) instead. In work funded by the EU through the South Pacific Regional Tuna Resource Assessment and Monitoring Project (SPRTRAMP), he successfully linked the population dynamics of skipjack tuna to changes in tuna habitat. Dr Lehodey is further refining the SEAPODYM model so that it will be a reliable tool in the management of Pacific tuna fisheries.

Jared Makaiau is Habitat Coordinator with the Western Pacific Fishery Management Council.

Russell Moffitt graduated from Dartmouth College with a BA in Environmental and Evolutionary Biology. Prior to coming to Marine Conservation Biology Institute (MCBI), he worked in Hawaii with the NOAA Pacific Islands Fisheries Science Center, conducting oceanographic and ecological research at coral reef ecosystems across the Pacific, including the Pacific Remote Island Areas and the Northwestern Hawaiian Islands. Mr Moffitt has also conducted reef biodiversity assessments with the Census of Marine Life (CoML) Census of Coral Reefs project and has recently been developing and deploying a standardized method to assess cryptic biodiversity on reefs worldwide using advanced molecular techniques. He is particularly interested in the interactions between marine biological communities and their physical environments and how those processes are affected by climate change. At MCBI, Mr Moffitt works on identifying biological hot spots on the high seas and other areas in need of protection and analyzes geospatial data, particularly on issues related to marine spatial planning.

Steven Murawski received his PhD from the University of Massachusetts–Amherst. He serves as Director of Scientific Programs and Chief Science Advisor for NOAA Fisheries, whose mission is to provide the scientific basis for conservation and management of living marine resources and their ecosystems. He is responsible for about 30 laboratories, eight offshore research vessels, and 1,400 staff throughout the United States. Prior to coming to NOAA Fisheries headquarters, he served as Chief Stock Assessment Scientist for the Northeast Fisheries Science Center in Woods Hole, Massachusetts (1990–2004). His research background is in fisheries biology and stock assessment. He has published in several journals, including the Canadian Journal of Fisheries and Aquatic Sciences, Marine Ecology Progress Series, Bulletin of Marine Science, and Ecological Applications. During his career, Dr Murawski has been a key representative on several committees and councils. His current roles include official US delegate to the International Council for the Exploration of the Sea, member on the Global Ocean Ecosystems Dynamics (GLOBEC) Program Steering Committee, and Project Manager for NOAA Fisheries' Ecosystem Management Pilot Projects.

Bryan P. Oles received his doctorate in Cultural Anthropology from the University of Pittsburgh. He is Senior Social Scientist at the MPA Science Institute, which is part of the NOAA National Marine Protected Areas Center. Dr Oles has extensive experience managing and conducting social science research projects in coastal communities in the United States and abroad. He has lived and worked with subsistence and small-scale commercial fishermen in the Federated States of Micronesia, investigating traditional marine tenure systems, human dimensions of local marine resource use, and socioeconomic impacts of global political and economic processes of change. More recently, he has been involved in research on the human dimensions of marine protected area management and coastal communities in the United States. Much of his work is concerned with clarifying theoretical and methodological issues related to the use of social science in support of marine resource management, such as modeling and implementing social impact assessments, incorporating local knowledge in management processes, and developing tools for effective participatory research and cooperative management.

Michael K. Orbach is Professor of Marine Affairs and Policy and Director of the Duke University Marine Laboratory and the Coastal Environmental Management Program in the School of the Environment at Duke University. His BA is in Economics from the University of California at Irvine, and his MA and PhD are in Cultural Anthropology from the University of California at San Diego. Dr Orbach has performed research on, and has been involved in, development and implementation of coastal and marine policy on all coasts of the United States and in Mexico, Central America, the Caribbean, Alaska, and the Pacific. He has published widely on social science and policy in coastal and marine environments.

Frank Parrish has been a fishery biologist with NOAA Fisheries Service, Honolulu Laboratory for over 15 years. His research focuses on benthic and demersal ecology, particularly as it relates to fishery and protected species. His publications include work on reef fish, deepwater snappers, sharks, lobster, monk seals, and diving technology. In recent years, his investigations have focused on identifying important foraging habitats of the endangered Hawaiian monk seal and studies of fish assemblages associated with deepwater corals. He serves as a member of the precious coral planning team for the Western Pacific Regional Fishery Management Council and is the NOAA diving supervisor for fisheries operations in Hawaii and the Western Pacific. He is currently leading development of the Hawaii Archipelago Ecosystem Research Plan.

John Petterson is President of Impact Assessment, Inc., a firm specializing in maritime social science around the United States and abroad. Dr Petterson has served as Principal Investigator for 24 US Department of the Interior studies conducted across coastal Alaska, throughout the Gulf of Mexico, and along the Mid-Atlantic coast since 1979. He has also completed a wide range of fisheries-specific social and economic studies for NOAA Fisheries Service; the Caribbean, Gulf of Mexico, and North Pacific Fishery Management Councils; and various state and local government agencies. Dr Petterson also specializes in social scientific assessment of oil spills and high-level nuclear waste storage. He recently assessed the effects of Hurricane Katrina on fishing communities along the Gulf of Mexico for NOAA Fisheries and completed a study designed to reverse trends of environmental degradation in the Lake Chad Basin for the United Nations Development Program and Lake Chad Basin Commission.

Richard Pollnac became interested in modeling human adaptation to large bodies of water while conducting research concerning intracultural variability in cognition on the shores of Lake Victoria, East Africa. He has since conducted research among coastal peoples in Africa, Southeast Asia, the Middle East, Europe, the Pacific Islands, the Caribbean, and North, Central, and South America. Recent research has been conducted in Southeast Asia and in Alaska, with involvement in projects investigating the success of MPAs and sustainability of integrated coastal management projects in the Philippines and Indonesia (2000–2003), conflicts between fisheries in Vietnam (2004), tsunami recovery in Thailand (2005), and responses of Native American communities in Norton Sound to the Community Development Quota (Fisheries) program (2005).

Jeffrey J. Polovina is Division Chief of NOAA Fisheries' Pacific Islands Fisheries Science Center. Dr Polovina is widely published, with extensive contributions in the biological oceanography of the Central and Western Pacific and with particular focus on the population dynamics of high trophic animals. His current research employs satellite telemetry and remotely sensed oceanographic data to investigate migratory behavior and ecosystem habitats of large pelagic animals, including turtles, tunas, whale sharks, and whales. He is also engaged in assessment of the impacts of interannual and decadal climatic variation on marine fisheries and ecosystems, and application of satellite remote sensing and ocean circulation models to fisheries and protected species research.

Samuel Pooley is Director of NOAA's Pacific Island Fisheries Science Center (PIFSC) in Honolulu, and US representative to the Governing Council of the North Pacific Marine Science Organization (PICES). Dr Pooley served for 20 years as the lead economist for NMFS Honolulu Laboratory, with responsibilities ranging from economic analysis of commercial fisheries to evaluation of the benefits of recreational fisheries and conservation of endangered species. He has published papers on bioeconomic analysis, alternative fishery management, and property rights regimes. He is also affiliate graduate faculty with several departments and programs at the University of Hawaii, as well as a member of the steering committee for the Joint Institute for Marine and Atmospheric Research (JIMAR). Dr Pooley received his doctorate in Political Science with a dissertation on economic decision-making from the University of Hawaii, and the masters in Economics from the University of Birmingham (UK).

Michael Quach is an IT specialist and program manager for PIFSC's Western Pacific Fisheries Information Network (WPacFIN). Michael Quach was acknowledged for designing and implementing a new system for efficiently scanning commercial fish catch records and other important documents and archiving images of them into a robust, easy-to-use database system. His system has been adopted by fisheries agencies in Hawaii, American Samoa, Guam, and the Commonwealth of the Northern Mariana Islands, all partners with the PIFSC in managing fisheries data in the Pacific Islands Region (PIR). Data collected by these agencies comprise the fisheries intelligence assets needed to effectively manage local fisheries and monitor fishery resource trends. The PIR is NOAA's largest marine resource management jurisdiction, making fishery data management a formidable challenge. The agencies coordinate data management through the Western Pacific Fisheries Information Network (WPacFIN), a PIFSC program managed by Quach. By using Quach's new scanning and archiving system, the agencies have enhanced their data handling procedures and virtually eliminated the risk of permanently losing valuable documents. Quach's new archiving system and centralized database have dramatically increased the ability of fisheries agencies in the PIR to secure and share scientific data.

Jesse Rosario is with the Office of the Dean of the College of Natural and Applied Sciences and the Office of the Director of the Agricultural Experiment Stations at the University of Guam. He is an indigenous fisherman from a long line of Guam fishermen and has a long-term understanding of marine ecosystems in the region. He has been involved in community-based fishery management on Guam for many years.

Craig Severance is a cultural anthropologist at the University of Hawaii at Hilo and a member of the Western Pacific Fishery Council Scientific and Statistical Committee. He also sits on the Council's Recreational Fisheries Data Task Force and Social Science Research and Planning Committee. He was a member of the National Research Council's Committee to evaluate the community development quota program in Alaska and its relevance to the Western Pacific. Dr Severance has done field research with fishermen and fishing communities in American Samoa, Commonwealth of the Northern Marianas Islands, Hawaii, Chuuk and Pohnpei. He has an interest in TEK, CMT, MMA, and SIA. He is a board member of Hilo Trollers and a part-time commercial, recreational, and subsistence troller-handliner.

Joeli Veitayaki is Director of the International Ocean Institute-Pacific Islands and Professor and Senior Lecturer in the marine affairs section of the School of Marine Studies at the University of the South Pacific in Suva, Fiji. He remains actively involved with the Locally Managed Marine Areas Network. This network involves the conduct of hands-on projects to facilitate effective management of marine resources throughout the region. Dr Veitayaki is coeditor of *Pacific Voices—Equity and Sustainability in the Pacific Islands Fisheries* and author of *Fisheries Development in Fiji: The Quest for Sustainability.* The latter is part of the larger project titled "Towards Sustainable Fisheries" funded by the Canada-South Pacific Ocean Development Program with collaboration from the University of Prince Edward Island's Institute of Island Studies (IIS), Australia National University (ANU), and the Secretariat of the Pacific Community (SPC).

Robert Wakeford joined Marine Resources Assessment Group (MRAG) Americas in 2006 as Technical Director, after 11 years working for MRAG Ltd. in London. He has a broad range of multidisciplinary skills within fisheries resource management and policy, including fish stock assessment, survey design, and analysis, statistical and empirical modeling, database design, and project management. He was awarded his doctorate degree from Imperial College London in 2000, following a biosocioeconomic study to develop alternative management strategies for the Seychelles artisanal fishery in the Indian Ocean. He has provided technical advice to the South Atlantic Fisheries Commission on the status of shared squid and finfish stocks between Argentina and the Falkland Islands and has contributed to the ICES and CCAMLR Fish Stock Working Groups. He has gained extensive experience in the design and implementation of a range of resource survey methods. These include baseline dive surveys, socioeconomic, ecological, and acoustic surveys for biological and fish stock assessment purposes. More recently, Dr Wakeford has developed a range of skills concerned with fisheries monitoring, control and surveillance, with particular emphasis on controlling illegal, unreported, and unregulated fishing.

Carl Walters is a Professor at the Fisheries Centre at the University of British Columbia. His areas of research include the development of rapid techniques for teaching systems analysis and mathematical modeling to biologists and resource managers. Dr Walters believes the heart of fisheries is how to manage harvest. A member of several of NSERC's grant committees since 1970, he has done extensive fisheries advisory work for public agencies and industrial groups. He has also conducted over two dozen 3- to 10-day workshops in the past decade, for the International Canadian

Fisheries Service, US Fish and Wildlife Service, and the International Institute for Applied Systems Analysis. In 1992, he gave the keynote address to the American Fisheries Society, titled: "Where Have All the Coho Gone?" Dr Walters is the editor of *The Open Fish Journal* and has been on the editorial boards of the *Journal of Applied Mathematics and Computation*, the *Northwest Environmental Journal*, the *Canadian Journal of Fisheries and Aquatic Sciences*, and *Marine and Coastal Fisheries*.

Peter C. Wiley is an Economist with NOAA's National Ocean Service. In his 16 years in this position, he has concentrated on developing a better understanding of the dynamic relationship between human society and coastal and ocean ecosystem resources. Mr Wiley received his bachelor's degree in economics from St. Mary's College of Maryland and his master's degree in economics from the George Washington University. His work has concentrated on the economic impact of management actions in National Marine Sanctuaries, strategic socioeconomic characterization and economic valuation of coastal and ocean resources, and estimation of the socioeconomic elements of marine-related outdoor recreation participation. Mr Wiley's recent work has included characterizing and valuing ecosystem services for NOAA's Ecosystem Goal Team.

David Witherell has served the council for over nine years as the staff biologist and plan coordinator for the fishery management plans for Salmon, Scallops, Bering Sea Groundfish, and Bering Sea Crab, as well as staff liaison for bycatch, ecosystem, and essential fish habitat issues. Prior to coming to Alaska in 1992, Mr Witherell worked with the Massachusetts Division of Marine Fisheries as a fisheries biologist, for the Massachusetts Cooperative Fishery Research Unit as a research assistant, and as a fisheries technician for the Connecticut Department of Environmental Protection.

Dirk Zeller is Senior Research Fellow and Project Manager of the *Sea Around Us* project (www.seaaroundus.org) at the Fisheries Centre, UBC. Dr Zeller leads international collaborations on catch data reconstruction dealing with illegal, unreported, and unregulated fishing, is responsible for global coral reef fisheries and global marine pollution modeling, engages in ocean governance and fisheries policy research, and collaborates with the Fisheries Economics Research Unit on issues in resource economics. Dr Zeller has a specialization in marine and fisheries ecology and has interests in marine resource policy, global ocean governance, marine reserves, coral reef ecology, and conservation. He also has expertise in biotelemetry and advanced tagging technology in quantitative ecology and fisheries research. Dr Zeller has investigated the effects of marine reserves and worked extensively in experimental field ecology, applied animal behavior, and the interactions between species, their distribution, and habitats. He also held a position as tenured Assistant Professor (Associate Lecturer) at James Cook University, contributing to and coordinating several courses in fisheries science, marine biology, and zoology. Over the last 25 years, he also conducted research on corals, sea turtles, and cephalopods. Dr Zeller has published widely, both in the primary literature (*Nature*, *Science*, *Marine Policy*, *Marine Ecology Progress Series*, *Environmental Health Perspectives*, etc.) and in dedicated research reports.

Appendix B
Select Demographic Tables

Table B.1. Select demographic conditions: Main Hawaiian Islands.

Factor	1990		2000	
	1,108,229		1,211,537	
Total population Ethnicity or race[a]	Number	Percent	Number	Percent
White	369,616	33.4	294,102	24.3
Black or African American	27,195	2.45	22,003	1.8
American Indian and Alaskan Native	5,099	.46	3,535	0.3
Eskimo	155	3.03	NA	NA
Aleut	206	4.04	NA	NA
Asian	685,236	61.83	503,868	41.6
Asian Indian	1,015	.15	1,441	0.1
Cambodian	119	<.01	NA	NA
Chinese	68,804	10.04	56,600	4.7
Filipino	168,682	24.62	170,635	14.1
Hmong	6	<.01	NA	NA
Japanese	247,486	36.15	201,764	16.7
Korean	24,454	3.57	23,537	1.9
Laotian	1,677	.24	NA	NA
Thai	1,220	.18	NA	NA
Vietnamese	5,468	.80	7,867	0.6
Other Asian	4,036	.59	42,024	3.5
Native Hawaiian/other Pacific Islander	162,269	14.64	113,539	9.4
Guamanian or Chomorro	2,120	1.31	1,663	0.1
Micronesian	NA	NA	NA	NA
Melanesian	291	.18	NA	NA
Native Hawaiian	138,742	85.50	80,137	6.6
Other Micronesian	1,848	1.14	NA	NA
Polynesian	NA	NA	NA	NA
Other Polynesian	885	.54	NA	NA
Samoan	15,034	9.26	16,166	1.3
Tongan	3,088	1.90	NA	NA
Other Pacific Islander	261	.16	15,573	1.3
Hispanic or Latino of any race	81,390	7.34	87,699	7.2
Cuban	NA	NA	711	0.1
Mexican	NA	NA	19,820	1.6
Puerto Rican	NA	NA	30,005	2.5
Other Hispanic or Latino	NA	NA	37,163	3.1

(continued)

Ecosystem-Based Fisheries Management in the Western Pacific, First Edition. Edward Glazier.

Table B.1. (*Continued*)

Factor	1990		2000	
Total population	1,108,229		1,211,537	
Ethnicity or race[a]	Number	Percent	Number	Percent
Some other race	21,083	1.90	47,603	3.9
Two or more races	NA	NA	259,343	21.4
Household income (median $)	38,829		49,820	
Poverty status (% of families below poverty level)	.06		7.6	

[a]The 2000 Census combines American, Indian and Alaskan Native. Persons of Cambodian, Hmong, Laotian, Thai, and Vietnamese ancestry now must report as "Other Asian." Consolidation of Native Hawaiian with other Pacific Islander categories eliminated reporting options for persons previously reporting under Micronesian, Melanesian, Other Micronesian, Polynesian, Other Polynesian, and Tongan options.

Table B.2. Select demographic conditions: American Samoa.

Factor	1990		2000	
Total population	46,773		57,291	
Ethnicity or race[a]	Number	Percent	Number	Percent
Single ethnic group	45,430	97.13	55,704	97.2
Samoan	41,444	88.61	50,545	88.2
Native Hawaiian	NA	NA	53,227	92.9
Niuean	51	.11	18	<.01
Tokelauan	68	.14	45	<.01
Tongan	1,726	3.69	1,598	2.8
Other Pacific Islander	265	.57	245	.4
Asian	824	1.76	1,647	2.9
Chinese	85	10.31	329	.6
Filipino	415	50.36	792	1.4
Korean	224	27.18	200	.3
Japanese	28	3.40	16	<.01
Other Asian	72	8.74	295	.5
White	903	1.93	682	1.2
Black	10	<.01	21	<.01
Other single ethnic group	139	.30	127	.2
Multiple ethnic groups	1,341	2.87	1,587	2.8
Samoan and other group(s)	1,196	89.19	NA	NA
Asian and other group(s)	373	27.81	NA	NA
White and other group(s)	391	29.15	NA	NA
Unspecified or unreported	2	<.01	NA	NA
Household income (median $)	16,114		18,219	
Poverty status (% of families below poverty level)	56.5		58.3	

[a]Changes in the 2000 Census preclude reporting of certain groups (N/A).

Table B.3. Select demographic conditions: Guam.

Factor	1990		2000	
Total population	133,152		154,805	
Ethnicity or race[a]	Number	Percent	Number	Percent
Single ethnic group	120,203	90.3	141,118	91.2
Chomorro	49,935	37.5	57,373	37.1
Carolinian	135	.10	123	<.01
Palauan	1,858	1.4	2,141	1.4
Chuukese	1,919	1.4	6,229	4.0
Kosraean	101	<.01	292	.2
Marshallese	71	<.01	257	.2
Native Hawaiian	NA	NA	75,851	48.99
Pohnpeian	589	.4	1,366	.9
Yapese	199	.1	686	.4
Other Pacific Islander	1,637	1.2	648	.4
Asian	39,281	29.5	51,106	33.0
Chinese	1,959	5.0	2,712	.2
Filipino	30,043	76.5	40,729	26.3
Japanese	2,244	5.7	2,093	1.4
Korean	3,931	10.0	3,816	2.4
Other Asian	1,104	2.8	1,002	.6
White	19,160	14.4	10,666	6.9
Black	3,158	2.4	1,585	1.0
Other single ethnic group	2,160	1.6	1,910	1.2
Multiple ethnic groups	12,877	9.7	13,687	8.8
Chomorro other group(s)	7,713	59.9	NA	NA
Asian and other group(s)	7,449	57.8	NA	NA
Unspecified or unreported	72	<.01	NA	NA
Household income (median $)	30,755		39,317	
Poverty status (percent of families below poverty level)	12.6		19.9	

[a]Changes in the 2000 Census preclude reporting of certain groups (N/A).

Table B.4. Select demographic conditions: CNMI.

Factor	1990		2000	
Total population	43,345		69,221	
Ethnicity or race[a]	Number	Percent	Number	Percent
Single ethnic group	40,990	94.6	65,888	95.2
Chomorro	12,555	28.9	14,754	21.3
Carolinian	2,348	5.4	2,652	3.8
Palauan	1,620	3.7	1,685	2.4
Chuukese	1,063	2.5	1,394	2.0
Kosroean	17	<.01	56	<.01
Marshallese	92	.2	112	.2
Native Hawaiian	NA	NA	25,127	36.3
Pohnpeian	522	1.2	640	.9
Yapese	152	.4	204	.3
Other Pacific Islander	197	.4	509	.7
Asian	21,332	49.2	38,616	55.8
Chinese	2,881	13.5	15,314	22.1
Filipino	14,160	66.4	18,141	26.2
Japanese	784	3.6	952	1.4
Korean	2,571	12.1	2,021	2.9
Other Asian	936	4.4	2,188	3.2
White	875	2.0	1,274	1.8
Black	24	<.01	43	<.01
Other single ethnic group	193	.4	491	.71
Multiple ethnic groups	2,354	5.4	3,333	4.8
Carolinian and other group(s)	639	27.2	NA	NA
Chomorro and other group(s)	1,639	69.6	NA	NA
Unspecified or unreported	1	<.01	NA	NA
Household income (median $)	20,644		22,898	
Poverty status (% of families below poverty level)	32.1		30.6	

[a]Changes in the 2000 Census preclude reporting of certain groups (N/A).

Index

Ecosystem-Based Fisheries Management in the Western Pacific, First Edition. Edward Glazier.

www.ingramcontent.com/pod-product-compliance
Lightning Source LLC
Chambersburg PA
CBHW081211180926
51613CB00003B/212
* 9 7 8 0 8 1 3 8 2 1 5 4 2 *